Charles Seale-Hayne Library
University of Plymouth
(01752) 588 588
LibraryandITenquiries@plymouth.ac.uk

History and Climate

History and Climate
Memories of the Future?

Edited by

P. D. Jones
Climatic Research Unit
University of East Anglia
Norwich, U.K.

A. E. J. Ogilvie
Institute of Arctic and Alpine Research
University of Colorado
Boulder, U.S.A.

T. D. Davies
School of Environmental Sciences
University of East Anglia
Norwich, U.K.

and

K. R. Briffa
Climatic Research Unit
University of East Anglia
Norwich, U.K.

Kluwer Academic / Plenum Publishers
New York, Boston, Dordrecht, London, Moscow

Library of Congress Cataloging-in-Publication Data

History and climate: memories of the future?/edited by P.D. Jones ... [et al.].
　　p.　cm.
　　Includes bibliographical references and index.
　　ISBN 0-306-46589-2
　　1. Climatic changes—Europe—History.　2. Climatic changes—Environmental
aspects—Europe—History.　I. Jones, P. D., 1952–

QC989.A1 H58 2001

2001025347

Front cover: The photograph on the front cover showing a ship under full sail
is from an illuminated Icelandic manuscript of ca. 1600. It comes from a chapter on maritime law
in the collection known as the Reykjabók (AM 354 fol.) and is reproduced by kind permission
of the Manuscript Institute in Reykjavik, Iceland (Stofnum Árna Magnússonar).

ISBN 0-306-46589-2

©2001, Kluwer Academic / Plenum Publishers, New York
233 Spring Street, New York, New York 10013

http://www.wkap.nl

10　9　8　7　6　5　4　3　2　1

A C.I.P. record for this book is available from the Library of Congress

Printed in the United States of America

THIS BOOK IS DEDICATED TO THE MEMORY OF

JEAN GROVE (1927–2001)

OUTSTANDING SCHOLAR OF HISTORY AND CLIMATE

PREFACE

This book owes its origins to the *Second International Climate and History Conference* held by the Climatic Research Unit at the University of East Anglia, Norwich, in September 1998. The conference was made possible thanks to financial support from the Department of the Environment, Transport and the Regions in the UK, and the World Meteorological Organization, as well as by a US National Science Foundation grant (OPP-9523529) awarded to Tom McGovern, Andy Dugmore and Astrid Ogilvie. We are indebted to many colleagues in the Climatic Research Unit, and the School of Environmental Sciences at the University of East Anglia, for their invaluable help in organizing the conference. In particular, we would like to thank Susan Boland and Don McKinlay.

The editors wish to thank all the authors of the chapters in this book, not only for their contribution, but also for their patience whilst their final chapters came together. We hope they are all pleased with the finished project and feel that the wait has been worthwhile.

All the chapters in this book have been reviewed both by the editors and by two specialists in the field, one of whom may not have attended the conference. In this regard, we are indebted particularly to the following: Lars Bärring, Reinhard Böhm, Keith Brander, Rudolf Brazdil, Paul Buckland, Gaston Demarée, Chris Folland, Povl Frich, Andy Haines, Trausti Jónsson, John Kington, Brian Luckman, Jürg Luterbacher, David Parker, Christian Pfister, Chet Ropelewski and Tom Wigley.

Most important and vital in bringing the book together has been Julie Burgess, who reformatted all the chapters and laid out all the text and diagrams into their final form.

The editors and authors also note their great sadness at the death of our colleague, Jean Grove, in January 2001. We are honoured to be able to dedicate this book to her memory.

P.D. Jones, A.E.J. Ogilvie, T.D. Davies and K.R. Briffa.
Norwich and Boulder, November, 2000

LIST OF CONTRIBUTORS

1. Dr Jürgen Alheit
   Baltic Sea Research Institute
   Seestraße 15
   D-18119 Rostock-Warnemünde
   Germany
   Tel. +49 381 51 97 208
   Fax. +49 381 51 97 440
   Email. juergen.alheit@io-warnemuende.de

2. Dr Ingeborg Auer
   Central Institute for Meteorology
     and Geodynamics
   Hohe Warte 38
   A-1190 Vienna
   Austria
   Tel. +43 1 36026 2206
   Fax. +43 1 36026 72
   Email. ingeborg.auer@zamg:ac.at

3. Dr Christoph Beck
   Geographisches Institute der
     Universität Würzburg
   Am Hubland
   D-97074 Würzburg
   Germany
   Tel. +49 931 888 4690
   Fax. +49 931 888 5544
   Email. christoph.beck@mail.uni-wuerzburg.de

4. Dr Reinhard Böhm
   Central Institute for Meteorology
     and Geodynamics
   Hohe Warte 38
   A-1190 Vienna
   Austria
   Tel. +43 1 36026 2203
   Fax. +43 1 36026 72
   Email. reinhard.boehm@zamg.ac.at

5. Dr Anita Bokwa
   Institute of Geography and
     Spatial Management
   Department of Climatology
   Jagiellonian University
   64 Grodzka Street
   PL-31-044, Kraców
   Poland
   Tel. +48 12 4223056
   Fax. +48 12 4225578
   Email. abokwa@grodzki.phils.uj.edu.pl

6. Dr Keith R. Briffa
   Climatic Research Unit
   University of East Anglia
   Norwich NR4 7TJ
   U.K.
   Tel. +44 1603 593909
   Fax. +44 1603 507784
   Email. k.briffa@uea.ac.uk

7. Drs J. Buisman
   KNMI, Climatological Services
   PO Box 201
   3730 AE De Bilt
   The Netherlands
   Tel. +31 30 220 65 22
   Fax. +31 30 220 46 14

8. Dr Christopher J. Caseldine
   Department of Geography
   School of Geography and Archaeology
   University of Exeter
   Amory Building, Rennes Drive
   Exeter EX4 4RJ
   U.K.
   Tel. +44 1392 263347
   Fax. +44 1392 263342
   Email. C.J.Caseldine@exeter.ac.uk

9. Professor Trevor D. Davies
    School of Environmental Sciences
    University of East Anglia
    Norwich NR4 7TJ
    U.K.
    Tel. +44 1603 592842
    Fax. +44 1603 507719
    Email. t.d.davies@uea.ac.uk

10. Dr Gaston R. Demarée
    Royal Meteorological Institute
     of Belgium
    Ringlaan, 3
    B-1180 Brussels
    Belgium
    Tel. +32 2 37 30 540
    Fax. +32 2 37 30 548
    Email. Gaston.Demaree@oma.be

11. Dr Henry F. Diaz
    NOAA/OAR/CDC
    325 Broadway
    Boulder, CO 80305
    USA
    Tel. + (303) 497 6649
    Fax. + (303) 497 6649
    Email. hfd@cdc.noaa.gov

12. Dr Jean M. Grove[†]
    Girton College
    Cambridge CB3 0JG
    U.K.

13. Dr Eberhard Hagen
    Baltic Sea Research Institute
    Seestraße 15
    18119 Warnemünde
    Germany
    Tel. +49 381 5197 150
    Fax. +49 381 5197 440
    Email. eberhard.hagen@io-
    warnemuende.de

14. Mr F. IJnsen
    KNMI, Climatological Services
    PO Box 201
    3730 AE De Bilt
    The Netherlands
    Tel. +31 30 220 65 22
    Fax. +31 30 220 46 14

15. Professor Jucundus Jacobeit
    Geographisches Institut
    Universität Würzburg
    Am Hubland
    D-97074 Würzburg
    Germany
    Tel. +49 931 888-5586
    Fax. +49 931 888-5544
    Email. jucundus.jacobeit@mail.uni-
    wuerzburg.de

16. Professor Phil Jones
    Climatic Research Unit
    University of East Anglia
    Norwich NR4 7TJ
    U.K.
    Tel. +44 1603 592090
    Fax. +44 1603 507784
    Email. p.jones@uea.ac.uk

17. Ms R.S. Kovats
    Department of Epidemiology
     and Population Health
    London School of Hygiene
     and Tropical Medicine
    Keppel Street
    London WC1E 7HT
    UK
    Tel. +44 20 7612 7844
    Fax. +44 20 7580 6897
    Email. s.kovats@lshtm.ac.uk

18. Ms Danuta Limanówka
    Institute of Meteorology and
     Water Management, Kraców Branch
    Borowego 14
    30 215 Kraców
    Poland
    Tel. +48 (12) 425 19 00 w.205
    Fax. +48 (12) 425 19 29
    Email. Danuta_Limanowka@imgw.pl

19. Dr Jürg Luterbacher
    Institute of Geography
    Climatology and Meteorology
    University of Bern
    Hallerstrasse 12, CH-3012 Bern
    Switzerland
    Tel. +41 31 631 8545
    Fax. +41 31 631 8511
    Email. juerg@giub.unibe.ch

20. Dr Anthony J. McMichael
Department of Epidemiology
  and Population Health
London School of Hygiene and
  Tropical Medicine
Keppel Street
London WC1E 7HT
U.K.
Tel. 020 76 12 7825
Fax. 020 7580 6897
Email. t.mcmichael@lshtm.ac.uk

21. Dr Axel Michaelowa
Hamburg Institute for International
Economics
Neuer Jungfernstieg 21
20347 Hamburg
Germany
Tel. +494042834309
Fax. +494042834451
Email. a-michaelowa@hwwa.de

22. Dr Neville Nicholls
Bureau of Meteorology Research
  Centre
PO Box 1289K
Melbourne 3001
Australia
Tel. +61 3 9669 4407
Fax. +61 3 9669 4660
Email. n.nicholls@bom.gov.au

23. Dr Astrid E.J. Ogilvie
Institute of Arctic and Alpine Research
Campus Box 450
University of Colorado
Boulder, CO 80309-0450
U.S.A.
Tel. +303 492 6072
Fax. +303 492 6388
Email. ogilvie@spot.Colorado.edu

24. Dr Wolfgang Schöner
Central Institute for Meteorology
  and Geodynamics
Hohe Warte 38
A-1190 Vienna
Austria
Tel. +43 1 36026 2203
Fax. +43 1 36026 72
Email. wolfgang.schoener@zamg.ac.at

25. Professor Dr Johann Stötter
Institut für Geographie
Universität Innsbruck
Innrain 52
A-6020 Innsbruck
Austria
Tel. +43 512 5403
Fax. +43 512 507 2895
Email. Hans.Stoetter@uibk.ac.at

26. Dr Aryan F.V. van Engelen
KNMI, Climatological Services
PO Box 201
3730 AE De Bilt
The Netherlands
Tel. +31 30 2206522
Fax. +31 30 220 46 14
Email. engelenv@knmi.nl

27. Dr Maria Wastl
Fachbereich 8
Abteilung Physiogeographie
Universität Bremen
Postfach 330440
D-28334 Bremen
Germany

and

Institut für Geographie
Universität Innsbruck
Innrain 52
A-6020 Innsbruck
Austria
Tel. +43 512 507 5415
Fax. +43 512 507 2895
Email. Maria. Wastl@uibk.ac.at

28. Dr Joanna Wibig
Department of Meteorology
  and Climatology
University of Lódz
Lipowa 81
90-568 Lódz
Poland
Tel. +48 42 6374547
Fax. +48 42 6376159
Email. zameteo@krysia.uni.lodz.pl

CONTENTS

INTRODUCTION

PRE-INSTRUMENTAL OBSERVATIONS

INSTRUMENTAL RECORDS

CLIMATE SYNTHESES

CLIMATE IMPACTS

UNLOCKING THE DOORS TO THE PAST: RECENT DEVELOPMENTS IN CLIMATE AND CLIMATE IMPACT RESEARCH

P.D. Jones[1], A.E.J. Ogilvie[2], T.D. Davies[3] and K.R. Briffa[1]

[1]Climatic Research Unit, University of East Anglia,
Norwich, NR4 7TJ, U.K.
[2]Institute of Arctic and Alpine Research, University of Colorado
at Boulder, CO 80309-0450, U.S.A.
[3]School of Environmental Sciences, University of East Anglia,
Norwich, NR4 7TJ, U.K.

INTRODUCTION

In 1979, the Climatic Research Unit at the University of East Anglia in Norwich held an international conference with the title "Climate and History". This landmark meeting was one of the first of its kind to bring together scholars from a variety of disciplines to focus on the issues of past climates and variability as well as the impacts of climate on societies. The conference resulted in an edited volume containing twenty-one peer-reviewed papers based on presentations at the meeting. This was published two years later as: Climate and History: Studies in past climates and their impact on Man (Wigley et al., 1981). With their interdisciplinary and multidisciplinary perspectives, the approaches of both the meeting and the book were highly innovative. Previously, for the most part, historians had tended to ignore the possible implications of climatic and environmental change for human history, and climatologists, with a few notable exceptions, had little interest in this field. The first meeting and the resulting volume also emphasised a further important development in climate research. Early investigations into past climate, dating from around the eighteenth century onwards, and, indeed, to as late as the mid-1970s, tended to assume that, on human timescales, climate was a relatively unchanging and constant factor. Certainly "weather" was acknowledged as a factor in certain historical events. The defeat of the Spanish Armada in 1588 is a well-known example; but subtler, longer timescale changes were not considered important by most historians and many climatologists. From the nineteenth century onwards, studies of climate based on instrumental records had been performed, but, in most cases, these were mere book-keeping exercises, concerned more with spatial rather than temporal variations of climate, and undertaken by small interest groups such as in a meteorological service or a university geography department.

The awakening of an interest in climatic change *per se* arose for two reasons, partly as the result of the commitment of a small group of far-sighted individuals, but also from the

History and Climate: Memories of the Future?
Edited by Jones et al., Kluwer Academic/Plenum Publishers, 2001

1

realisation gained by the fledgling climate modelling community that human factors could, in fact, modify the climate system and lead to changes in our supposedly benign climate. The modern understanding of the concept of climatic change was born out of the realisation that, in order to determine whether anthropogenic influences were acting to change our climate, we needed to increase our knowledge concerning both climatic variations that have occurred in the past, and the natural factors that influence the climate system as a whole.

RECENT DEVELOPMENTS IN CLIMATE RESEARCH

A detailed discussion of the developments in palaeoclimatology, historical climatology, climate modelling, and the multitude of related studies during the 21 years since the first "Climate and History" conference is clearly beyond the scope of these introductory remarks. However, it is appropriate to draw together a few of the strands in this interesting story. Much current research into palaeoclimate focuses on the use of high-resolution proxy data; a variety of natural and human "data banks". These include information drawn from the fields of: dendroclimatology; marine and lake sediment cores; ice-core studies; corals; glaciology; and documentary historical records. In the late 1960s and early 1970s, a number of seminal papers were published by the pioneers in these fields. Two particular developments may be detected from around this time: one is the evolution and refinement of separate disciplines within palaeoclimatology; the second is the realisation of the importance of interdisciplinary and multidisciplinary research. Of interest here are the early papers written by Hubert Lamb, the first Director of the Climatic Research Unit (see, e.g., Lamb, 1977, and references cited therein). A climatologist and meteorologist by training, he had become convinced early in his career that climate was an ever-changing element, not a constant factor. He also showed considerable foresight in his championship of the importance of interdisciplinarity.

Lamb had a particular interest in historical climatology, and one of the first major projects fostered by him in the Climatic Research Unit involved the use of historical documentary evidence to assess changes in past climate. A vitally important development in this field was the realisation of the necessity for careful analysis of all historical evidence to be used for climate reconstruction in order to ensure its reliability (Bell and Ogilvie, 1978; Ingram *et al.*, 1981). Further to this, painstaking research and care in locating and assessing early meteorological observations and accounts has led to a steadily increasing knowledge of the climate of recent centuries (see, e.g., Kington, 1997; Demarée *et al.*, 1998; Jónsson, 1998). In tandem with this has been the important work on the establishment of data banks of documentary climate information by researchers such as Christian Pfister and colleagues (see, e.g., Pfister, 1985).

Parallel with these developments in historical climatology has been the increasing refinement of quantitative methods in climate reconstruction. By the early 1970s, it was becoming clear that qualitative interpretations of proxy climate data were no longer acceptable; rigorous statistical analyses were increasingly being applied, particularly in the areas of tree-ring, pollen, and fossil insect data (see, e.g., La Marche, 1974; Fritts, 1976; Berglund, 1986; Atkinson *et al.*, 1987).

At the same time, the need for the development of a quality-controlled, global-scale instrumental temperature data set was perceived. After much exhaustive data collection and homogenization by many individuals, a standard land-based, gridded temperature data set was established (Jones *et al.*, 1982, 1986a,b; Kelly *et al.*, 1982; Raper *et al.*, 1984). In the mid 1980s, the analysis was extended to the marine sector and the first ever synthesis of land and marine data was produced in 1986 (Jones *et al.*, 1986c). This work has continued to be updated (Jones *et al.*, 1999).

The firm foundation laid in the areas of historical climatology and palaeoclimatology has also been of value to other disciplines. One example of this is its use in the fields of archaeology and history. The work of Thomas McGovern and colleagues in the North

Atlantic area may be cited here (see, e.g., Barlow *et al.*, 1997). Historical and palaeoclimate studies have, more generally, given valuable insights for the development of climate-impact studies (see, e.g., Delano Smith and Parry, 1981; Wigley *et al.*, 1981; Mörner and Karlén, 1984; Pfister and Brázdil, 1999)

The progress made in the areas discussed above undoubtedly also played a significant part in one of the most exciting and important developments in the history of climate research; studies aimed at the detection of the anthropogenic signal in the global warming of the twentieth century (Wigley and Barnett, 1990; Santer *et al.*, 1995, 1996a,b). This work has had a profound influence on climate-related policy, and provides part of the raison d'être for a growing emphasis on the human dimensions of global climate change (see, e.g., ARCUS, 1997).

THE CLIMATE OF THE LAST 1000 YEARS

In 1979, as now, the two most widely recognised climate epochs of the past millennium were the so-called "Medieval Warm Period" (MWP ~900-1300) and the "Little Ice Age" (LIA ~1450-1850). Although the years shown here correspond broadly to the views of a number of researchers, the fact remains that attempting to assign hard and fast dates to these hypothesized climatic periods is contentious (Ogilvie and Jónsson, 2001). This is because different researchers have varying opinions concerning their meaning and, indeed, their usefulness. Also, as information regarding the climate of the past thousand years has grown, the complexity of the situation has become more apparent. Certainly a major difference between now (2000) and 21 years ago (1979) is that the amount of proxy climatic information available for the past millennium has increased by at least two orders of magnitude. More importantly, spatial climatic patterns of the millennium are beginning to be mapped for large parts of the Earth's surface. In 1979, evidence was limited to Europe, eastern North America and the Far East, and it was tempting to make broad generalizations from these limited data. The last few years have seen the development of multiproxy averages (the combination of different types of proxy evidence), providing year-by-year estimates of average temperatures for the Northern Hemisphere for both annual and extended summer seasons (Jones *et al.*, 1998; Mann *et al.*, 1998, 1999; Briffa, 2000; Crowley and Lowery, 2000). There were pioneering attempts in the early 1980s (e.g. Williams and Wigley, 1983), but the growth of tree-ring evidence from the northern high latitudes (Briffa, 2000) and the emergence of highly resolved information from corals in the tropics (see, e.g., Cole, 1996) has dramatically improved our knowledge base.

The new series (see Figure 1) are challenging the concepts embodied in the terms "Medieval Warm Period" and "Little Ice Age". Both periods are barely recognizable in the few Southern Hemisphere reconstructions available, and although both find some support in Figure 1 their amplitudes are relatively small (as, indeed, one might expect when signal and noise aspects are considered; see, e.g. Wigley and Raper, 1995). Does this mean that we should abandon or redefine the terms "Medieval Warm Period" and "Little Ice Age"? Redefining terms is not new in science: indeed, it is often a sign of positive progress in understanding; in this case of what has happened in the past, and the reasons why. The "Medieval Warm Period" and "Little Ice Age" are most clearly recognized as European phenomena. However, as we gain more evidence of the timing and scale of climatic changes in the tropics and the Southern Hemisphere, the appropriateness of these terms as defining distinct climatic episodes becomes clouded. Analysis of the last 150 years of the instrumental period also clearly shows that, for the understanding of the climate system as a whole, no one geographical area is more important than any other. Furthermore, just as it was necessary to incorporate temperature records from as many regions of the world as possible in order to develop global average series for the instrumental period, so it will be necessary to include proxy evidence from many regions in order to better define pre-instrumental changes. As well as the increasing availability of palaeoclimatic data, enhancements in computing power have also fostered more complex, and hence more realistic, pictures of past climate. Figure 1 also shows that the twentieth century is clearly the warmest century of the millennium; a fact that provides important support to the idea that human influences on climate are significant.

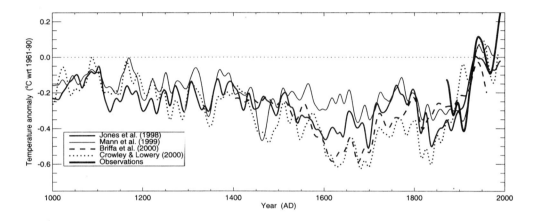

Figure 1. Reconstructions of Northern Hemisphere temperature for the last 1000 years. All series have been smoothed to highlight variations on the 50-year timescale. Some reconstructions are of annual conditions, some for the growing season (April to September or June to August). Instrumental temperatures for the Northern Hemisphere for the April to September season are also shown, similarly filtered. Further details concerning the various series are given in the original publications and in Briffa *et al.*, 2000.

THE PRESENT VOLUME: DATA ANALYSES AND CLIMATE-IMPACT STUDIES

This volume is the product of the "Second International Climate and History Conference", organised by the Climatic Research Unit in 1998, and held, as the first one, at the University of East Anglia. The volume presents new evidence from several specific regions, including the Netherlands (van Engelen *et al.*); Poland (Bokwa *et al.*); England and France (Michaelowa); Austria (Auer *et al.*); Iceland (Wastl *et al.*) as well as studies ranging over a wider European area (Grove; Jacobeit *et al.*, Luterbacher) and the eastern North Atlantic (Alheit and Hagen). The paper by Demarée and Ogilvie focuses primarily on Europe as a whole, while Diaz *et al.*, have a spatially extensive approach including both Eastern and Western Hemispheres. The twelve chapters, written by leading authors (names emboldened later in this section) in their fields, cover a number of different aspects of climate research. Of these, the two main subject areas covered are: i) the elucidation of changes in the climate of the past millennium, incorporating discussions of both sources of climate data and the methodologies used; and ii) past and present climate impacts.

Since the time of the 1979 'Climate and History' conference, the research effort expended on the study of documentary sources compared to natural archives (trees, ice cores, corals, etc.) has possibly declined, despite the continuing efforts of Pfister and Brazdil (1999), Ogilvie (1992) and others. This change may partly reflect the time it takes to become proficient in the languages of the past and in acquiring the experience to interpret the significance of historical records in a correct social and contemporary context. In this volume, Bokwa *et al.*, discuss the information content in diaries from sixteenth-century Krakow. Diaries provide extremely fine resolution information about extremes of weather throughout the year, in stark contrast to the one value per year represented in many natural archives. Also based on the use of documentary evidence, the paper by Luterbacher relates many of the indices derived by Pfister and colleagues for central Europe to atmospheric circulation patterns, considering how altered circulation states might have explained the differences in temperature and precipitation indices. One of the main difficulties with documentary and diary information in Europe is that, after the development of meteorological instruments in the seventeenth and eighteenth centuries, such historical sources become much harder to find. This makes the earlier records harder to interpret, because there are fewer (compared with the case for natural archives) opportunities for calibrating the diary/documentary information against instrumental data.

Early instrumental data provide climatic information directly, but considerable effort and care are necessary to derive homogeneous time series. Europe has a wealth of early data and Jones lists locations and analyses of some of the long series. The most intriguing result of this work is that, while western Europe has experienced long-term warming in annual series, this has only occurred during the winter half of the year. Summers were milder in the 1750-1850 period than in much of the twentieth century, particularly the 1760s and 1820s. By considering possible circulation changes since 1780, Jacobeit *et al.*, attempt to explain these seasonal differences.

All authors clearly discuss their methods and potential, as well as the limitations of their data. Thus, van Engelen *et al.*, present time series of indices for over a thousand years of Dutch temperatures based on instrumental and documentary sources. Their methods of estimation of seasonal temperatures are carefully explained. This can be contrasted with the Central England temperature series developed by Manley (1974). Although the Manley series is the best known, longest, and most extensively analysed instrumental series available, the original papers give few details about how it was produced. The development and availability of early instrumental series varies throughout Europe. In some countries, many series are available, while in others there are few. Austria, for example, has been fortunate to have had a national programme which has led to the production of many long instrumental time series. These results are discussed by Auer *et al.* This type of work in other European countries is to be encouraged. There is also potential to extend existing instrumental records in other continents, notably North and South America and parts of southern Asia.

The advances and retreats of glaciers have been used for the last 80 years as evidence of past climate change. Indeed, it was a glaciologist (Matthes, 1939, 1940) who first coined the phrase 'Little Ice Age' to identify the period of the most recent major advance of glaciers since the Last Ice Age. It is worth noting that his definition of the term applied to the entire period of the past 4,000 years. In her paper, Grove considers the differences in timing of advances in diverse alpine regions. Strictly defining the term as referring specifically to glaciation and not temperature, she explores the evidence for the timing of 'Little Ice Age' glaciation and concludes that it may have begun as early as the fourteenth century in some regions. Wastl *et al.*, focus on glacier-sea-ice-climate relationships in northern Iceland, and possible analogues for climate variations during the Holocene. They focus on the period from the second half of the nineteenth century when glaciers in northern Iceland reached a maximum extent, and suggest a termination date of ca. 1920 for the "Little Ice Age" in Iceland.

Assessing the potential impacts on various sectors as a result of future anthropogenic climate change has become a virtual growth industry since 1979. This work, which is generally quantitative and model-based, has helped identify some of the problems inherent in past impact studies. The economic costs/benefits of future climate change need to be considered alongside a whole range of other non-climatic elements. Climate change is clearly only one of the factors which will contribute in the future. Nevertheless, studies of the past provide valuable insights. Here, the effects of climate on agriculture in eighteenth-century England and France are discussed by Michaelowa. The influences of climate in these two regions might be expected to be similar, but the impacts were quite different. In England, poor seasons led to innovation; but in France, a succession of poor harvests eventually led to revolution. Climate was undoubtedly a factor in these changes but it cannot be considered in isolation. Equally, however, it would be wrong for historians to ignore the effects of climate on past societies.

The paper by Demarée and Ogilvie concerns the effects in Europe and elsewhere of the *Lakagígar* (Laki fissure eruption) which occurred in Iceland in the years 1783-1784. The haze caused by the eruption was seen in many parts of the Northern Hemisphere and came to be called the "great dry fog". In this paper, a number of the climatological, phenological, environmental and human dimensions impacts of this fog are presented on the basis of many new findings in hitherto unused historical documents.

Analysing present-day climate impacts on fisheries, Alheit and Hagen evaluate the effects that changes in climate have had on pelagic fish stocks throughout the world, basing their study particularly on the long records from the eastern North Atlantic. Today, conservation of fish stocks is an urgent necessity, but it is also important to establish how fish stocks have varied naturally in the past.

The chapter by Diaz et al., considers the issue most likely to rally public opinion behind moves to reduce emissions: the impact of climate on health. Indeed, dramatic changes to regions affected by vector-borne diseases are suggested by the latest climate scenarios. Diaz et al., consider health impacts through climate over the last 150 years, assessing what can be learnt from a number of case studies. Climate has clearly affected human health, but the scale of impacts is very dependent on the knowledge base of the society. Climate is an important factor, but just as it is not the sole factor, its influence cannot be ignored.

It is clear today, that, over the past twenty years or so, great progress has been made in our understanding of the climates of the past and the present. Nevertheless, there is still an urgent need for the further refinement and development of individual disciplines within climatology. In addition to this, the increasing realisation of the value, and indeed the necessity, of an interdisciplinary approach should continue to be built upon. By interweaving the many elements that make up climate research, and by continuing to compare and correlate different proxy indicators, our understanding of the climates of the past and the present will continue to illuminate and instruct us, and help us prepare for the challenges of future climates. In this spirit, the present volume of studies focusing on the climate of the past millennium is offered as a contribution to current research in climate and climate-impact studies.

REFERENCES

ARCUS, 1997, People in the Arctic. A Prospectus for Research on the Human Dimensions of the Arctic System (HARC) for the National Science Foundation Arctic System Science Program, The Arctic Research Consortium of the United States (ARCUS), Fairbanks.

Atkinson, T., Briffa, K.R., and Coope, R., 1987, Seasonal temperatures in Britain during the past 22,000 years reconstructed using beetle remains, *Nature* 325:587-592.

Barlow, L.K., Sadler, J.P., Ogilvie, A.E.J., Buckland, P.C., Amorosi, T., Ingimundarson, J.H., Dugmore, A.J., and McGovern, T.H., 1997, Interdisciplinary investigations of the end of the Norse Western Settlement in Greenland, *The Holocene* 7:489-499.

Bell, W.T., and Ogilvie, A.E.J., 1978, Weather compilations as a source of data for the reconstruction of European climate during the medieval period, *Climatic Change* 1:331-348.

Berglund, B.E. (Ed.), 1986, *Handbook of Holocene Palaeoecology and Paleohydrology*, Wiley, Chichester: 869pp.

Briffa, K.R., 2000, Annual climate variability in the Holocene: interpreting the message of ancient trees, *Quat. Sci. Rev.* 19:87-105.

Briffa, K.R., Osborn, T.J., Schweingruber, F.H., Harris, I.C., Jones, P.D., Shiyatov, S.G., and Vaganov, E.A., 2000, Low-frequency temperature variations from a northern tree-ring density network, *J. Geophys. Res.* (in press).

Cole, J., 1996, Coral records of climatic change: understanding past variability in the tropical ocean-atmosphere, in: *Climatic Variations and Forcing Mechanisms of the last 2000 years*, P.D. Jones, R.S. Bradley and J. Jouzel, eds., Springer, Berlin:331-353.

Crowley, T.J., and Lowery, T.S., 2000, How warm was the Medieval warm period? A comment on 'Man-made versus Natural Climate Change', *Ambio* 39:51-54.

Delano Smith, C., and Parry, M., eds, 1981, *Consequences of Climatic Change*, Department of Geography, University of Nottingham.

Demarée, G.R., Ogilvie, A.E.J., and Zhang Deíer, 1998, Comment on Stothers, R.B. 'The great dry fog of 1783' (Climatic Change 32:1996): Further documentary evidence of Northern Hemispheric coverage of the Great Dry Fog of 1783, *Climatic Change* 39:727-730.

Fritts, H.C., 1976, *Tree Rings and Climate*, Academic Press, London, New York, San Francisco:567pp.

Ingram, M.J., Farmer, G. and Wigley, T.M.L., 1981, Past climates and their impact on Man: a review, in: *Climate and History: Studies in past climates and their impact on Man*, T.M.L. Wigley, M.J. Ingram, and G. Farmer, eds., Cambridge University Press, Cambridge:3-50.

Jones, P.D., Wigley, T.M.L., and Kelly, P.M., 1982, Variations in surface air temperatures, Part 1: Northern Hemisphere, 1881-1980, *Mon.Wea. Rev.* 110:59-70.

Jones, P.D., Raper, S.C.B., Bradley, R.S., Diaz, H.F., Kelly, P.M., and Wigley, T.M.L., 1986a, Northern Hemisphere surface air temperature variations: 1851-1984, *J. Clim. Appl. Met.* 25:161-179.

Jones, P.D., Raper, S.C.B., and Wigley, T.M.L., 1986b, Southern Hemisphere surface air temperature variations: 1851-1984, *J. Clim. Appl. Met.* 25:1213-1230.

Jones, P.D., Wigley, T.M.L., and Wright, P.B., 1986c, Global temperature variations, 1861-1984, *Nature* 322:430-434.

Jones, P.D., Briffa, K.R., Barnett, T.P., and Tett, S.F.B., 1998, High-resolution palaeoclimatic records for the last millennium: Integration, interpretation and comparison with General Circulation Model control run temperatures, *The Holocene* 8:455-471.

Jones, P.D., New, M., Parker, D.E., Martin, S., and Rigor, I.G., 1999, Surface air temperature and its changes over the past 150 years, *Revs. Geophys.* 37:173-199.

Jónsson, T., 1998, Reconstructing the temperature in Iceland from early instrumental observations: data availability and a status report, in: *Documentary Climatic Evidence for 1750-1850 and the 14th Century*, Palæoklimaforschung 23, B. Frenzel, ed., Special Issue: ESF Project European Palaeoclimate and Man 15, European Science Foundation, Strasbourg, Akademie der Wissenschaften und der Literatur, Mainz:87-98.

Kelly, P.M., Jones, P.D., Sear, C.B., Cherry, B.S.G., and Tavakol, R.K., 1982, Variations in surface air temperatures, Part 2: Arctic regions, 1881-1980, *Mon. Wea. Rev.* 110:71-83.

Kington, J., 1997, Observing and measuring the weather: a brief history, in: *Climates of the British Isles*, M. Hulme and E. Barrow, eds, Routledge, London and New York:137-152.

La Marche, V.C., 1974, Paleoclimatic inferences from long tree-ring records, *Science* 183:1043-1048.

Lamb, H.H., 1977, *Climate Present, Past and Future. Volume 2 Climatic History and the Future*, Methuen & Co., London:835pp.

Manley, G., 1974, Central England temperatures: monthly means 1659 to 1973, *Quat. J. Roy. Met. Soc.* 100:389-405.

Mann, M.E., Bradley, R.S., and Hughes, M.K., 1998: Global-scale temperature patterns and climate forcing over the past six centuries, *Nature* 392:779-787.

Mann, M.E., Bradley, R.S., and Hughes, M.K., 1999, Northern Hemisphere temperatures during the past millennium: inferences, uncertainties and limitations, *Geophys. Res. Letts.* 26:759-762.

Matthes, F., 1939, Report of Committee on Glaciers, *Trans. Amer. Geophys. Union* 20:518-23.

Matthes, F., 1940, Committee on Glaciers, 1939-40, *Trans. Amer. Geophys. Union* 21:396-406.

Mörner, N.-A., and Karlén, W., eds, 1984, *Climate Changes on a Yearly to Millennial Basis*, Reidel, Dordrecht:667pp.

Ogilvie, A.E.J., 1992, Documentary evidence for changes in climate in Iceland, in: *Climate since A.D. 1500*, R.S. Bradley and P.D. Jones, eds, Routledge, London:92-117.

Ogilvie, A.E.J., and Jónsson, T., 2001, 'Little Ice Age' research: a perspective from Iceland, in: The Iceberg in the Mist: Northern Research in Pursuit of a "Little Ice Age", A.E.J. Ogilvie and T. Jónsson, eds, *Climatic Change*, in press.

Pfister, C., 1985, *CLIMHIST - a weather data bank for Central Europe 1525 to 1863*. May be ordered from METEOTEST, Fabrikstr. 29a, CH 3012 Berne.

Pfister, C., and Brázdil, R., 1999, Climatic variability in sixteenth century Europe and its social dimension: A synthesis, *Climatic Change* 43:5-53.

Raper, S.C.B., Wigley, T.M.L., Mayes, P.R., Jones, P.D. and Salinger, M.J., 1984, Variations in surface air temperature, Part 3: The Antarctic, 1957-82, *Mon. Wea. Rev.* 112:1341-1353.

Santer, B.D., Taylor, K.E., Wigley, T.M.L., Penner, J.E., Jones, P.D., and Cubasch, U., 1995, Towards the detection and attribution of an anthropogenic effect on climate, *Climate Dynamics* 12:77-100.

Santer, B.D., Taylor, K.E., Wigley, T.M.L., Johns, T.C., Jones, P.D., Karoly, D.J., Mitchell, J.F.B., Oort, A.H., Penner, J.E., Ramaswamy, V., Schwarzkopf, M.D., Stouffer, R.J., and Tett, S.F.B., 1996a, A search for human influences on the thermal structure of the atmosphere, *Nature* 382:39-46.

Santer, B.D., Wigley, T.M.L., Barnett, T.P., and Anyamba, E., 1996b, Detection of climate change and attribution of causes, in: *Climate Change 1995: The Science of Climate Change*, J.T. Houghton, L.G. Meira Filho, B.A. Callander, N. Harris, A. Kattenberg and K. Maskell, Eds., Cambridge University Press, Cambridge:407-443.

Wigley, T.M.L., and Raper, S.C.B., 1995, Modelling low-frequency climate variability: the greenhouse effect and the interpretation of paleoclimatic data, in: *Natural Climate on Decade to Century Time Scales*, D.G. Martinson, K. Bryan, M. Ghil, M.M. Hall, T.R. Karl, E.S. Sarachik, S. Sorooshian, and L.D. Talley, eds., National Academy Press, Washington, D.C., 169-174.

Wigley, T.M.L., and Barnett, T.P., 1990, Detection of the greenhouse effect in the observations, in: *Climate Change: The IPCC Scientific Assessment*, J.T. Houghton, G.J. Jenkins and J.J. Ephraums, eds., Cambridge University Press, Cambridge:239-255.

Wigley, T.M.L., Ingram, M.J., and Farmer, G., eds, 1981, *Climate and History: Studies in Past Climates and their Impact on Man*, Cambridge University Press, Cambridge, 530pp.

Williams, L.D., and Wigley, T.M.L., 1983, A comparison of evidence for the late Holocene summer temperature variations in the Northern Hemisphere, *Quaternary Res.* 20:286-307.

PRE-INSTRUMENTAL WEATHER OBSERVATIONS IN POLAND IN THE 16th AND 17th CENTURIES

Anita Bokwa,[1] Danuta Limanówka,[2] and Joanna Wibig[3]

[1]Institute of Geography, Department of Climatology, Jagiellonian University, Poland
[2]Institute of Meteorology and Water Management, Branch Kraków, Poland
[3]Department of Meteorology and Climatology, University of Łódź, Poland

1. INTRODUCTION

A few remarkable episodes can be distinguished in the European climate of the last millennium. The first is the Medieval Warm Epoch (MWE). The culmination of the MWE occurred in the 12th and 13th centuries in the period 1150-1300 (Lamb, 1982). This warm period was followed by a colder one which lasted to the second half of the 19th century and is known as the so-called Little Ice Age (LIA). The climate during the LIA was not constantly cold. The recoveries to warmer conditions in the first half of the 15th century and around 1700-1750 are evident (Lamb, 1982). They were followed by reversions to colder conditions. Some authors stress the coincidence of the two greatest periods of cooling with the two prolonged minima of sunspot activity, the Spörer Minimum from 1400 to 1510 and the Maunder Minimum from 1645 to 1715 (Schuurmans,1981; Pfister, 1994b). An increase of volcanic activity is also mentioned as a reason for the LIA. Since the second part of the 19th century the temperature has been rising although some reversions to colder conditions are also present. This warming is a global feature (Jones *et al.*, 1986; Jones, 1994) and can be linked to an anthropogenic increase of CO_2 and other greenhouse gas concentrations as well as to solar and volcanic activity variations (Schönwiese, 1984).

The climatic fluctuations during the last millennium can be determined by means of instrumental observations, written historical documents and proxy data. Instrumental data are available only for the last part of this period. The first written weather report from the territory of Poland comes from the winter of 940/941 and informs us that this winter was relatively cold (Polaczkówna, 1925, after *Annalium Cracoviensium brevium Complementum*). The notes from the 10th century are extremely scarce but the frequency of information increases with time. These singular descriptions usually concern remarkable individual climatological or hydrological events, i.e. they characterise some extreme weather phenomena. Often they provide evidence of average weather conditions over a longer period such as a month or season. On that basis the weather journals are

History and Climate: Memories of the Future?
Edited by Jones *et al.*, Kluwer Academic/Plenum Publishers, 2001

9

distinguishable by the regularity of descriptions which were being kept for a longer time as marginal notes in astronomical and ecclesiastical calendars or diaries.

A few such weather diaries are known from the Polish territories. The first record consists of meteorological observations made in Cracow in the 16th century by professors from the Academy of Cracow. Marcin Biem of Olkusz systematically observed weather conditions during the period 1502-1540 and his notes form the most important data-base for those years. Weather notes of other professors complete this file.

Hellmann (1883) also described the 62-year record of observations made in Oleśnica near Wrocław from 1536. Oleśnica was then a German city and the original manuscript, if it exists, is probably in German archives. Johann Kepler made non-instrumental meteorological observations at Żagań in south-western Poland from 1628 to 1630. He described results in his papers edited by Frisch (1858-1870).

Friedrich Buethner kept systematic observations of weather phenomena in the period 1655-1699. He was a mathematician, astronomer and astrologer, professor and rector of a Gymnasium in Gdańsk. The correspondence he carried out with Hevelius indicates that his observations were used for the verification of some astronomical calculations. Hellmann (1883) mentioned that the manuscript of Buethner's observations titled *"Observationes meteorol. singulis diebus Calendarii annotatae ab a. 1655 ad a. 1699"* was in the Library of Gdańsk.

The first meteorological observations near Szczecin were made by Abraham Rockenbachs in Nowogard during the period 1561-1564 (Miętus, 1994). Weather phenomena were also noted in Szczecin by Wawrzyniec Eichstadt from 1 January 1635 to 31 December 1638. These observations were in the form of a calendar. Eichstadt was a physician, mathematician and astronomer. He probably made his observations for seven years because he wanted to prove a proposed theory about the recurrence of weather according to a cyclic influence of the seven planets. Such ideas were common at that time. The notes from the first year of observations were edited together with notes of landgrave Herman Fourth of Haga under the title *"Uranophilus Cyriandus"*. Other notes were published in his own work *"Ephemeris parva Uraniburgo, seali Astronomiae instauratae accomodata, atq.; in tali forma constructa, ut Astrophili cursum Planetarum,...Una cum Appendice Status aeris in Stetino Vet. Pomeranorum M. DC XXXVIII. Autore Laurentio Eistadio, Med, D & Physico Ordinario in Veteri Sedino Pomeranorum Statini, Anno, 1637"*. A copy of this work is in the British Museum Library in London (Klemm, 1976).

In the second half of the 17th century, Jan Antoni Chrapowicki, a Polish nobleman, kept a private diary in which he noted weather almost every day. His diary covers the period 1656-1685. At the beginning of this period Chrapowicki lived in north-eastern Poland and then moved east to the territory of present-day Byelorussia.

The main objective of this paper is to present the climatic conditions in selected periods of the 16th and 17th centuries in Poland on the basis of the observations of professors at the Academy of Cracow (16th century) and Jan Antoni Chrapowicki (17th century). Before this a short review is presented of weather conditions in Europe during these periods.

2.   WEATHER CONDITIONS IN EUROPE IN SELECTED PERIODS

2.1 The First Half of the 16th Century

The first half of the 16th century seems to have been relatively warm, although it belongs to the middle phase of LIA. Lamb (1982) has indicated that in England the temperature was higher and it was generally drier than in the previous century as well as in the latter half of the 16th century (Fig. 1-2). He suggested that the warmth was caused by

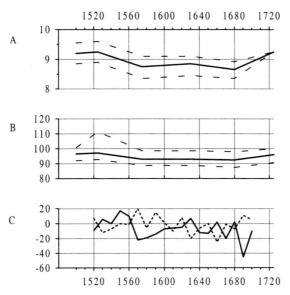

Figure 1. Estimated annual temperature and precipitation conditions at selected locations in Europe: (a) Probable fifty-year averages of annual temperature (solid line) with confidence limits (dashed lines) according to Lamb (1982), (b) probable fifty-year averages of annual precipitation in percentages of 20th century values (solid line) with confidence limits (dashed lines) according to Lamb (1982), (c) Ten-year temperature (solid line) and precipitation (dashed line) annual indices for Switzerland according to (Lamb, 1982 after Pfister). Indices are defined by a numbers of warm (wet) minus cold (dry) months indicated by documentary records.

relatively frequent anticyclones in the zone 45-50°N and westerly winds over northern Europe, whereas earlier and later the anticyclones dominated the zone north of 60°N. Pfister (1995) has shown that the weather conditions in Switzerland during the period 1530-1560 were favourable (Fig. 1-3). Though winters and springs were slightly colder than in the present century optimum, summers, autumns and years as a whole were warmer and it was generally drier. According to the findings of Borisenkov (1995) the first half of the 16th century was also relatively warm in central Russia (Fig. 2-3). The statistics of extreme precipitation seasons indicate that there was not any perceptible dryness or wetness during this period.

2.2 The Second Half of the 17th Century

During the end of the 17th century the culmination of the LIA seems to have occurred, simultaneous with the period of the Maunder Minimum of solar activity (1645-1715). According to Lamb (1982) the temperature in England was the lowest in the second millennium A.D. but just before the end of the century it rose considerably. The instrumental record of temperature in central England (Manley, 1974) indicates that the first part of Maunder Minimum period (i.e. 1660s) was relatively warm (Fig. 4), with springs slightly cooler than during the period 1901-1960, summers slightly warmer and other seasons oscillating near the mean value. According to Lamb it was also the driest period in the millennium in all seasons but summer. Summers in the 17th century were generally extremely wet. In Switzerland (Pfister, 1995) it was relatively warm in the 1650s and in the 1660s but the temperature then fall to the minimum value in the 1690s. Simultaneously, winters and springs were dry. The summer and autumn precipitation levels were close to the normal values with drier conditions in the first two decades (the 1650s and the 1660s) and wetter in the next three. The thirty-year seasonal mean temperature for central Russia shows that during the period 1650-1680 it was extremely cold during winter, colder than usual in spring and autumn but the summer temperature was slightly above normal. At the same time, the number of extremely dry seasons was considerably higher than those with wet conditions (Borisenkov, 1995).

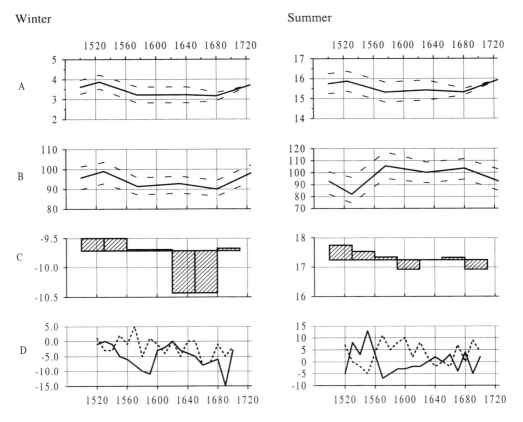

Figure 2. Estimated temperature and precipitation conditions at selected locations in Europe during winter and summer: (a) Probable fifty-year averages of seasonal temperature (solid line) with confidence limits (dashed lines) according to Lamb (1982), (b) probable fifty-year averages of seasonal precipitation in percentages of 20th century values (solid line) with confidence limits (dashed lines) according to Lamb (1982), (c) Thirty-year averages of seasonal temperature for central Russia compared with the long-term mean value (Borisenkov 1995 after Lyakhov), (d) Ten-year temperature (solid line) and precipitation (dashed line) seasonal indices for Switzerland according to (Lamb, 1982 after Pfister).

3.   16th CENTURY WEATHER DIARIES IN CRACOW

Cracow has one of the oldest universities in Europe, the Jagiellonian University (formerly the Academy of Cracow), which was founded in 1364. It is assumed that the systematic comparison of astrological prognostics and observed weather was developed at the school of mathematics and astrology at the Academy of Cracow which, at that time, was one of the leading centres in this field. It was the Academy of Cracow where Nicolaus Copernicus studied in the period 1491-1495. The practice of keeping weather diaries spread from Cracow to Ingolstadt (Germany) and from there to other universities. The keeping of

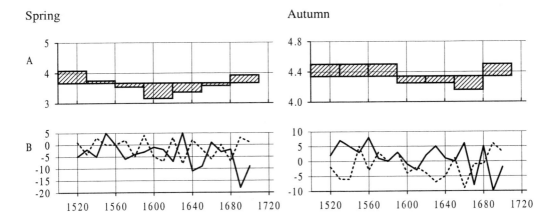

Figure 3. Estimated temperature and precipitation conditions at selected locations in Europe during spring and autumn: (a) Thirty-year averages of seasonal temperature for central Russia compared with the long-term mean value (Borisenkov, 1995 after Lyakhov), (b) Ten-year temperature (solid line) and precipitation (dashed line) seasonal indices for Switzerland according to (Lamb, 1982 after Pfister).

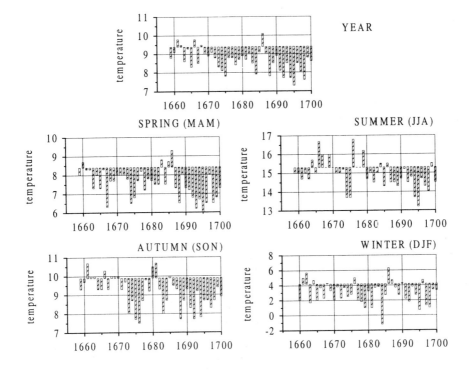

Figure 4. Seasonal and annual temperature in central England as departures of 1901-1960 mean values according to Manley (1974).

Figure 5. Marcin Biem (Museum of the Jagiellonian University in Cracow)

daily weather observations was promoted to improve and test the reliability of astro-meteorological predictions. The mass publication of astronomical calendars (ephemerides) from the late 15th century provided a suitable medium by which such observations could be recorded and some space was left empty at the end of every line. It could be used to record personal notes such as observations of the weather. Because the space was restricted weather diarists had to reduce their notes to a few keywords. The oldest known copy of ephemerides containing meteorological entries is deposited in the Jagiellonian University

Figure 6. The oldest panorama of Cracow (from the Historical Museum of Cracow, Firlet, 1998).

Library of Cracow. Under the date of April, 9, 1468, we read an anonymous note: "9... *in sabbato dominice Palmarum A.D. millesimo CCCCLXVIII frigore temperato*", i.e. on Palm Day 1468 it was moderately cold. This entry was ephemeral; it was not followed by a systematic effort. All the weather notes recorded by many of the professors in the Cracow ephemerides were written in Latin. They refer usually to single days, although sometimes there are descriptions of seasons or years. The entries comprise one or two descriptive terms (e.g. "*dies clara et calida*", sunny and hot day), sometimes preceded by a specification of the time of the day or followed by a term that expresses the magnitude or the intensity of the phenomenon (e.g. "*pluvia in nocte copiosa*", during the night much rain) (Pfister *et al.*, 1999).

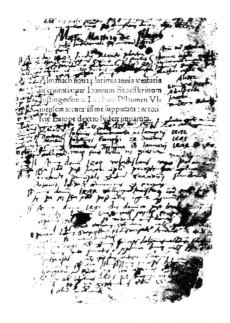

Figure 7. Front page of J. Stöffler and J. Pflaum *Almanach Nova* (1499-1531, Ulm. 1499) containing written weather records made by professors of the Cracow Academy (from the Jagiellonian University Library).

Marcin Biem (c. 1470-1540), astronomer and theologian, was the most outstanding among the weather observers in Cracow at that time (Fig.5-6). He was born in Olkusz near Cracow. In the winter term of 1486/87 he was admitted to the Artistic Faculty of the Academy of Cracow. On 6 March 1500 Biem was appointed to the position of Dean of the Collegium Maius. During the period 1507-1513 he stayed in Olkusz but, in 1513 he returned to Cracow. On 22 January 1517 he obtained a Doctorial Degree. From 1521 to 1528/29 he held the position of Dean of the Theological Faculty. In 1529 he was elected Rector of the Academy. He died on 9 November 1540. From the spring of 1487 Biem began to record occasional notes on daily weather conditions. From February, 1488, his observations became somewhat more systematic, but they were laid down in the common, academic astronomical calendars (Fig.7). Biem's notes – apart from the gaps - form one of the longest series of this kind known to exist for that period (*Polski Słownik Biograficzny*, 1960).

Biem's own, private observations are recorded in two ephemerides: *Almanach Nova 1499-1531* (Stöffler, Pflaum, 1499) and *Ephemerides 1534-1551* (Gauricus, 1533). Weather

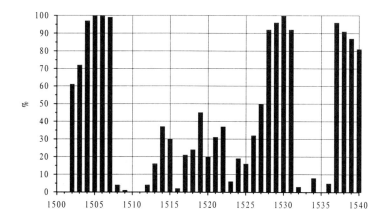

Figure 8. Percentages of daily weather observations by Marcin Biem of Olkusz. Data have been converted to the Gregorian style.

observations were regularly and carefully noted on a day to day basis for no less than 682 months and contain 5915 entries. There are three phases: 1502 to 1507, 1524 to 1531 and 1535 to 1540, during which the notes are almost continuous (Fig.8). In the intermediate periods they are intermittent (Pfister *et al.*, 1999). Biem observed not only the weather but also astronomical phenomena, for example, the solar eclipses on 1 October, 1502, 8 June, 1518 and 28 March, 1530 and comets in the years 1506, 1531 and 1539. Biem corresponded with Copernicus and probably made the observations to provide him with the material for comparisons. When making the astronomical observations Biem also noted especially carefully the weather conditions (Polski Słownik Biograficzny, 1960). Besides Biem there were a few other professors of the Academy whose weather observations might be considered significant, namely, Michał of Wiślica (1699 entries in the period 1527-1555); Bernard of Biskupie (1685 entries; 1510-1531); Jan Muscenius (1649 entries; 1555-1568); and Mikołaj Sokolnicki (934 entries; 1521-1531). Their notes complete the records of Biem. The whole data set was incorporated within the international data base of the research programme EURO-CLIMHIST, supervised by Pfister at the University of Bern, Switzerland (Pfister *et al.*, 1994a). The programme allowed us to calculate numbers of significant weather days, e.g. days with frost, days with precipitation, etc. for different places in Europe and then reconstruct the climatic situations and analyse their synoptic patterns. The results of the programme are presented in Glaser *et al.* (1999).

Another possible investigation of the data by Biem is the analysis of the additional marginal notes and descriptions he made (so-called "marginalia"), concerning information on flood events of the Vistula River. In the 16th century many more floods occurred in Cracow during the summer season than in spring. During the first half of the century there were 19 flood events and only 5 of them were noted for winter months; the floods occurred in: August, 1502, June, 1504, April-September, 1505, May-September, December, 1515, May, July, 1518, August, 1520, June-July, 1522, June, December, 1523, March, 1524, May, September, 1526, July, 1527, June-July, 1528, February, 1529, March, 1530, May, 1531, January, July, 1533, April-July, 1534, September, 1538, September, December, 1539 (Limanówka, 1999). Especially disastrous were the inundations of 1505, 1515, 1528 with overflowing rivers. The inundations of 1515 were classified by other authors as the greatest of the century, with 6 freshets and inundations of the Vistula River. The detailed weather

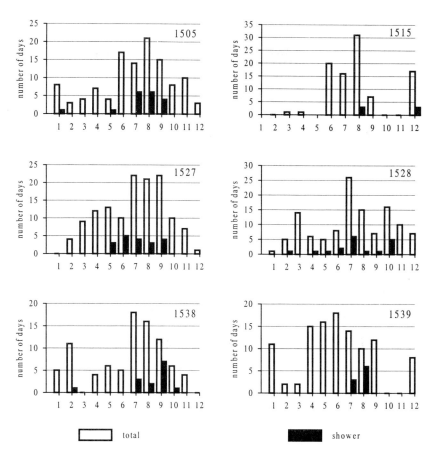

Figure 9. Frequencies of days with precipitation in Cracow in selected years, when some of the mentioned floods occurred.

descriptions of Marcin Biem indicate that the floods of 1528 were more dangerous and catastrophic: in the notes for February he wrote: "on the Christmas Eve there was a great flood", on the 14 of June "there was a great flood of the Vistula River which destroyed the fields and meadows", from 27 to 29 June "a great flood caused by constant rains, the water in the Vistula River rose suddenly and caused much damage around", in July "in this month there was the greatest flood which caused damage even to fields and meadows located far away from the river, houses were destroyed and a bridge was swept away by the river, many people were killed", on the 19 July "there was a great flood and the whole Bernardine monastery was flooded", in August "the pest occurred in Cracow". Figure 9 presents the number of days with precipitation in Cracow in selected years, when some of the above-mentioned floods occurred.

If we consider the three periods for which Biem's notes are most complete (1502-1507, 1527-1531 and 1535-1540) it is possible to investigate the frequency of days with chosen phenomena. Figures 10-12 present the courses of frequency of days with frost, thunderstorm and snowfall in Cracow for these periods. It turns out that the most frosty was the period 1527-1531 and the most frosty month was November 1539 (21 days with frost). The

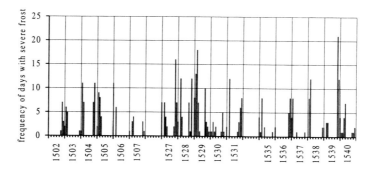

Figure 10. Monthly frequencies of days with severe frost in the periods 1502-1507, 1527-1531, 1535-1540.

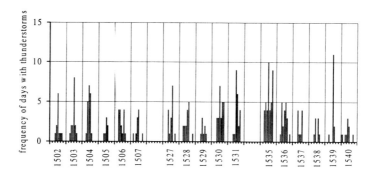

Figure 11. Monthly frequencies of days with thunderstorm in the periods 1502-1507, 1527-1531, 1535-1540.

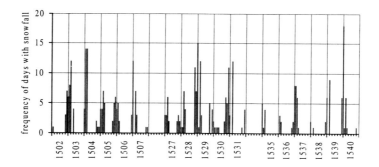

Figure 12. Monthly frequencies of days with snowfall in the periods 1502-1507, 1527-1531, 1535-1540.

highest frequency of days with thunderstorms was in the period 1535-1540 and for the month of July in all periods. The annual frequency of days with snowfall ranged from 5 in 1538 to 40 in 1529. The snowfall occurred from September (1530) to May (1503 and 1530). The periods 1502-1507 and 1527-1531 were more snowy than 1535-1540. For the days with precipitation (rainfall plus snowfall) we can compare with mean monthly number of

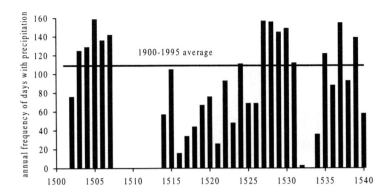

Figure 13. Annual frequency of days with precipitation in Cracow in 16th century compared with the average value for the period 1900-1995

days of precipitation and annual number of days with precipitation in Cracow for the period 1900-1995 from the Climatological Stations of the Department of Climatology, Jagiellonian University in Cracow (Fig. 13, 14). The data from the 16th and 20th century (Fig. 14) show no large difference in the mean annual numbers of days with precipitation, but it is clear that in the 16th century the frequency of days with precipitation was higher during summer and lower in winter. A serious environmental problem for Poland today is a growing water deficit in some regions, but this is not due to a decrease in the number of days with precipitation but rather the fact that precipitation totals change. Unfortunately, we shall probably never be able to compare them with the historical data that only record the number of rain days and not the total amount of rain.

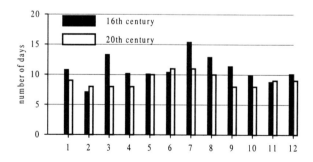

Figure 14. Mean monthly numbers of days with precipitation in Cracow in selected periods of the 16th century and in the 20th century.

4. DIARY OF JAN ANTONI CHRAPOWICKI

Jan Antoni Chrapowicki (Fig. 15) was born on 11 October 1612 as a third or fourth son of Krzysztof Chrapowicki. His elder brothers enlisted in the army or became priests, and he was destined for a political career. Jan Antoni Chrapowicki spent his school years at the

Figure 15. Jan Antoni Chrapowicki (Chrapowicki, J.A., 1978)

Nowodworski College in Cracow and then began his study at the Cracow Academy. He also continued studies in western Europe, firstly in the Netherlands, at Leyden and Amsterdam, where he studied fortification skills. Then he spent a few months at the Sorbonne in Paris. At the end of this period he went to Padua, where he studied for a few years. In 1638 he returned to Poland. So Jan Antoni Chrapowicki was quite well educated especially in comparison with the average representatives of noblemen in 17th century Poland.

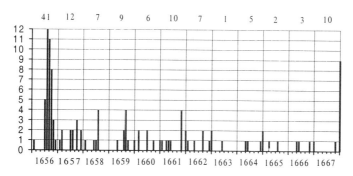

Figure 16. Monthly frequencies of days without weather descriptions in the period 1656-1667. Numbers above the graph indicate the annual values.

He began keeping his diary in 1656 and continued it up to his death on 3 November 1685. In this diary he noted many of his activities, letters he had received and written, people who visited him, journeys, expenses, family events and weather. He was really very enthusiastic about the weather and made notes about it every day and sometimes even several times per day. He also wrote about special astronomical phenomena he had observed, such as comets, eclipses, etc. He wrote seven volumes of this diary. Unfortunately many of them have been lost and only copies are available. These copies were made at the end of the 18th century by order of Chrapowicki's successors. The notes from 1656 to December 1667 are retained in the original manuscript and those in copies are very close to the original version. However, the copier of the latter part of the diary was apparently not very interested in weather and only gave laconic comments on it. Therefore, only the notes from the period 1656-1667 are taken into consideration here (Chrapowicki, 1978; 1988).

Chrapowicki was very meticulous in his weather descriptions. There were only 112 days without notes (i.e. 2.5%) during the twelve years mentioned above. Figure 16 shows the distribution of these days. There is a major problem with homogeneity of data, because, as a politician, member of parliament and a person accompanying the Polish king, Jan Kazimierz in his peace negotiations with Russia, Chrapowicki travelled a great deal. The parliamentary sessions took place in Warsaw and lasted for two or three weeks, and the travelling took up another two weeks. So each year he spent at least five weeks away from his home. The negotiations with Russia were conducted in 1664 and 1665 and he spent a few months in the eastern part of contemporary Poland attending upon Jan Kazimierz. So it is impossible to link his notes with a specific place. They describe rather the weather conditions in north-eastern Poland.

As mentioned above, Chrapowicki described weather usually once but occasionally several times per day. He noted either the day was clear or cloudy, so it is possible to obtain some information about the state of the sky. Precipitation events were also reported: rain, showers, thunderstorms, snowfall and hail. Temperature was characterised by statements such as frost, severe frost, slight or night frost, cold, cool, warm or hot. It is impossible to translate these descriptions into temperature degrees, especially during summer, but they do give information about the lengths of seasons with frost, and temperature features of the seasons. There are also notes about snow cover, ice and thaws. On that basis some ideas concerning circulation can be obtained.

For example, long snow cover periods with frost suggests blocking situations, whilst a high frequency of days described as warm with thaws can be linked to zonal circulation patterns. During the warm season the mixture of precipitation and temperature data also gives a picture of circulation. For instance, warm days with showers indicate anticyclonic weather whereas continuous rain indicates cyclonic conditions. There are also some direct descriptions of wind that contain information about its strength, temperature features and sometimes also direction.

An example of the record of weather variations in January 1660 is quoted below.

1.1.   Cloudy day, thaw in afternoon
2.1.   Cloudy day, quite a heavy snowfall in the morning
3.1.   Humid day with cold yet westerly wind
4.1.   Snowing in the morning, severe snowstorm, then fine weather, it began to freeze in the evening
5.1.   Humid day with cold wind, quite heavy snow in the evening
6.1.   Severe frost in the night and morning, then wind and fine rain in the afternoon, yet very cold
7.1.   Humid day with westerly wind
8.1.   Cloudy in the morning, then severe frost and fine weather with strong wind
9.1.   Frosty in the morning, then thawed a great deal
10.1. It was freezing in the morning, then thawed, quite heavy snowfall in the night, snowstorm with the wind
11.1. Frosty day with westerly wind and snowfall
12.1. Frosty day, snowing for a while

13.1. Cloudy day with severe frost and wind, in the evening severe gale with snowstorm
       and dry snow
14.1. During the night and in the morning a severe snowstorm with frost, then cleared up
       for some time and overcast once more
15.1. Severe frost and northerly wind all day
16.1. Frosty day, eased in the morning
17.1. Cloudy and cold day
18.1. Hard frost and clouds
19.1. Fine weather, a frosty day with a wind
20.1. Frosty day, strong wind, calm evening
21.1. A little more settled, calm day
22.1. Cloudy morning with strong wind, then fine weather, calm yet cold
23.1. Cloudy day and thawed, fall of snow
24.1. Cloudy day with wind and severe frost, a heavy snowstorm
25.1. Fine weather, frosty
26.1. Fine weather, strong frost, calm
27.1. Dark and cloud all day, very frosty in the morning, then a fall of rime
28.1. Very cold day with wind, cloudy
29.1. Frosty and cloudy day, snow at times
30.1. Fine weather day, wind
31.1. Cold and cloudy day, strong and cold wind, decreasing in the evening, light fall of
       snow

4.1 Temperature Conditions in Poland

To characterise the weather in winter the frequency of days with frost ($t_{max} < 0°C$) was
calculated. Figure 17 shows the course of frosty days during the analysed period. Frosty
days occurred from October to April but generally they were more frequent from November
to March. The earliest frost appeared on 19 October 1665 and the latest day with frost in
spring was 24 April 1665. Similar values are characteristic for north-eastern Poland today.
The frequency of days with frost varied from 46 in 1661/1662 to 109 in 1666/1667 around
the mean value of 75.3. In comparison with present-day 63 frost days in Suwałki (*Climatic
Atlas of Poland*, 1971), it is evident that the winters in the period 1656-1667 were more

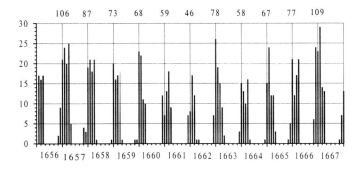

Figure 17. Monthly frequencies of days with frost in the period 1656-1667. Numbers above the graph indicate
the annual values.

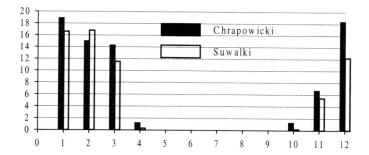

Figure 18. Annual course of frosty day frequencies in comparison with 1951-1960 mean values for Suwałki in north-eastern Poland.

severe than today. The frequencies of frosty days per month in the Chrapowicki data are also higher than in Suwałki in all months except February (Fig. 18). This corresponds with the results of Borisenkov (1995) that the mean winter temperature was lower than at present and also with the Pfister assumption that during the Maunder Minimum of solar activity there was a cooling in Europe. But it should be noted that an extremely long period of ice on the Baltic near Riga in the winter of 1658/1659 (Lamb, 1982) occurred during quite an average winter in north-eastern Poland.

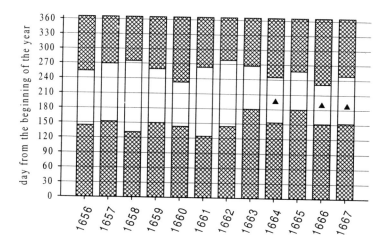

Figure 19. The dates of the first and last occurrences of slight frost. Triangles indicate the cases of slight frost during summer.

The analysis of days with slight frost indicates that springs were colder than at present (Fig. 19). The comparison of dates of the first and last occurrence of days with slight frost with the record from Suwałki in north-eastern Poland for the period 1963-1980 (*Climatic Atlas of Poland*, 1990) shows that in the 17th century there were many such days in spring as well as in autumn and they occurred later in spring and earlier in autumn, so the length of the frost-free period was evidently shorter, particularly during the last four years of the analysed period. This agrees with the findings of Borisenkov (1995), who reported that in

European Russia frost was recorded in the midsummer of 1666 (Chrapowicki noted frost on the 9th of July and from the 20th to 24th of August) and in late spring and early autumn of 1667 (Chrapowicki noted the dates 1 June, 10 July, and six other days in September).

In Switzerland the temperature was below normal in winter and close to normal in the other seasons (Pfister, 1995). The ten-year frequencies of extreme temperature seasons in northern Bohemia (Brázdil, 1994) show that in the 1650s and the 1660s extremely warm winters were more frequent than extremely cold ones.

It is more difficult to compare summer temperatures, because it is impossible to translate the statements "cold", "cool", "warm" or "hot" into degrees of temperature. But some comparisons can be made. Borisenkov (1995) reported a cold and rainy summer in 1660. In north-eastern Poland summer 1660 was very variable. In June whilst six days were reported as hot, nine days were described as cold or extremely cold. Cold and rainy days dominated also in July (11 cold and 21 rainy days) and in the first decade of August. On 8 August Chrapowicki recorded a flood, because of continuous rains on previous days. The rest of August was warm or hot. Borisenkov also reported a poor harvest in 1661 because of cold weather in early spring and snowfall throughout the season. Chrapowicki noted frost in late spring and a poor harvest because of heavy rains in August.

Hot summers in England occurred in 1665 and 1666. In July 1665 Chrapowicki reported six hot days, ten warm, two sultry and only one as cool. In July 1666 there were ten hot days, eleven warm days and only two cool. But one day with slight frost and ice was also noted. The other summer months of these seasons were more variable.

4.2 Precipitation in Poland

Figure 20 presents the monthly frequencies of days with precipitation in the period 1656-1667: graph (a) shows days with precipitation, (b) days with snowfall and (c) days with thunderstorms. Snowfall occurred from October to May, but it was most frequent from November to March. The days with thunderstorms were probably underestimated. Their frequency was twice as low as at present, though there was no evidence of such a dramatic change of frequency of anticyclonic weather situations. There were no perceptible variations in the total number of days with precipitation; Chrapowicki reported an annual average of 157.4 such days, which is very close to 158.0 days at Suwałki in the period, 1951-1960. But it is evident from Figure 21 that the annual course of these frequencies changes dramatically. During 1656-1667 the number of days with precipitation was higher in summer and lower in winter. The number of snowfall days was also lower. The result was probably a smaller annual total of precipitation because the average sum of precipitation on a summer day is much higher than on a winter day in Poland. This corresponds with the variation of rainfall over England and Wales estimated by Lamb (1982). He stated that winters were drier and summers were wetter (Fig.1-2) compared with 20th century values. Simultaneously, the annual totals were higher. In England and Wales autumn and winter precipitation dominates so that the winter tendencies determine the annual totals, whereas in Poland the summer precipitation dominates so the annual tendency is determined by variations during the warm season. Winter and spring precipitation was also lower in that period in Switzerland (Pfister, 1995), but conditions there in summer and autumn were very close to present day ones. According to Brázdil (1994) anomalies in the Czech lands hardly occurred in spring and autumn, but it was a little wetter there in winter and drier in summer, that is, just the opposite to north-eastern Poland.

In the first two years of Chrapowicki's record the frequency of days with precipitation was evidently smaller than in the latter part of the series. This corresponds with Borisenkov results that in western Russia droughts were common in the period 1640-1659 (11 dry years). In June 1659 the prolonged drought destroyed crops.

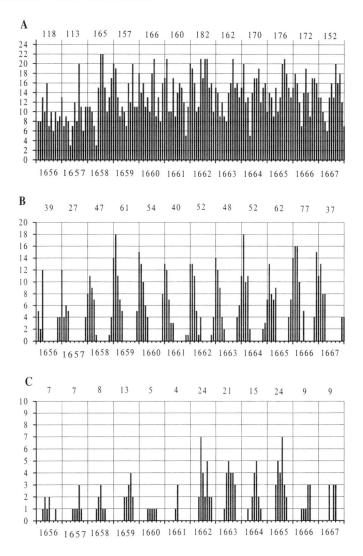

Figure 20. Monthly frequencies of days with (a) precipitation, (b) snowfall, (c) thunderstorms in the period, 1656-1667. Numbers above the graph indicate the annual values.

Heavy rains were reported a few times by Chrapowicki. Torrential floods occurred in June 1658, August 1660, June 1664 and September 1667. Prolonged rains destroyed rye crops in August 1661. The same summer was reported by Borisenkov (1994) as cold and rainy with a poor harvest. Prolonged rains prevented harvesting also in September 1662 and heavy rains damaged crops in June 1667. Twice in the analysed period snowmelt floods occurred: in March 1659 and in January 1661.

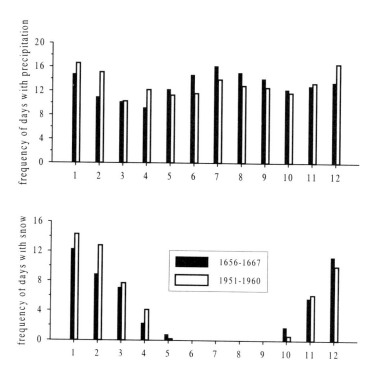

Figure 21. Annual course of days with (a) precipitation, (b) snowfall in comparison with 1951-1960 mean values for Suwałki in north-eastern Poland.

5.   SUMMARY

This short study shows that historical data although non-instrumental and scarce allow climatic conditions in the past to be determined. The more complete comparison of weather in the first half of the 16th century in Cracow and other European locations was made by Pfister at al. (1999), but even from this short review it is possible to infer, on the basis of the change in the annual course of days with precipitation, that it was a little wetter than today.

It is more difficult to compare the data contained in the diary of Chrapowicki with present-day values because they were made at different places, so they characterise the general conditions in north-eastern Poland rather than those at an individual location. But it is clear that winters were colder and drier, springs were later and summers were wetter. The weather anomalies in north-eastern Poland were more similar to those of western Russia than to those of central or western Europe.

There are many other written reports from the Polish territory and it seems that a careful analysis of these data will allow us to reconstruct the variability of climate in the region during the last millennium.

Acknowledgements

John Kington, Climatic Research Unit, is thanked for reading the manuscript and correcting the text.

REFERENCES

Borisenkov, E.P., 1994, Climatic and other natural extremes in the European territory of Russia in the late Maunder Minimum (1675-1715), in: *Climatic Trends and Anomalies in Europe 1675-1715*, ed. Burkhard Frenzel, Special Issue: ESF Project, European Paleoclimate and Man 8, Gustav Fischer Verlag.

Borisenkov, E.P., 1995, Documentary evidence from the U.S.S.R. in: *Climate since A.D. 1500*, R.S. Bradley and P.D. Jones, ed., Routledge, London and New York.

Brázdil, R., 1994, Climatic Fluctuations in the Czech Lands during the Last Millennium, *GeoJournal* 32:199-205.

Chrapowicki, J.A., 1978, *Diariusz, cz. I: lata 1656-1664*, Instytut Wydawniczy PAX, Warszawa.

Chrapowicki, J.A., 1988, *Diariusz, cz. II: lata 1665-1669*, Instytut Wydawniczy PAX, Warszawa.

Climatic Atlas of Poland, 1971, Institute of Meteorology and Water Management, ed. Wyd. Kom. i Łącz., Warszawa.

Climatic Atlas of Damaging for Agriculture Elements and Phenomena in Poland, 1990, Institute of Soil Science and Plant Cultivation in Puławy and Agriculture University of Szczecin.

Firlet, E.M., 1998, Najstarsza panorama Krakowa, Muzeum Historyczne miasta Krakowa.

Frisch, C., ed., 1858-1870, Joannis Kepleri astronomi opera omnia Volumines I et VII, Frankfurt am Main.

Gauricus, L., 1533, *Ephemerides 1534-1551*, Venezia (in the archives of the Jagiellonian University Library, Cracow, catalogue number BJ Cim.5521).

Glaser, R., Brázdil, R., Pfister, C., Dobrovolny, P., Barriendos Vallve, M., Bokwa, A., Camuffo, D., Kotyza, O., Limanówka, D., Rácz, L., and Rodrigo, F.S., 1999, Seasonal temperature and precipitation fluctuations in selected parts of Europe during the sixteenth century, *Climatic Change* 43:169-200.

Hellmann, G.,1883, *Repertorium der Deutschen Meteorologie*. Leipzig

Jones, P.D., 1994, Recent warming in global temperature series, *Geoph. Res. Letters* 21:1149-1152.

Jones, P.D., Raper, S.C.B., Bradley, R.S., Diaz, H.F., Kelly, P.M., and Wigley, T.M.L., 1986, Northern Hemisphere surface air temperature variations: 1851-1984, *Jour. of Clim. and Appl. Met.* 25:161-179.

Klemm, F., 1976, Die Entwicklung der meteorologischen Beobachtungen in Nord- und Mitteldeutschland bis 1700, *Annalen der Meteorologie (neue Folge)* 10:37-43.

Lamb, H.H., 1982, *Climate, History and the Modern World*, Methuen, London and New York.

Limanówka, D., 1999, Powodzie w Krakowie w świetle zapisków pogodowych z XVI wieku, ed., Interdyscyplinarność w badaniach dorzecza, Kraków-Dobczyce, 21-23 May 1999 (in Polish).

Manley, G., 1974, Central England temperatures: monthly means 1659 to 1973, *Quart. J. Roy. Met. Soc.* 100:389-405.

Miętus, M., Wielbińska, D., and Owczarek, M., 1994, History of meteorological observations at some stations at the Polish coast, *Reports of Institute of Meteorology and Water Management* 42/2:41-64 (in Polish).

Pfister, C., 1995, Monthly temperature and precipitation in central Europe from 1525-1979: quantifying documentary evidence on weather and its effects. in: *Climate since A.D. 1500*, R.S. Bradley and P.D. Jones, eds., Routledge, London and New York.

Pfister, C., Kington, J., Kleinlogel, G., Schüle, H., and Siffert, E., 1994a, High resolution spatio-temporal reconstructions of past climate from direct meteorological observations and proxy data, in: *Climatic Trends and Anomalies in Europe 1675-1715*, ed. Burkhard Frenzel, Special Issue: ESF Project, European Paleoclimate and Man 8, Gustav Fischer Verlag.

Pfister, C., Yan, Z., and Schüle, H., 1994b, Climatic variations in western Europe and China, AD 1645-1715: a preliminary continental-scale comparison of documentary evidence, *The Holocene* 4:206-211.

Pfister, C., Brázdil, R., Glaser, R., Bokwa, A., Holawe, F., Limanówka, D., Kotyza, O., Munzar, J., Rácz, L., Strömmer, E., and Schwarz-Zanetti, G., Daily Weather Observations in Sixteenth-Century Europe, *Climatic Change* 43:111-150.

Polaczkówna, M., 1925, Climatic variations in Poland during the Middle Ages (in Polish and English), *Travaux Geographiques publiés sous la direction de E. Romer*, Książnica-Atlas, Lwów-Warszawa.

Polski Słownik Biograficzny, PWN, Kraków, 1935-1971.

Schönwiese, Ch.-D., 1984, Northern Hemisphere temperature statistics and forcing. Part B: 1579-1980 AD, *Arch. Met. Geoph. Biocl., Ser., B* 35:155-178.

Schüle, H., Data handling and process structure in the EURO-CLIMHIST Data Bank, in: *Climatic Trends and Anomalies in Europe 1675-1715*, ed. Burkhard Frenzel, Special Issue: ESF Project, European Paleoclimate and Man 8, Gustav Fischer Verlag.

Schuurmans, C.J.E., 1981, Climate of the last 1000 years. in: *Climatic Variations and Variability: Facts and Theories*, A. Berger, ed., D. Reidel Publ. Comp., Dordrecht - Boston - London.

Stöffler, J., and Pflaum, J., 1499, *Almanach Nova 1499-1531*, Ulm (in the archives of the Jagiellonian University Library, Cracow, catalogue number BJ Inc. 2697).

THE LATE MAUNDER MINIMUM (1675-1715) - CLIMAX OF THE 'LITTLE ICE AGE' IN EUROPE

Jürg Luterbacher

Institute of Geography
Climatology and Meteorology
University of Bern
Hallerstr. 12, CH-3012 Bern, Switzerland
E-mail: juerg@giub.unibe.ch

1. INTRODUCTION

The Maunder Minimum (MM; 1645-1715) delineates the coldest phase of the so-called 'Little Ice Age' (LIA; variously assessed as ~AD 1300 to 1900; Holzhauser, 1997; Pfister et al., 1998; Wanner et al., 2000) with marked climatic variability over wide parts of Europe. This period coincides with an enhanced volcanic (Briffa et al., 1998) and a reduced solar activity, as well as a low number of sunspots (Spörer, 1887; Maunder, 1922; Eddy, 1976) and an increase in atmospheric ^{14}C (Stuiver and Braziunas, 1993). Estimates of the total radiative solar output changes for the MM are in the order of 0.2 to 0.4% relative to present levels (Hoyt and Schatten, 1993; Nesmes-Ribes et al., 1993; Lean et al., 1995; Reid, 1997; Lean and Rind, 1998; 1999). Solar activity during the MM was near its lowest levels within the past 8000 years (Lean and Rind, 1999) and the UV (200-300 nm) irradiance was also lower (Lean et al., 1995). This in turn could have had an influence on stratospheric chemistry (ozone) and dynamics (absorption). The reduced solar activity might have resulted in a decrease of the stratospheric ozone content as proposed by Wuebbles et al. (1998). In agreement with this proposal, levels of δ^{14}C and δ^{10}Be cosmogenic isotopes in tree-rings and ice cores were found to be elevated (Eddy, 1976; Stuiver and Braziunas, 1993). However, several authors (i.e. Landsberg, 1980; Cullen, 1980; Xu et al., 2000) believe that a decline in solar activity may not have been the cause of the climate severity during the LMM, since evidence from numerous local histories, especially from east Asia, suggest that sunspots were not rare in the seventeenth century. Mann et al. (1998) have found lower annual Northern Hemisphere (NH) mean surface temperatures with decreases between 0.2° and 0.4°C compared to the reference period of 1902-1980. Jones et al. (1998) report of a decrease of the NH April to September temperatures in the order of around 0.3°-0.6°C compared to the reference period of 1961-1990. However, there is no evidence of an advance of European alpine glaciers. The Great Aletsch and the Lower Grindelwald Glaciers show a series of years with a nearly stable or even a negative mass balance (Wanner et al., 2000).

History and Climate: Memories of the Future?
Edited by Jones et al., Kluwer Academic/Plenum Publishers, 2001

Within the MM, the Late Maunder Minimum (LMM; 1675-1715) appears to be of particular interest from a climatological point of view since it is one of the few very cold periods in recent centuries that persisted over decades. The annual Central England Temperature (CET; Manley, 1974; Parker *et al.*, 1992) was 0.75°C warmer in 1961-1990 than during the LMM. In addition, a huge amount and a broad spectrum of high-resolution multi-proxy data (monthly, seasonal) and instrumental data for Europe are available for this period (Wanner *et al.*, 1995; Luterbacher *et al.*, 1999; Pfister, 1999; Luterbacher *et al.*, 2000a,b). During the LMM, European winters and springs were characterised by a much higher frequency of severe climatic conditions in comparison with the twentieth century. Wide parts of Europe experienced lower winter mean temperatures of the order of 0.7°C to 1.5°C (Pfister, 1999). According to Pfister (1994; 1999), the wintertime cooling began in the mid-1670s over the British Isles and the westernmost part of the continent. It then spread to central Europe and 10 years later to eastern Europe (Hungary). The climax of the LIA was reached in the 1690s. The year 1695 was the fifth coldest over the last millennium with a decrease of 1°C over the whole Northern Hemisphere (Jones *et al.*, 1998). Mann *et al.* (1998) found the coldest decade to be 1696-1705 over the NH with 0.35° lower NH annual temperatures relative to the 1902-1980 mean. The four years 1692, 1694, 1695, 1698 were among the coldest in the CET series with a drop of mean temperature of the order of 1.5°C to 2°C compared to the period 1961-1990. In central and eastern Europe especially, the winters were extremely dry, with recurrent long-lasting and strong advection of continental air from northeastern Europe (Borisenkov, 1994; Brázdil *et al.*, 1994; Pfister, 1994; 1999; Wanner *et al.*, 1995; Kington, 1995; 1997; 1999; Luterbacher *et al.*, 2000a,b). Koslowski and Glaser (1999) found an increase of winter ice severity in the Western Baltic from 1655 till 1710 with a maximum during the 1690s which in turn contributed to the coldness over central and eastern Europe. Xoplaki *et al.* (2000) report slightly wetter and cooler conditions over the southern Balkans and eastern Mediterranean, but with higher variability than in the recent century. They argue that the persistency and intensity of the extensive high-pressure systems over Scandinavia and northeastern Europe explain the more severe winters over the southeastern Balkans. In combination with low pressure over the Mediterranean, these conditions could account for a higher frequency and quantity of snow in Greece and in the eastern Mediterranean. Similar climatic conditions were prevalent in the western Mediterranean (Barriendos, 1997). According to Alcoforado *et al.* (2000) cold relapses occurred more frequently in the LMM in Portugal than during the last decades of the twentieth century.

In northwestern, central and eastern Europe, the springs from 1687 to 1717 were the most severe of the last 500 years (Pfister, 1999) with often a decrease of 2°C in mean temperature compared to the reference period 1901-1960. No spring season exceeded the 1901-1960 average temperature either in England or in Switzerland between 1695 and 1703 (Pfister, 1994; Luterbacher *et al.*, 2000b). In general, LMM summers, generally, were wetter and cooler than the twentieth century in western, central and eastern Europe. Apart from the exception of the 1680s, the LMM summer NH temperatures were reduced by 0.1°C to 0.4°C compared with the recent century's conditions (Briffa *et al.*, 1998). Tree-ring evidence from northern Fennoscandia also indicates distinctly cooler summers, with some of them among the coldest over the past 6 centuries (Briffa *et al.*, 1998). Jones *et al.* (1998) report that the summers from 1691 to 1700 were the coldest of the present millennium over the NH (0.7°C below from the 1961-1990 mean). The autumns were in general colder in many parts of Europe (Pfister, 1999) but showed the least deviations from the twentieth-century climatic conditions. Warming began in 1699, first over the British Isles (Kington, 1999), and by 1704 it had spread to western central Europe and reached eastern Europe by the end of the LMM (Pfister, 1999). However, the temperature increase is not visible in the Portuguese data (Alcoforado *et al.*, 2000).

The long-lasting climatic anomalies over decades with persistent low temperatures over wide parts of Europe must have had their origin in a marked change of the atmospheric circulation. One of the chief objectives of the EU-funded ADVICE (Annual to Decadal Variability In Climate in Europe) research project (Jones *et al.*, 1999; Luterbacher *et al.*, 2000a,b) was to improve understanding of the climate variability during this period. Due to the virtually complete absence of human influence, the LMM is an important period in which to study the different processes leading to natural climate variability and extended cooling. Particular focus was placed on investigating the LMM atmospheric circulation dynamics. Luterbacher *et al.* (2000a) used statistical methods to reconstruct North Atlantic European monthly mean Sea Level Pressure (SLP) fields for the LMM. The reconstructed SLP fields consist of gridded (5°x5° latitude by longitude grid containing 96 grid points) values from 25°W to 30°E and 35°N to 70°N. Depending on the season, 18 (autumn), 20 (spring and summer) to 21 (winter) partly measured (i.e. Central England Temperature (CET; Manley, 1974)), monthly mean wind directions from Øresund (Denmark) obtained from ship's logs (Frydendahl *et al.*, 1992), Paris station pressure and temperature (Legrand and Le Goff, 1992)) as well as indexed time series (relative temperature and precipitation values reconstructed from documentary proxy evidence, see below) from several European sites and the Western Baltic Sea-Ice severity index (only for winter; Koslowski and Glaser, 1995; 1999) were available for the LMM (Luterbacher *et al.*, 1999; 2000a). These time series formed the predictors for the SLP reconstructions.

The authors followed the common approach for reconstructing past climate namely to establish the quantitative relationship between the proxy and measured data and the twentieth century instrumental data which are derived for a 'calibration' period, in our case from 1901-1960. The reconstructions are based on a canonical correlation analysis (CCA), using a limited number of canonical pairs sharing 95% of variance (without a EOF truncation) with the standardised station data as predictors and the SLP pressure fields as predictands. CCA identifies pairs of patterns and associated new, optimally correlated variables, from two multivariate data sets. The new variables are obtained by projecting the original data onto the respective patterns, i.e. CCA accounts for correlations in the predictors and/or predictands. The established relationships can be used in a linear regression model to simultaneously estimate all predictands from the predictors (Barnett and Preisendorfer, 1987; von Storch and Zwiers, 1999). Thus, CCA optimally summarises the linear relationship between the large-scale patterns of the atmospheric circulation (gridded SLP) over the eastern North Atlantic and Europe and the proxy and measured data. The statistical relationships were then verified with independent data (1961-1990) in order to assess the model performance outside of the calibration period. In all seasons, the overall model performance (assessed by calculating the fraction of variance shared across the whole 96 grid-point region) and the spatial model performance (expressed as the shared variance over each grid point separately) did not reveal any systematic deficiencies and the regression equations developed for the majority of the grid points contained good predictive skill (Luterbacher *et al.*, 2000a). The overall performance expressed by the average shared variance over all 96 grid points is best in winter, with 61%, followed by spring and autumn (43%) and summer (33%), respectively. However, considerably poorer reconstruction skill at individual grid points was obvious for regions with little or no data, especially during summer time. In general, the predicted SLP fields for the period 1961-1990 are in good agreement (in terms of the location of the pressure centres and to a lesser extent the prevailing pressure gradient) with the observed SLP distribution (Luterbacher *et al.*, 2000a). Based on a backward elimination technique it was shown that Paris station pressure is the most important predictor sharing a large amount of SLP variability. Assuming stationarity of the statistical and climatological relationships, the calibrated statistical model for each season can then be applied to the data available during the LMM in order to estimate monthly mean gridded pressure fields (Luterbacher *et al.*, 2000a). These LMM SLP reconstructions are synoptically compared with the climate normal period of 1961-1990 which itself is the independent period in the twentieth century. This offers a unique chance to obtain a better insight into the feedback mechanisms between ocean, sea ice and the atmosphere.

The paper is structured as follows: Section two shortly discusses how temperature and precipitation indices are obtained and validated and introduces the objective statistical techniques to compare the climatic conditions and synoptic situations during the LMM with those of the period 1961-1990. In section three the seasonal distribution of selected indexed temperature and precipitation time series from southern Portugal, Switzerland and Hungary is investigated in order to test the reliability of documentary proxy data in reconstructing past climate conditions. Additionally, the highlights of the unique climate character of the LMM within the last 500 years are shown, using Switzerland as an example. The synoptic part presents the seasonal differences in atmospheric circulation between the LMM and the 1961-1990 period based on average seasonal mean SLP difference charts and objective SLP pattern classification. A short section shows the SLP distribution of the coldest seasons within the LMM according to the CET series (winter 1684, spring 1695, summer 1695 and autumn 1676) and their departures with respect to the period 1961-1990. The discussion relates to possible significant processes in the lower atmosphere of the eastern Atlantic-European area and to the influence of forcing factors. The conclusions are presented in the last section.

2. DATA AND METHODS

The reconstructed monthly SLP patterns of the LMM have been taken from Luterbacher *et al.* (2000a). As mentioned in the introduction, the period 1961-1990 is considered as the independent period since it was not used for the LMM reconstructions. The observed gridded monthly SLP data (NCAR, 1997) from this period were used for the synoptic comparison with the LMM.

Except for CET (Manley, 1974; Parker *et al.* 1992), Paris station pressure (Legrand and Le Goff, 1992) and mean wind direction from Øresund (Frydendahl *et al.*, 1992), only indexed temperature and precipitation indices from different European areas which have been prepared within ADVICE are available. It needs to be briefly outlined how temperature and precipitation indices are obtained and validated. Documentary data have a couple of strengths: high (daily to seasonal) time resolution; coverage of all months of the year; separation of the effects of temperature and precipitation and a high sensitivity to anomalies and natural hazards. However, many types of descriptive documentary proxy data are discontinuous and heterogeneous (Pfister *et al.*, 1994) and, as a consequence, they must be combined using relatively simple and robust mathematical techniques (Banzon *et al.*, 1992).

The main emphasis when assessing unequivocal climatic tendencies in terms of indices involves comparing and cross-checking different types of concurrent, high-resolution natural and documentary proxy data. The derivation of climatic indices is a first step towards transforming different kinds of documentary evidence into time series (Glaser *et al.*, 1999). Indices cannot be directly calibrated with instrumental series, because there is hardly any overlapping period between quality documentary observations and early instrumental measurements. As soon as meteorological instruments became available, the keenest and most reliable weather diarists switched to instrumental observations. Thus, another approach of validating index series was developed from the example of the sixteenth century (Glaser *et al.*, 1999). As a first step, indices for different countries (Switzerland, Germany, Hungary, Czech Republic) were derived from instrumental data. For the second step, these 'instrumental' indices were compared to those obtained from documentary proxy data for the same countries. It turned out that 'pre-instrumental indices' were correlated at almost the same level as the 'instrumental' ones (Glaser *et al.*, 1999). The indices are integer values between +3 and −3, where values of +3 and −3 are applied to anomalies that are 'extreme' by twentieth century standards, i.e. greater than two standard deviations from the mean of the reference period 1901-1960. Values of +2 and –2 are given to months with less extreme deviations that are 1.41-2.0 standard deviations from the mean of the period 1901-1960. Values of +1 and –1 are applied to cases which deviate between 0.7 and 1.4 standard deviations from the mean of the reference period, or are poorly documented. Values of 0 are given to months that correspond to the average climate of the reference period or to missing data (Pfister *et al.*, 1994).

The three indexed seasonal temperature and precipitation time series for Lisbon (Portugal; Alcoforado *et al.*, 2000), Zurich (Switzerland; Pfister, 1999) and Budapest (Hungary; Rácz, 1999) were selected to investigate the seasonal statistical distribution of the temperature and precipitation indices both for the LMM and the period 1961-1990 (see section 3.1). Adequate scaling of the seven point Pfister index (-3 to +3) is based on discrete intervals of 0.7 standard deviation units, based on the 1901-1960 period (Pfister 1994; see above). By means of a one-tailed X^2-test, it was decided whether the observed frequencies of the Pfister indices for temperature and precipitation during the LMM and the period 1961-1990 differ from reference values for 1901-1960. Based on these results, some highlights of the unique climatic character of the LMM within the last 500 years are shown for Switzerland.

Several approaches are known for investigating differences in atmospheric circulation between two periods. In this paper, two were applied.

- The differences in mean SLP during the LMM and the period 1961-1990 were analyzed with average seasonal mean SLP difference charts (average seasonal means of LMM minus average seasonal means for 1961-1990). The regions with statistically significant SLP differences according to Student's t-test are also given. In addition, the SLP distribution of the coldest season during the LMM according to the CET series are given together with the SLP departures from the 1961-1990 mean.

- The other approach involved a classification of atmospheric surface-pressure patterns. The underlying idea here was to identify major geographical modes of variation of atmospheric pressure over the North Atlantic-European area (Huth, 1996). A correlation-based classification is used here and is briefly described. For more details see Yarnal (1993), Huth (1996), Yarnal and Frakes (1997), Blair (1998) and Kaufmann *et al.* (1999). One of the primary goals of classification is to maximise between-class distance and minimise within-class variance. The first step of the classification procedure is the standardisation of the SLP grids for both periods. Standardisation removes the seasonal influence on absolute pressure and also eliminates the seasonal impact on pressure-pattern intensity. Thus, only the generalised map pattern (grid) remains, and pressure configurations of various months are comparable (Yarnal, 1993). The similarity between the standardised SLP grids is expressed in terms of a pattern correlation. A correlation coefficient is calculated for every possible grid-pair for the two periods separately. The correlation coefficient and a sums-of-squares value (Kirchhofer technique) are equivalent in classifying patterns. Based on this correlation, the algorithm identifies groups of months with similar pressure patterns. In this study, a correlation coefficient threshold of 0.7 was used. This means that in each group the correlation coefficient between each of the members is higher than 0.7. It is the best compromise between the percentage of the classified months and the number of groups (Yarnal, 1993). In addition, in order to ensure pattern similarity in all areas of the grids, sub-grid scores were calculated for columns and rows respectively, with data normalised using the appropriate row and column averages and standard deviations (Blair, 1998). We used a positive correlation coefficient for rows and columns. Then, based on these values, the computer algorithm identifies groups of months with similar air pressure patterns. The month with the largest number of correlations exceeding the given pre-set threshold is denoted as a key pattern ('key month'). All the patterns that correlate better than the given overall correlation threshold of 0.7 and the row and column correlations with this key month are removed from the analysis. The next key month is found using the same procedure among the remaining patterns, again with the largest number of correlations exceeding the given correlation threshold. After the selection of all key patterns, all months are reclassified by looking at their highest correlation with one of the selected key patterns. Months which do not have a correlation with any of the key patterns higher than the threshold value, are not classified. All the months from the two periods belonging to the respective classes are then averaged and plotted together with their standard deviations and their relative frequencies.

3.   RESULTS

3.1 A comparison of station temperature and precipitation for the LMM and the 1961-1990 period

This section focuses on some special features of the characteristic differences between the moisture and thermal regimes during the LMM and the period 1961-1990 for some selected European areas. The seasonal observed and the expected reference period frequencies (1901-1960) of the Pfister temperature and precipitation Indices (+3 to -3) for the two periods for each station (Lisbon, Portugal; Zurich, Switzerland; Budapest, Hungary) are presented in Figures 1 to 3. All seasonal frequency distributions which differ at a statistically significant (99% level) from the normal distribution are marked with an asterisk. The lack of an asterisk means that the observed frequency is not significantly different from the 1901-1960 reference distribution, which itself is the basis for the Pfister indices. If the respective seasonal curve for each station is close to the zero-line, then the observed relative frequency is more likely to be normally distributed. The more pronounced the deviations from the zero-line, the more distinct is the seasonal anomaly of the respective Pfister indices.

The comparison with the seasonal curves between the two periods show marked differences. A typical feature for Lisbon, or generally for southern Portugal, is the distinctly higher frequency of the Pfister index zero during the LMM. This distribution suggests, in general, that over southern Portugal 'normal' conditions were experienced more in the LMM than during the twentieth century. However, cooler and drier summers and autumns (Pfister Indices -1) were more frequent than the 1901-1960 reference distribution. Slightly colder winters (Pfister Indices –1 to –3) during the LMM go along with a slightly enhanced frequency for very dry conditions (Pfister Index of –2). By contrast, warm conditions (Pfister Indices of +1 to +3) in Lisbon, were generally less frequent during the LMM than expected by the 1901-1960 reference distribution.

Zurich shows a higher frequency of very dry winters (Pfister Index –2) during the LMM, a situation not supported by Budapest. On the other hand, a tendency to wetter conditions during the LMM is supported by both Zurich (summer, Pfister Index +2) and Budapest (in all seasons, Pfister Index +1). More precipitation in summer implies a southward shift of the storm tracks during this season, and thus lower pressure in the mid-latitudes. This is only partly supported by the seasonal pressures measured at Paris (Legrand and Le Goff, 1992) (not shown). This fact can be explained only by a higher intensity of precipitation events probably connected with frontal thunderstorms, processes which cannot be resolved in a seasonal analysis. Moreover, the above normal frequency of higher pressure in Paris during summer promotes northwesterly flow over central Europe (especially when Paris is situated on the edge at the eastern border of the surface high) with wetter than normal conditions.

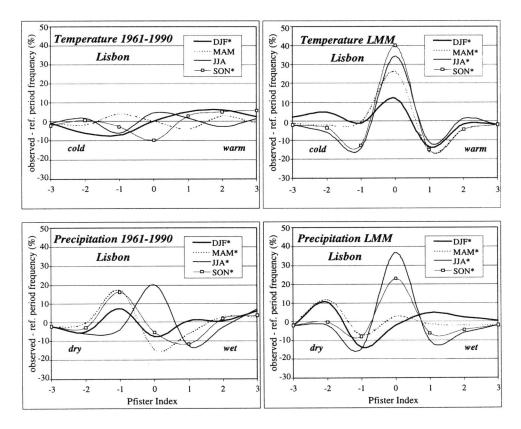

Figure 1. Observed and reference period frequencies (1901-1960) of the Pfister Indices (+3 to -3) (temperature and precipitation) for the LMM (right of the respective figures) and 1961-1990 (left of the respective figures) for Lisbon (Portugal) for each season separately. All seasonal frequency distributions which differ statistically significantly (99% level) from the normal distribution are marked with an asterisk (*). The lack of an asterisk means that the observed frequency is not significantly different from the 1901-1960 reference distribution, which is itself the basis for the Pfister indices. If the respective seasonal curve for each station is close to the zero-line, then the chance that the observed relative frequency is normally distributed is more probable. The more pronounced the deviations from the zero-line, the more distinct is the seasonal anomaly of the respective Pfister indices.

Zurich and Budapest temperatures show a similar development in the observed values during both periods. Both stations have in common a significantly different behaviour of winter and spring conditions during the LMM compared to the 1901-1960 reference distribution. During these seasons more negative Pfister temperature indices are prevalent, an indication of more frequent cold air outbreaks from the Baltic and Scandinavian regions connected with dry air advection.

Some doubts appear considering the seasonal positive bias (except for spring) for positive Pfister temperature indices (+1) for Budapest. Zurich does not support this. One cannot definitively say, if this is real or an artefact of the construction. Budapest is closer to continental air masses than Zurich. Persistent inversions at the southern border of an anticyclone over central Europe or at the southwestern border of a northeast European anticyclone promote negative temperature values. This must be the result of cold air advection and radiative cooling although mass divergence and strong subsidence also promotes significant warming during winter months. It is therefore questionable as to why

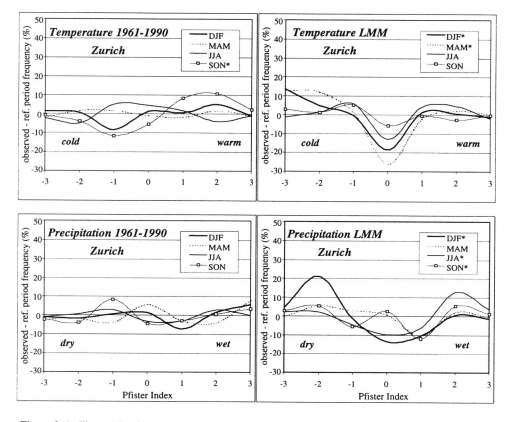

Figure 2. As Figure 1 but for Zurich (Switzerland)

Budapest shows positive temperature deviations and also wetter conditions during all seasons. Recurrent cold air outbreaks from the northeast cannot have led to such situations.

Similar analyses for the area of Greece revealed slightly wetter and colder conditions during the LMM but with higher variability in temperature and precipitation than during the recent century (Xoplaki *et al.,* 2000).

3.2 The Late Maunder Minimum in Switzerland in the context of the last 500 years

The results for Switzerland are interpreted in the context of the last five hundred years. Figure 4 displays the number of extreme (+3 and -3) monthly temperature anomalies per decade for the winter half year (October to March). In a second step, the monthly anomalies were classified according to precipitation. For the instrumental sub-period (from 1864 onwards) the classification draws on an aggregate of four precipitation series from the main parts of the Swiss Plateau (Neuchâtel, Bern, Zurich and St. Gallen) and is based on departures from the 1901-1960 average of this aggregate series. Months with a precipitation index of ≥ +2 are designated 'wet', those with an index of ≤ -2 are designated 'dry'. Months within these thresholds are not designated. This combined temperature-precipitation classification thus yields six types of anomalies: Warm-wet, warm-dry, warm; cold-wet, cold-dry and cold (Pfister, 1999; Heino *et al.*, 1999).

The two decades 1676 to 1695 witnessed by far the largest number of cold extremes in the winter half year within the last five centuries (Figure 4). Whereas the number of cold

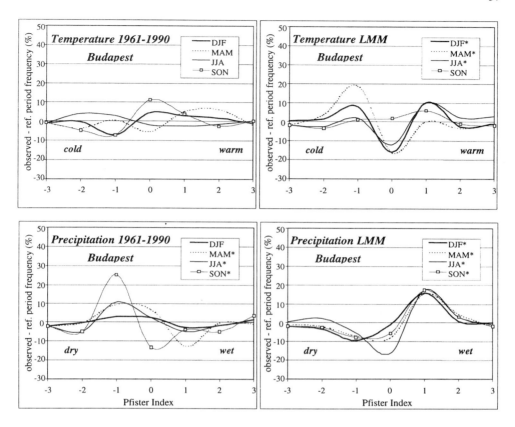

Figure 3. As Figure 1 but for Budapest (Hungary)

and dry months connected to more or less permanent flow of continental air to central Europe was rather small in the first decade, it became substantial in the decade 1686-1695. Considering the absolute number of extremes in the winter half-year, the decade 1676 to 1685 was the most variable within the last five hundred years. Specifically, this refers to the fact that the number of extremely warm months was far above average whereas the number of severe months was at a maximum. This variability is also supported by the reconstructed SLP monthly fields during this decade (Luterbacher *et al.*, 2000a) and from a wealth of anomalous circulation patterns (zonal, blocking) over western, central and eastern Europe.

3.3   The coldest seasons within the LMM and their SLP distribution

Figure 4 clearly showed that the 20 years from 1676 to 1695 revealed by far the largest number of cold extremes in the winter half-year in Switzerland within the last five centuries. In this subsection the SLP distribution of the coldest seasons within the LMM according to the CET series (Manley, 1974), all appearing within the two mentioned decades, are presented together with the SLP departures from the 1961-1990 long-term mean (Figure 5) in order to gain some idea about the atmospheric circulation during these seasons in comparison with the recent period.

The coldest winter in the CET series which goes back to 1659 (Manley, 1974) was in 1684. The mean winter temperature was -1.2°C, thus 5.3°C below the long-term mean from 1961-1990. It was also a very cold winter over central and eastern Europe (in Switzerland a

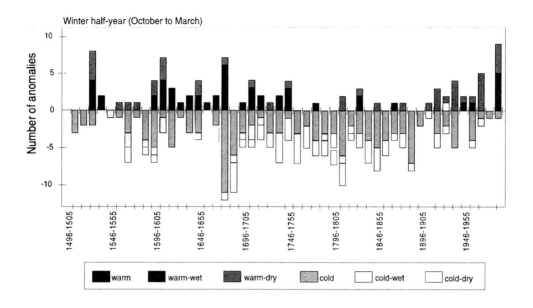

Figure 4. Number of combined temperature and precipitation extreme (+3 and -3) monthly anomalies in the winter half year (October to March) per decade 1496-1995 for the northern part of Switzerland (Pfister, 1999)

Pfister Index of –2,67; Pfister, 1999; in Hungary a Pfister Index of –2; Rácz, 1999). In addition, severe ice conditions were experienced in the western Baltic area (Koslowski and Glaser, 1999) and North of Iceland (Ogilvie, 1996). The corresponding sea-level pressure distribution shows a strong high over northeastern Europe and an extended low pressure system reaching from the eastern North Atlantic to the central Mediterranean. Persistent cold air advection from the east was responsible for the severity over wide parts of Europe. The anomaly chart (difference between winter SLP 1684 and winter SLP 1961-1990; Figure 5 bottom) indicates positive SLP anomalies up to 16 hPa over Finland and negative departures in the order of 10 hPa over Spain, thus strong anomalous easterly flow over Europe. In central and eastern Europe it was dry, but in the Mediterranean area above normal precipitation was experienced due to an enhanced cyclonic situation (Xoplaki et al., 2000).

The lowest LMM mean spring CET with 6.0°C was in 1695 (Manley, 1974). The departure from the 1961-1990 mean is 2.3°C. It was the second coldest in the whole CET series. In contrast, in Switzerland, the springs of 1701 and 1714 (Pfister Index of -2,3) were the coldest during the LMM. The Azores high was little developed and extended low pressure covered large parts of Europe. Cold air was advected from the polar region towards the British Isles and then diverted towards central Europe. Also the southern part of Portugal experienced a very wet spring (Alcoforado et al., 2000). The anomaly SLP pattern indicates anomalous high pressure over Iceland and negative values over most of the European continent.

The mean summer CET in 1695 was 13.2°C (Manley, 1974), representing a lowering of 2.2°C in comparison with the long-term mean from 1961-1990. It was the second coldest in the CET series. The SLP distribution for summer 1695 (Fig. 5) shows high pressure over

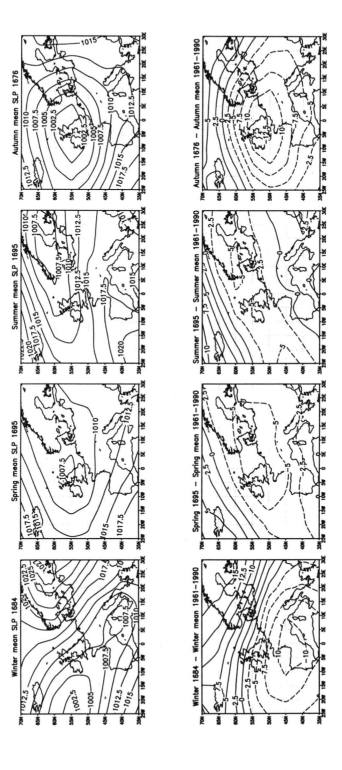

Figure 5. Sea level pressure (SLP) distribution of the coldest seasons within the LMM according to the CET series (top) and difference patterns (bottom). Continuous lines mark positive deviations, and dashed lines negative deviations. The contours are drawn at 2.5 hPa intervals.

Iceland and a low pressure system extending from the North Atlantic over central Europe towards western Russia - a pattern similar to the preceding spring. Recurrent low pressure activity connected with westerly flow brought cooler and wetter conditions over northern, central and eastern Europe. No indication of a cooling has been found in the Mediterranean area. The SLP difference pattern looks similar to that for spring 1695, with a negative anomaly stretching from the Azores over the British Isles towards Scandinavia, whereas positive SLP anomalies prevailed northwest of Scotland.

The coldest LMM autumn in the whole CET series were 1676, 1740 and 1786 with a mean temperature of 7.5°C (Manley, 1974), thus a departure of minus 2.8°C in relation to the 1961-1990 mean. For Switzerland, November 1676 was the coldest autumn month for the last 500 years with a reduction of 6 to 7°C compared to the 1961-1990 period (Pfister, 1999). On the basis of the reconstructed SLP pattern for autumn 1676, the coldness cannot be explained. Lower autumn temperatures over the British Isles and also over Europe are mainly the result of cold air advection from the North (polar air) and also from the east during November. A look at the reconstructed monthly pressure patterns (not shown) shows no indication of northerly flow at the eastern flank of a high pressure system located over the eastern Atlantic, which might explain such conditions. The SLP maps and the SLP difference maps support moist air advection associated with marginally lower temperatures towards the British Isles and central Europe. Similarly reconstructed 500 hPa height and 500-1000 hPa thickness fields revealed that the coldness can be explained by cold air advection in the mid-troposphere connected with a strong northwesterly component over the British Isles and central Europe (not shown).

3.4 Seasonal sea level pressure (SLP) differences between the LMM and the period 1961-1990

The seasonal mean SLP fields for the two periods are presented in Figure 6 (top and middle). The differences of the mean SLP patterns are analysed by average seasonal mean SLP difference charts (LMM minus 1961-1990). These patterns indicate persistence or changes in the mean circulation pattern which implies redistribution of mass in the troposphere (depletion, accumulation). These difference maps together with the regions with statistically significant SLP differences (95% significance levels according to Student's t test) are presented in Figure 6 (bottom). They show pronounced differences for all seasons and are very similar to the difference patterns presented in Figure 5, though the amplitude of the pressure anomalies is smaller. The most striking feature of the LMM is significantly positive pressure anomalies in winter, up to 5 hPa over northern Europe (northern Scandinavia and the Baltic area) connected with more blocking conditions during the LMM and anomalous cold air advection towards central and eastern Europe. Significant negative winter pressure anomalies are prevalent over central Europe and the western and central Mediterranean area. The positive winter anomalies move to a belt between Iceland and northern Scandinavia from spring to autumn connected with a weakening of the Icelandic low. Together with significantly negative SLP anomalies in the mid-latitudes and the Mediterranean area (spring to autumn) this led to a reduced pressure difference between the Azores and Iceland, and thus to negative NAOIs which was also shown by Luterbacher *et al.* (1999). Thus, there is evidence for a southward shift of the storm tracks. Tests of the model performance (Luterbacher *et al.*, 2000a) indicate that the shape of these anomalies is very unlikely to be an artefact of the reconstruction technique.

3.5 Comparison of correlation based map pattern classification of surface pressure

Simplifying the atmospheric circulation of the two periods (1675-1715 and 1961-1990) into a finite number of representative map patterns enables a comparison between the two periods. The correlation-based classification method (section 2) was applied to both periods

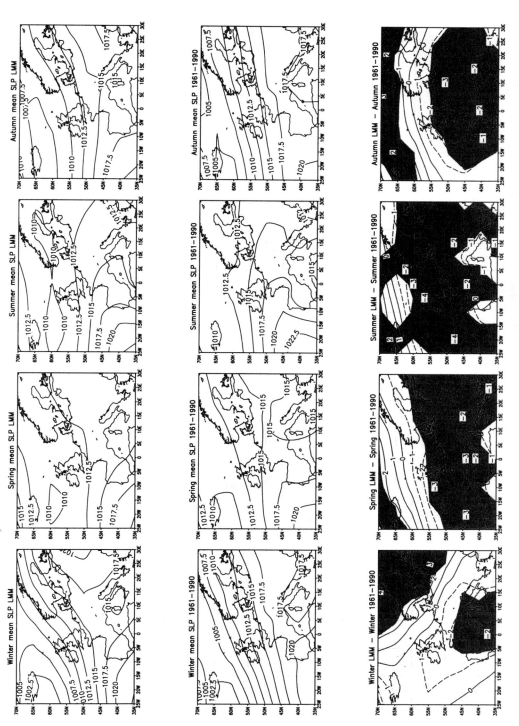

Figure 6. Average seasonal mean SLP distribution for the LMM (top) and the period 1961-1990 (middle) and their difference patterns (LMM minus analysis 1961-1990) (bottom). The contours of the top and middle charts are drawn at 2.5 hPa intervals, the contours of the bottom charts at 1.0 hPa, respectively. Continuous lines mark positive deviations, dashed lines negative deviations. The areas within grey contours (bottom charts) exceed the 95% confidence level of non-zero correlation using a 2-tailed t-test.

Figure 7. Average mean sea level pressure (SLP) fields (solid lines), their standard deviation (dashed lines) and relative frequencies (%) (in brackets) of classes 1 to 10 for the LMM based on a Correlation-based classification.

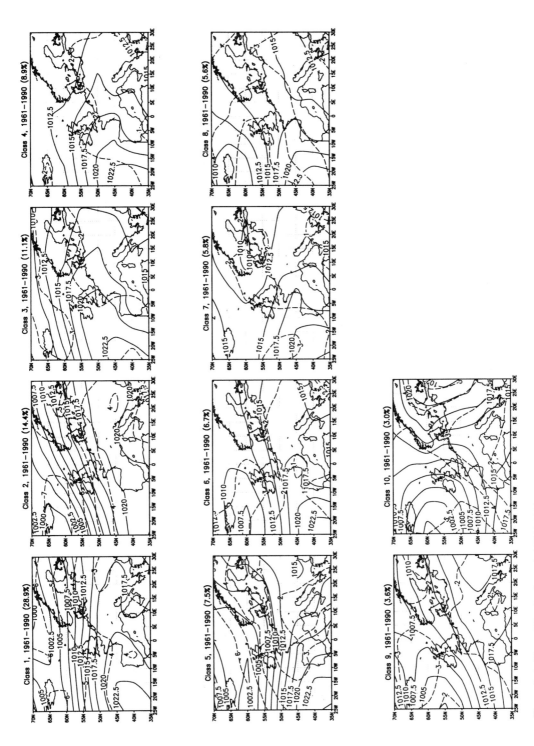

Figure 8. As Figure 7 but for the period 1961-1990.

separately. All the monthly pressure fields were standardised in order to remove the seasonal influence on absolute pressure as well as the seasonal impact on pressure-pattern intensity. A correlation threshold of 0.7 is used in this study. This means that in each class the correlation coefficient between each of the class members and the key pattern is higher than 0.7. In addition, the correlations of the columns and rows of the pressure grid respectively are positive in order to ensure pattern similarity for all areas of the grids. In this study we have chosen 10 classes as appropriate for both periods. 15 months in the LMM and 5 months in the period 1961-1990 could not be classified according to these subjective settings. The months within each class have been averaged and plotted together with their standard deviations and relative frequency. The charts are given in Figures 7 and 8.

The four most frequent classes in the 1961-1990 period have a cumulative frequency of around 63%, those during the LMM 58%. The most common classes during the LMM and the second and ninth important classes during the recent three decades, can be found in winter and autumn. The SLP distribution for months belonging to these classes are connected with high pressure stretching from the Azores over central Europe to western Russia. The storm tracks are situated over the British Isles and Scandinavia. The second most important class during the LMM is similar to the dominant pattern 1961-1990. It is a typical winter pattern showing strong flow with a west-southwest component in the central and northern part of Europe, based on the dominant pressure systems over the eastern North Atlantic: the Icelandic Low and the Azores High (Schmutz and Wanner, 1998). However, the pressure gradient over northern Europe is lower during the LMM compared to 1961-1990 and the relative frequency is around 10% less. This class is quite uniform with little overall variability. Classes 3 and 4 during the period 1961-1990 can be considered as typical summer patterns. The Azores high is strong and has moved northward and reaches far into the continent. The Icelandic low is clearly weaker. The cumulative frequency of these classes is around 20%. The corresponding classes in the LMM are numbers 5 and 8, occurring 13% of the time. More important for the LMM late spring and summer circulation is class 7 (7,5%) which is not among the 10 most important in the 1961-1990 period. It is the most frequent class for LMM spring months (Luterbacher *et al.*, 2000b) with a relative frequency of 35% (not shown). Only a few months from 1961-1990 are classified in this class. It shows a remarkable southeastward shift of the North Atlantic low pressure system. Its centre is in the vicinity of the British Isles, whereas in the last 30 years the centre remained over the North Atlantic. The resulting cyclonic or cyclonic easterly flow over the British Isles leads to unsettled, rather cold, windy and rainy weather. Nevertheless, the standard deviation for this circulation type is very high and implies high variability and thus different positions of the low. Classes 3 and 6 are typical winter patterns for the LMM with a cumulative frequency of 21%. Their appearance in early spring is a signal of the continental cold air anticyclone over eastern Europe and western Russia (Schmutz and Wanner, 1998; Luterbacher *et al.*, 2000b) connected with stronger blocking over central and eastern Europe and thus drier and cold conditions. These classes underline the severe climate conditions during the LMM. Other studies (Schmutz and Wanner, 1998) have demonstrated that these pressure patterns influenced station temperatures in central Europe despite their relatively low frequency.

4. DISCUSSION

The results section clearly showed some significant climatic and atmospheric circulation differences during the LMM in comparison with the period 1961-1990, which is the independent reference period in the twentieth-century as our LMM reconstructions are based on the 1901-1960 period. The period 1961-1990 displays two clearly different circulation regimes over Europe in winter: a more meridional period with mostly negative NAO indices between 1961 and the early 1970s and a zonal phase with positive NAO indices since about 1974 (Hurrell, 1995; Hurrell and van Loon, 1997). However, the three decades from 1961-1990 do not all show the different seasonal climatic conditions.

The differences between the observed and the expected reference period frequencies (1901-1960) of the Pfister Indices (+3 to -3) for the three stations (Lisbon, Zurich, Budapest) revealed differences between the LMM and the 1961-1990 period: during the LMM, the observed frequencies differ much more from the expected values with respect to the period 1901-1960 and therefore more seasons show statistical significance (cf. Figures 1 to 3). The overall balance during the LMM shows more monthly deviations of one Pfister score for Zurich and Budapest than expected by the reference period distribution and more negative than positive values. This leads to the significantly colder climate for which the LMM is well known. This is in contrast to the situation during the period 1961-1990, which shows more maritime behaviour and becomes especially evident during wintertime when a strongly maritime regime supports higher monthly mean temperature. The colder the month the less the influence of maritime air masses. The total amount of precipitable water in wintertime clouds is smaller in continental air masses than in oceanic. Apart from some exceptional events (warm air advection from the Adriatic Sea around a Mediterranean low / cut off low dynamics) cold air advection from the east is mostly accompanied by subsidence in the middle troposphere. Subsidence effectively damps precipitation mechanisms and increases atmospheric dryness due to widening the difference between the absolute temperature and the dew point. Cold air advection from the northeast becomes more important as do local and regional radiative forcing. This explains why alpine glaciers did not advance during the LMM (Wanner *et al.*, 1995).

A typical feature for Lisbon and also for other Mediterranean stations (Xoplaki *et al.*, 2000), is the fact that the frequency of mean LMM seasonal precipitation and temperature between -0.7 and +0.7 standard deviations (Pfister index 0) is distinctly above the expected value from the 1901-1960 reference period. Due to the lack of weather information for some LMM months for southern Portugal, as well as Greece (Xoplaki *et al.*, 2000) and Spain (Barriendos, 1997), the precipitation and temperature index value of zero was given. Existing gaps could mean that the climatic conditions were normal compared to the reference period or the lack of differentiation between average conditions and missing data can lead to an impression of much greater incidence of average conditions than actually occurred (Xoplaki *et al.*, 2000). Thus, the results for southern Portugal should be interpreted cautiously.

The comparison of SLP patterns between the LMM and the period 1961-1990 based upon correlation-based classification revealed some significant changes in atmospheric circulation. However, despite the positive attributes concerning the correlation-based classification, the considerable disadvantages of this method should not be overlooked. Several studies have shown that the results of correlation-based classification are subjective and that the number of synoptic types and months vary with the number of investigator-controlled variables (Yarnal, 1993; Huth, 1996; Yarnal and Frakes, 1997; Blair, 1998). Furthermore, relevant patterns are usually missed and 'important climate months' are not always classified (such as the cold January 1963 with a reversal in SLP), and the knowledge of the investigator cannot be included in the classification process (Yarnal and Frakes, 1997). In addition, the number and size of classes identified by this method depends on the threshold of the correlation coefficient and minimum group size. As a result, one cannot be sure whether the synoptic patterns identified by this technique represent groups that are generated by a meaningful meteorological phenomenon or whether they emerge randomly (Kaufmann *et al.*, 1999). In addition, it is worth mentioning that quite similar patterns may result as different types. Thus, synoptic classification is a method of obtaining a rough overview of the differences in mean atmospheric circulation. This method should be used together with other procedures to help establish any significant changes in circulation, since different classification approaches can lead to different results and thus are not stable and do not permit clear statements about differences between two periods (Huth, 1996).

In this context, a joint correlation-based classification (rather than conducting the classification separately for each period it was based on) has been performed. The joint approach was used for characterising systematic differences, since the major changes in atmospheric circulation were expected in the mean state. The classes for each period were in agreement with the correlation-based classification for each period, though the relative frequencies for each class revealed slightly different results (not shown). In another study by Luterbacher *et al.* (2000b) a joint classification based on a principal component analysis (PCA) (S-mode which concerns SLP varying over space) with varimax rotation and a subsequent 'k-means' clustering of the principal components scores was performed. A comparison between the different approaches revealed that slightly different frequencies of the classes and also the dominant pressure patterns appeared (Luterbacher *et al.*, 2000b). The main findings support the results obtained here with an increased frequency of months connected with persistent cold air advection from the east (at the southwestern edge of the strong cold high) towards central Europe. Such winter pressure patterns seem to be much more frequent during the LMM (especially the first few winters, 1684, the 1690s and 1709). During the whole twentieth century no winter was similarly outstanding as regards the climatic influence on the whole of Europe as were 1684, 1695 and 1709 (Luterbacher *et al.*, 2000b). Wanner *et al.* (1995) pointed out that the winter period from 1675-1704 was characterised by strong SLP reversals, with high sea level pressure over northern or northwestern Europe and below normal pressure over southwestern Europe (negative phase of the North Atlantic Oscillation) connected with large outbreaks of cold continental air from the northeast. This is in agreement with Kington (1995; 1997; 1999), who found an increased frequency of blocking and meridional patterns over the British Isles during this period. However we could not find any single month with a pressure reversal. In this sense their results should be interpreted as anomalous pressure distributions as presented in Figure 6 rather than absolute pressure patterns.

The SLP difference charts for spring (Figure 6) indicated significantly higher SLP during the LMM over Iceland and lower pressure south of 55°N. The dominant class for the LMM (class 7) clearly shows the southeastward shift of the Icelandic low towards the British Isles leading to unsettled, cold and rainy (1693) or snowy weather over central Europe. No spring season exceeded the 1901-1960 average temperature both in England and in Switzerland between 1695 and 1703 (Pfister 1994; Luterbacher *et al.*, 2000b). The SLP distribution of the months belonging to this class indicate reduced zonality (negative NAOI) derived by the simple pressure difference between the Azores and Iceland.

The main differences during summertime can be shown by the average SLP distribution and of classes 3 and 4 for the period 1961-1990 and the classes 4, 5 and 8 during the LMM. The LMM can be characterized by a much higher relative frequency of situations with the Azores high extending to 50° to 55°N, allowing low pressure systems from the Atlantic to reach the continent further south than normal. These situations are connected with more frequent spells of wet, cool and windy, but not necessarily cold westerly conditions over western and central Europe. The worst summer of this type was 1692, leading to famine in the subsequent year in many parts of western and central Europe (e.g. Lachiver, 1991). In upland parishes of Scotland the harvests (largely oats) failed in seven out of eight years between 1693 and 1700 which caused a mass emigration (Lamb, 1982). Around 75% of the LMM summer months correspond to classes 4,5 and 8 which can be attributed to a southward shift of the mid-latitude storm tracks. In Switzerland, no single summer was warm and dry from 1695 to 1705 (Pfister, 1994; 1999). In contrast, during the period 1961-1990 the subtropical high in classes 4 and 5, where the majority of the summer months from 1961-1990 are classified, ridges more towards the northeast, leading to stable, warm and rather dry conditions (Luterbacher *et al.*, 2000b).

During LMM autumns significantly higher pressure was experienced north of around 60°N, whereas lower SLP was prevalent over much of continental Europe and the Mediterranean (Figure 6). This suggests, that the change from summer to winter circulation was somewhat advanced (Luterbacher *et al.*, 2000b). The below normal pressure during the LMM can be explained by a higher frequency of the southward expansion of the Icelandic low (Luterbacher *et al.*, 2000b). Causality in the differences in atmospheric circulation is difficult to derive since the eastern North Atlantic and Europe represent only a sector of the Northern Hemisphere.

Which mechanisms drive the different climates of the LMM? Three possible mechanisms might be responsible for the anomalous climate: Are they only connected to external changes in the solar activity (leading to changes in solar radiation output)? Do large volcanic eruptions act additionally? What 'role' did the internal variability of the atmosphere-ocean system in the North Atlantic play? However, it has to be pointed out that the differences between the two periods might also be partly due to the modern anthropogenic influence which led to some specific patterns within the period 1961-1990.

4.1 Volcanic influence on atmospheric circulation

Volcanic eruptions are able to force rapid climate change and to change climate on longer time scales (Zielinski, 2000; Robock, 2000). Volcanic eruptions that inject large quantities of sulphur-rich gases into the stratosphere have the capability of cooling global climate by 0.2-0.3°C for several years (the complete impact last for up to 4 years) after the eruption (Zielinski, 2000). Recent studies (Robock and Free, 1995; Free and Robock, 1999; D'Arrigo *et al.*, 1999; Crowley and Kim, 1999; Zielinski, 2000; Robock, 2000; Hyde and Crowley, 2000) suggest a significant role of volcanism in decadal-scale climate fluctuations, with eruptions between 1400 and 1850 explaining 18%-25% of the decadal-scale variance in Northern Hemisphere mean temperature. The mid-latitudes of the NH seem to be particularly sensitive, experiencing winter warming but summer cooling for at least the following year (Robock and Free, 1995; Robock and Mao, 1995; Robock, 2000). The aerosol cloud after a large explosive volcanic eruption produces heating in the stratosphere but cooling at the surface (Robock, 2000). For tropical eruptions this warming is larger in the tropics than in the high latitudes. The heating of the tropical lower stratosphere by absorption of terrestrial and solar radiation expands that layer and results in an enhanced equator-to-pole temperature gradient (Robock, 2000). In the NH winter, a stronger polar vortex traps the wave energy of the tropospheric circulation, and the stationary wave pattern, known as the NAO (stronger westerlies) dominates the winter circulation, producing the winter warming of NH continents (Robock and Mao, 1995; Free and Robock, 1999; Robock, 2000).

In the mid-latitude summers, the cooling is mainly due to having more sunlight to block. The direct radiative forcing of the surface, with a reduction of total downward radiation leads to a cooling of the surface. Transfer functions have been developed to quantify the cooling in summers following volcanic eruptions. Briffa *et al.* (1998) acknowledge that there are other forcing factors that can lead to cool summers, but the coincidence of the lowest densities of tree-rings in several parts of Europe strongly indicates that volcanic eruptions are most likely the cause of the cool summers that produce these characteristics in the tree-ring data. Briffa *et al.* (1998) investigated the evidence for volcanic forcing of NH summer temperatures for the last six centuries. Apart from Gamkonora (Indonesia) in 1673 (Volcanic Explosivity Index (VEI) 5?), the large eruption of Long Island (New Guinea) (VEI 6) may have happened between 1666 and 1675 (Briffa *et al.*, 1998). Concluding from 1675 which, in central Europe, was another year without a summer (besides 1816; Robock, 1994; Harrington, 1992) the years 1673 and 1674 are the most likely candidates for the eruption. Indeed, our reconstructed winter SLP patterns from 1675 to 1679, generally, indicate an enhanced westerly flow over the eastern North Atlantic and western Europe, which supports the expected circulation after large volcanic eruptions. However, a strong high-pressure system was situated over Scandinavia and western Russia during these winters, leading to cold air advection towards central Europe. Apart from western Russia (Lyakhov, 1987; Borisenkov, 1994) and Romania (Boroneant and Chiper, 1998), Central England (except for 1676), Switzerland and Hungary experienced cold conditions for the winters from 1675 to 1679. There are thus inconsistencies with other eruptions of the last 200 years, which indicated a general warming over Eurasia in the winters following.

For the years mentioned, the summers were cool and wet in western and central Europe. As in winter the SLP patterns were connected with a rather strong westerly circulation, which is expected after large volcanic eruptions. Increased levels of sulphate in polar ice-cores suggest that the cool summers of 1695 and 1698 might also have volcanic causes (Zielinski, 1995) in response to Hekla (Iceland) and Serua (Indonesia) in 1693 and Komaga-Take (Japan) and Aboina (Indonesia) in 1694. Additionally, Jones *et al.* (1995) and D'Arrigo *et al.* (1999) suggest another likely, but previously undetected event. Our reconstructed SLP patterns for these summers also indicate a reduced influence of the Azores High with a westerly to northwesterly circulation over western and central Europe, connected with somewhat cooler and wetter conditions over the British Isles, France, Germany, Switzerland and the eastern Mediterranean. These conditions are in agreement with the climatic expectations after large volcanic eruptions. However, the winter surface pressure patterns indicate a reduced zonal circulation over Europe connected with coldness, and therefore are not in agreement with the atmospheric circulation after large eruptions.

Robock (2000) reported about positive height anomalies (500 hPa) over Europe and negative anomalies over the Near and Middle East (corresponding to warm and cold anomalies over the same locations) for the 1991/1992 winter following the 1991 mount Pinatubo eruption. According to our 500 hPa monthly reconstructions (not shown) we did not find such anomaly patterns for the above mentioned LMM winters which is not surprising since the temperature conditions were different for the LMM than for the winter after the Pinatubo eruption. These results may support the findings of Zielinski (2000) who reports that an eruption that occurs during a cooler climatic mode may enhance or extend those cooler conditions (due to other forcing factors). This suggests, that certain eruptions of the past may have enhanced and likely extended the cool climate (such as the LMM) existing at the time of the eruption (Zielinski, 2000).

4.2 Solar activity, influence on temperature and atmospheric circulation change

Volcanic eruptions and solar variations were both important causes of climate change in the 'Little Ice Age' (Briffa *et al.*, 1998; Bertrand *et al.*, 1999; Crowley and Kim, 1999; Free and Robock, 1999; D'Arrigo *et al.*, 1999; Beer *et al.*, 2000, Robock, 2000; Zielinski, 2000). The discussion concerning the solar influence on the earth's climate is of much interest, since accurate assessment of anthropogenic impacts on global change requires reliable estimates of the magnitudes of natural forcing mechanisms (Mann *et al.*, 1998; Rind *et al.*, 1999). Several recent studies (Lean *et al.*, 1995; Mann *et al.*, 1998; Lean and Rind, 1998; 1999) pointed to the fact that the paleo-reconstruction of NH surface temperature correlates highly with historical solar irradiance during the pre-industrial period from 1610 to 1800 implying a predominant solar influence during the MM. They report an empirical sun-climate relationship for the period prior to 1800 with a 0.16°C surface reduction per 0.1% irradiance decrease. Similar results have been found through equilibrium simulations of climate response to changing solar radiation using different GCMs (Rind and Overpeck, 1993; Wuebbles *et al.*, 1998; Lean and Rind, 1999) with a global average cooling of around 0.5°C based on a lowering of 0.24% in solar total irradiance during the MM. Haigh (1994; 1996) performed simulation studies to investigate the relationship between the 11-year solar activity cycles, ozone production and climatic change. She found a warming of the lower stratosphere by the absorption of more sunlight at solar maximum, one to two percent more ozone, a strengthening of the stratospheric winds and a poleward shift of the mid-latitude storm tracks. Shindell *et al.* (1999) report about zonal wind changes between solar maximum and solar minimum with an associated decrease of ~1.1 hPa in the northern latitudes and a zonal mean SLP increase of ~0.7 hPa from 30°N to 45°N. The opposite effect of Haigh's (1996) and Shindell's *et al.* (1999) simulations may have played a role in the climate during the MM. The reduced solar activity might have resulted in a decrease of the stratospheric ozone content as suggested by Wuebbles *et al.* (1998). Assuming that this decrease leads to the opposite effect of that simulated by Haigh (1996) and Shindell *et al.* (1999), a decrease of the latitudinal extent of the Hadley cell-circulation follows, connected with an expansion of the Polar cells and a southward shift of the storm tracks or a general reduction of SLP which goes along with the wetter and cooler conditions during LMM spring and summertime. Our average seasonal mean difference patterns of SLP (LMM minus analysis 1961-1990; see Figure 6) indicate significant lower pressure in the mid-latitudes and thus a slight southward shift of the mean polar front axes. However, contrary results have been proposed by Tinsley (1988) and Brown and Johns (1979) who found that on a longer time-scale the climatic changes of the MM and other periods of low solar activity can be interpreted as a poleward shift of storm tracks and reduced heat exchange between low and high latitudes.

4.3 Internal variability of the atmosphere-ocean system in the North Atlantic area

It is likely that the climate severity during the LMM can also be partly attributed to the internal variability of the atmosphere-ocean system in the North Atlantic. Evidence comes from millennial North Atlantic SST variations of a type quite at variance with the so-called Bond cycles (Bond *et al.*, 1997). The phasing of these cycles would tentatively suggest a lower SST in the mid-to-high latitudes of the North Atlantic at the MM. Bond *et al.* (1997) suggest these might be associated with thermohaline circulation changes, possibly coincidental with lower solar output at the MM. However, these ideas are controversial and are variously discussed. Lamb (1979) suggested much below normal SSTs between the Faeroes and southeast Iceland for the period 1675-1705. In addition, Stuiver and Braziunas (1993) and Rind *et al.* (1999) hypothesised that the North Atlantic region experienced colder conditions. According to Rind *et al.* (1999) the temperatures are lower over land than over the ocean due to the land's smaller heat capacity. The temperatures over land peak with a 0-10 year phase lag, while over the ocean, the peak has a 0-15 year time lag. The land temperatures are affected by ocean temperature changes, and both are affected by the lag induced by the feedback amplification (Rind *et al.*, 1999). Thus the study of Rind *et al.* (1999) indicates a rather simultaneous decrease of the temperature over land and ocean (Atlantic Ocean) in the case of reduced solar irradiance. However, recent modelling studies of Trudinger *et al.* (1999) indicate that the observed decrease in CO_2 is consistent with the effect of decreased temperature on either terrestrial or oceanic exchange, however, the increase in $\delta^{13}C$ favours a terrestrial response to cooling. This is in agreement with Etheridge *et al.* (1998) who conclude that lower levels of CH_4 (methane) for the same period were linked to cooling of the northern hemisphere land surfaces rather than the oceans. In addition, these authors reported that global SSTs were not significantly different for the 1550-1800 period. Luterbacher *et al.* (2000b) discussed the possibility of higher SSTs during the LMM compared to 1961-1990 mainly based on simple analogous studies with observational data from the twentieth century and other evidence and independent data. Kushnir (1994) and Kushnir and Held (1996) determined the differences in surface atmospheric conditions connected with warm and cold sea surface temperatures (SSTs) in the northern North Atlantic during the twentieth century. The SLP difference fields (warm SST years minus cold SST years) for the cold and warm season revealed a positive SLP anomaly from Iceland to northern Scandinavia and anomalous low pressure in the middle of the North Atlantic. Similar anomaly charts have been found by Peng and Mysak (1993) who investigated the teleconnection of interannual SST fluctuations (positive SSTs minus negative SSTs) in the northern North Atlantic and precipitation and runoff over western

Siberia. These composite SLP patterns are both similar to our SLP difference charts (LMM minus 1961-1990, see Figure 6), thus SSTs in parts of the northern North Atlantic during the LMM might also be above those of 1961-1990 which itself has experienced large SST fluctuations in the North Atlantic (such as the freshening and cooling, the Great Salinity Anomaly (GSA) (Dickson *et al.*, 1988) in the 1970s and the 'smaller GSA' (Reverdin *et al.*, 1997) in the early 1980s). The controversial discussions concerning the state of the North Atlantic will require further coupled GCM studies in combination with better volcanic and solar chronologies in order to obtain additional insight into the cause of the extreme periods occurring in the 'Little Ice Age' such as the Maunder Minimum.

5.   CONCLUSIONS AND FUTURE WORK

During the Late Maunder Minimum (LMM; 1675-1715) the 'Little Ice Age' in Europe was at its climax. A combined temperature-precipitation classification for Switzerland indicated that the two decades 1676 to 1695 witnessed by far the largest number of cold extremes in the winter half-year within the last five centuries. The anomalous coldness is less pronounced in the Mediterranean area, though the climate variability in respect to temperature and precipitation was higher compared to recent decades.

Synoptic comparison between the LMM and the period 1961-1990 revealed some significant differences in atmospheric circulation. For winter, significant higher pressure was prevalent over Scandinavia and the Baltic countries, but below normal SLP over the central and the western Mediterranean. The SLP classification results indicate a much higher frequency of months with a westward extension of the cold Scandinavian high connected with very cold and dry polar or even arctic air advection towards eastern and central Europe, thus less maritime influence. Springs were generally cold and summers rather wet. The evidence is basically in favour of a weaker Azores high and of the on average lower latitude in the LMM for the jet stream over the eastern North Atlantic and Europe. A higher frequency of dry and cold air advection in March and April might be the reason for the regular delay in the change over from winter to summer circulation. Late Maunder Minimum autumns indicated cooler and significantly higher SLP mainly north of around 65°N and lower pressure over continental Europe and the Mediterranean area, an indication of an advanced change from the summer to the winter circulation.

The results of the correlation-based SLP classifications revealed that the dominant pressure patterns in the LMM are also prevalent in the last 30 years and show similar relative frequencies and seasonal distribution. However, some classes during the LMM are unique and appear in winter and spring connected with strong cold air outbreaks from east to northeast and wetter and cooler conditions in summer, respectively. These classes are responsible for the climate variability within the LMM.

The outstanding climate during the LMM can possibly be explained by external forcing factors such as solar variability, volcanic impact and internal oscillation in the North Atlantic. The SLP fields with below normal pressure in the mid-latitudes from spring to autumn for the first years of the LMM as well as some years of the 1690s (1695-1699) might be attributed to a number of explosive volcanic events. In agreement with several modelling studies the reduced solar activity during the LMM might be a reason for the southward shift of the mid-latitude storm tracks. The 'role' of the internal variability of the atmosphere-ocean system in the North Atlantic is not yet clear. More coupled GCM studies are needed in order to get more insight into the dynamics and physics of the North Atlantic area during the LMM.

Further studies will include the analysis of the low-frequency variability of the atmospheric circulation (SLP and geopotential height fields) during the LMM in comparison with the periods of 1500 to 1674 and of 1716 to 1779, the EIP (Early Instrumental Period; 1780-1860) and RIP (Recent Instrumental Period) performed on a decadal basis and with respect to the four seasons. Both classification techniques will be included and the differences analysed and interpreted.

The SLP reconstructions are a step on the way towards having variables with more substantial information about the monthly mean circulation regime, such as reconstructed geopotential height fields and the deduction of the 500-1000 hPa thickness patterns. The upper level wave pattern gives an indication of where to expect recurrent divergence / convergence, lifting and sinking, probably cut off low dynamics, as well as mid-level cloudiness and areas where precipitation might be expected to increase. In addition, the 500-1000 hPa thickness is strongly correlated with the 300 hPa height fields and also the 850 hPa temperature. This means that the shape of the thickness patterns provide an impression of the polar front ridges, troughs and meanders, as well as about the thermal regime. Additionally, further studies will include the analysis of the low frequency variability of the atmospheric circulation including regime studies over the past 500 years for which continuous reconstructions are now available.

Acknowledgements

This work is part of the ADVICE project, funded by the European Commission under contract ENV4-CT95-0129. The Swiss project was funded by the Bundesamt für Bildung und Wissenschaft (BBW) under contract 95.0401. Part of the research was also funded by the Swiss National Science Foundation (Priority Programme Environment, No. 5001, 34888). The author would like to thank Prof. C. Pfister (University of Bern) for his contribution to an earlier version of this manuscript and his fruitful comments. Prof. P.D. Jones (Climatic Research Unit, Norwich), E. Xoplaki (University of Bern), Prof. Jucundus Jacobeit (University of Würzburg) and anonymous reviewers are acknowledged for their corrections, comments and suggestions which improved this paper. Thanks go also to Mary Brown (Meteotest) and Chris Sidle (University of Bern) for proof-reading the English text.

REFERENCES

Alcoforado, M.J., Nunes, M.F., Garcia, J.C., and Taborda J.P., 2000, Temperature and precipitation reconstructions in southern Portugal during the Late Maunder Minimum (1675 to 1715), *The Holocene* 10, in press.

Banzon, V., de Franceschi, G, and Gregori, G.P., 1992, The mathematical handling and analysis of non homogeneous and incomplete multivariate historical data series, in: European Climate Reconstructed from Documentary Data: methods and results, B. Frenzel, C. Pfister, eds., ESF, Stuttgart, 137-151.

Barnett, T., and Preisendorfer, R., 1987, Origins and levels of monthly and seasonal forecasts skill for United States surface air temperature determined by Canonical Correlation Analysis, *Mon. Wea. Rev.* 115:1825-1850.

Barriendos, M., 1997, Climatic variations in the Iberian peninsula during the Late Maunder Minimum (AD 1675-1715): An analysis of data from rogation ceremonies, *The Holocene* 7:105-111.

Beer, J., Mende, W., and Stellmacher, R., 2000, The role of the sun in climate forcing, *Quat. Sci. Rev.* 19:403-415.

Bertrand, C., van Ypersele J.-P., and Berger, A., 1999, Volcanic and solar impacts on climate since 1700, *Clim. Dyn.* 15:355-367.

Blair, D., 1998, The Kirchhofer technique of synoptic typing revisited, *Int. J. Climatol.* 18:1625-1635.

Bond, G., Showers, W., Cheseby, M., Lotti, R., Almasi, P., de Menocal, P., Priore, P., Cullen, H., Hajdas, I., and Bonani, B., 1997, A pervasive millennial-scale cycle in North Atlantic Holocene and glacial climates, *Science* 278:1257-1266.

Borisenkov, Y.P., 1994, Climatic and other natural extremes in the European territory of Russia in the Late Maunder Minimum (1675-1715), in: *Climatic Trends and Anomalies in Europe 1675-1715*, B. Frenzel, C. Pfister, and B. Glaeser B., eds., Gustav Fischer Verlag, Stuttgart, Jena, New York, 83-94.

Boroneant, C., and Chiper, M., 1998, Climatic anomalies in Romanian territory compared to the climate of Europe in the years of Maunder Minimum (1675-1715), *Second International Climate and History Conference*, Norwich, U.K., September 1998, p. 7.

Brázdil, R., Dobrovolny, P., Chocholác, B., and Munzar, J. 1994, Climatic and other natural extremes in the European territory of Russia in the Late Maunder Minimum (1675-1715), in: *Climatic Trends and Anomalies in Europe 1675-1715,* B. Frenzel, C. Pfister, and B. Glaeser B., eds., Gustav Fischer Verlag, Stuttgart, Jena, New York, 83-94.

Briffa, K.R., Jones, P.D., Schweingruber F.H., and Osborn, T.J., 1998, Influence of volcanic eruptions on northern hemisphere summer temperature over the past 600 years, *Nature* 393:450-455.

Brown, G.M., and Johns, J.I., 1979, Solar cycle influences in tropospheric circulation, *J. Atmos. Terr. Phys.* 41: 43-52.

Crowley, T.J., and Kim, K.-Y., 1999, Modeling the temperature response to forced climate change over the past six centuries, *Geophys. Res. Lett.* 26:1901-1904.

Cullen, C., 1980, Was there a Maunder Minimum? *Nature* 283:427-428.

D'Arrigo, R.D., Jacoby, G.C., Free, M., and Robock, A., 1999, Northern hemisphere temperature variability for the past three centuries: tree-ring and model estimates, *Clim. Change* 42:663-675.

Dickson, R.R., Meincke, J., Malmberg, S.-A., and Lee, A. J., 1988, The "Great Salinity Anomaly" in the northern North Atlantic 1968-1982, *Prog. in Oceanogr.* 20:103-151.

Eddy, J.A., 1976, The Maunder Minimum, *Science* 192:1189-1202.

Etheridge, D.M., Steele, L.P., Francey, R.J., and Langenfelds, R., 1998, Atmospheric methane between 1000 AD and present: evidence for anthropogenic emissions and climate variability, *J. Geophys. Res.* 103:15979.

Free, M., and Robock, A., 1999, Global warming in the context of the Little Ice Age, *J. Geophys. Res.* 104:19057-19070.

Frydendahl, K., Frich, P., and Hansen, C., 1992, Danish weather observations 1675-1715 (DMI Technical Report 92-3, Danish Meteorological Institute (DMI), Denmark-2100 Copenhagen, p. 23.

Glaser, R., Brázdil, R., Pfister, C., Dobrovolný, P., Barriendos Vallve, M., Bokwa, A., Camuffo, D., Kotyza, O., Limanówka, D., Rácz, L., and Rodrigo, F.S., 1999, Seasonal temperature and precipitation fluctuations in selected parts of Europe during the sixteenth century, *Clim. Change* 43: 169-200.

Haigh, J.D., 1994, The role of stratospheric ozone in modulating the solar radiative forcing of climate, *Nature* 370:544-546.

Haigh, J.D., 1996, The impact of solar variability on climate, *Science* 272:981-984.

Harrington, C.D. (ed.), 1992, *The Year without a Summer. Word Climate in 1816*, Canadian Museum of Nature, Ottawa, 576 pp.

Heino, R., Brázdil, R., Førland, E., Tuomenvirta, H., Alexandersson, H., Beniston, M., Pfister, C., Rebetez, M., Rosenhagen, G., Rösner, S., and Wibig, J., 1999, Progress in the study of climatic extremes in northern and central Europe, *Clim. Change* 42:151:181.

Holzhauser, H.P., 1997, Fluctuations of the grosser Aletsch glacier and the Gorner glacier during the last 3200 years: new results, in: *Paläoklimaforschung/Paleoclimate Research* 24:35-58.

Hoyt, D.V., and Schatten, K.H., 1993, A discussion of plausible solar irradiance variations, 1700-1992, *J. Geophys. Res.* 98:18895-18906.

Hurrell, J.W., 1995, Decadal trends in the North Atlantic Oscillation: regional temperatures and precipitation, *Science* 269:676-679.

Hurrell, J.W., and van Loon, H., 1997, Decadal variations in climate associated with the North Atlantic Oscillation. *Clim. Change* 36:301-326.

Huth, R., 1996, An intercomparison of computer-assisted circulation classification methods, *Int. J. Climatol.* 16:893-922.

Hyde, W.T., and Crowley, T.J., 2000, Probability of future climatically significant volcanic eruptions. *J. Climate (Letters)* 13:1445-1450.

Jones, P.D., Briffa, K.R., and Schweingruber, F.H., 1995, Tree-ring evidence of the widespread effects of explosive volcanic eruptions, *Geophys. Res. Lett.* 22:1333-1336.

Jones, P.D., Briffa, K.R., Barnett, T.P., and Tett, S.F.B., 1998, High-resolution palaeoclimatic records for the last millennium: interpretation, integration and comparison with general circulation model control-run temperatures, *The Holocene* 8:455-471.

Jones, P.D., Davies, T.D., Lister, D.H., Slonosky, V., Jónsson, T., Bärring, L., Jönsson P., Maheras, P., Kolyva-Machera, F., Barriendos, M., Martin-Vide, J., Rodriguez, R., Alcoforado, M.J., Wanner, H., Pfister, C., Rickli, R., Luterbacher, J., Schüpbach, E., Kaas, E., Schmith, T., Jacobeit, J., and Beck, C., 1999, Monthly mean pressure reconstruction for Europe for the 1780 – 1995 period, *Int. J. Climatol.* 19:347-364.

Kaufmann, R.K., Snell, S.E., Gopal, S., and Dezzani, R., 1999, The significance of synoptic patterns identified by the Kirchhofer technique: A Monte Carlo approach, *Int. J. Climatol.* 19:619-626.

Kington, J., 1995, The severe winter of 1694/95, *Weather* 50:160-163.

Kington, J., 1997, The severe winter of 1696/97, *Weather* 52:386-391.

Kington, J., 1999, The severe winter of 1697/98, *Weather* 54:43-49.

Koslowski, G., and Glaser, R., 1995, Reconstruction of the ice winter severity since 1701 in the western Baltic, *Clim. Change* 31:79-98.

Koslowski, G., and Glaser, R., 1999, Variations in reconstructed ice winter severity in the western Baltic from 1501 to 1995, and their implications for the North Atlantic Oscillation, *Clim. Change* 41:175-191.

Kushnir, Y., 1994, Interdecadal variations in North Atlantic sea surface temperature and associated atmospheric conditions, *J. Climate* 7:142-157.

Kushnir, Y., and Held, I.M., 1996, Equilibrium atmospheric response to North Atlantic SST anomalies', *J. Climate* 9:1208-1220.

Lachiver, M., 1991, *Les Années de Misère*, Fayard, Paris.

Lamb, H.H., 1979, Climatic variations and changes in the wind and ocean circulation. The Little Ice Age in the northeast Atlantic, *Quaternary Res.* 11:1-20.

Lamb, H.H., 1982, *Climate, History and the Modern World*, Methuen and Co Ltd., London.

Landsberg, H.E., 1980, Variable solar emissions, the Maunder Minimum and climatic temperature fluctuations, *Arch. Meteor. Geophys. Bioklim.* B28:181.

Lean, J., and Rind, D., 1998, Climate Forcing by Changing Solar Radiation, *J. Climate* 11:3069–3094.

Lean, J., and Rind, D., 1999, Evaluating sun-climate relationships since the little ice age, *J. Atmos. Sol.-Terr. Phys.* 61:25-36.

Lean, J., Beer, J. and Bradley, R.S., 1995, Reconstruction of solar irradiance since 1610: Implications for climate change, *Geophys. Res. Lett.* 22:3195-3198.

Legrand, J.-P., and Le Goff, M., 1992, Les observations météorologiques de Louis Morin entre 1670 et 1713, in: *Direction de la Météorologie Nationale*, Monographie Nr. 6, Météo-France, Trappes.

Luterbacher, J., Schmutz, C., Gyalistras, D., Xoplaki, E., and Wanner, H., 1999, Reconstruction of monthly NAO and EU indices back to 1675, *Geophy. Res. Lett.* 26:2745-2748.

Luterbacher, J., and 33 co-authors: 2000a, Reconstruction of monthly mean pressure over Europe for the Late Maunder Minimum period (1675-1715) based on canonical correlation analysis', *Int. J. Climatol., in press.*

Luterbacher, J., Rickli, R., Xoplaki, E., Tinguely, C. Beck, C., Pfister, C., and Wanner, H., 2000b, The Late Maunder Minimum (1675-1715) – a key period for studying decadal scale climatic change in Europe. *Clim. Change, in press.*

Lyakhov, M., 1987, Years with extreme climatic conditions, in: *Data of Meteorological Studies No. 13.* Institute of Geography, Academy of Sciences of the USSR: Moscow (in Russian), 119-178.

Manley, G., 1974, Central England temperatures: monthly means 1659 to 1973, *Quart. J. Roy. Met. Soc.* 100:389-405.

Mann, M.E., Bradley, R.S., and Hughes, M.K., 1998, Global-scale temperature patterns and climate forcing over the past six centuries, *Nature* 392:779-787.

Maunder, E.W., 1922, The prolonged sunspot minimum 1675-1715, *Brit. Astron. Ass. J.* 32:140-145.

NCAR, 1997, Trenberth's northern hemispheric sea level pressure, 5°x5°, monthly. DSS/D/DS010.1 dataset, Boulder, Colorado.

Nesmes-Ribes, E., Ferreira, E.N., Sadourny, R., Le Treut, H., and Li, Z.X., 1993, Solar dynamics and its impact on solar irradiance and the terrestrial climate, *J. Geophys. Res.* 98:18923-18935.

Ogilvie, A.E.J., 1996, Sea ice conditions off the coasts Iceland AD 1601-1850 with special reference to part of the Maunder Minimum period (1675-1715), *AmS-Varia* 25, Archaeological Museum of Stavanger, Norway, 9-12.

Parker, D.E., Legg, T.P., and Folland, C.K., 1992, A new daily central England temperature series, 1772-1991, *Int. J. Climatol.* 12:317-342.

Peng, S., and Mysak, L.A, 1993, A teleconnection study of interannual sea surface temperature fluctuations in the North Atlantic and precipitation and runoff over western Siberia, *J. Climate* 6:876-885.

Pfister, C., 1994, Spatial patterns of climatic change in Europe 1675-1715, in: *Climatic Trends and Anomalies in Europe 1675-1715,* B. Frenzel, C. Pfister, and B. Glaeser B, eds., Gustav Fischer Verlag, Stuttgart, Jena, New York, 287-317.

Pfister, C., 1999, *Wetternachhersage. 500 Jahre Klimavariationen und Naturkatastrophen 1496-1995*, Paul Haupt Verlag, Bern, Stuttgart, Wien.

Pfister, C., Kington, J., Kleinlogel, G., Schüle, H. and Siffert, E., 1994, High resolution spatio-temporal reconstructions of past climate from direct meteorological observations and proxy data, in: *Climatic Trends and Anomalies in Europe 1675-1715,* B. Frenzel, C. Pfister, and B. Glaeser B., eds., Gustav Fischer Verlag, Stuttgart, Jena, New York, 329-376.

Pfister, C., Luterbacher, J., Schwarz-Zanetti, G., and Wegmann, M., 1998, Winter air temperature variations in western Europe during the early and high middle ages (AD 750-1300), *The Holocene* 8:535-552.

Rácz, L., 1999, *Climate History of Hungary since 16th Century: Past, Present and Future*, Pécs, p. 160. Zoltán Gál, Pécs.

Reid, G.C., 1997, Solar forcing of global climate change since the mid-17th century, *Clim. Change* 37:391-405.

Reverdin, G., Cayan, D.R., and Kushnir, Y., 1997, Decadal variability of hydrography in the upper northern North Atlantic, 1948-1990, *J. Geophys. Res.* 102:8505-8532.

Rind, D., and Overpeck, J., 1993, Hypothesised causes of decade-to-century climate variability: climate model results, *Quat. Sci. Rev.* 12:357-374.

Rind, D., Lean, J., and Healy, R., 1999, Simulated time-dependent climate response to solar radiative forcing since 1600', *J. Geophys. Res.* 104:1973-1990.

Robock, A., 1994: Review of year without a summer? World climate in 1816. *Clim. Change* 26:105-108.

Robock, A., 2000: Volcanic eruptions and climate, *Rev. Geophys.* 38:191-219.

Robock, A., and Free, M.P., 1995, Ice cores as an index of global volcanism from 1850 to the present, *J. Geophys. Res.* 100:11549-11567.

Robock, A., and Mao, J., 1995, The volcanic signal in surface temperature observations, *J. Climate* 8:1086-1103.

Schmutz, C., and Wanner, H., 1998, Low frequency variability of atmospheric circulation over Europe, *Erdkunde (Earth Science)* 52:81-94.

Shindell, D., Rind, D., Balachandran, N., Lean, J., and Lonergan, P., 1999, Solar Cycle variability, Ozone, and Climate, *Science* 284:305-308.

Spörer, F.W.G., 1887, Über die Periodizität der Sonnenflecken seit dem Jahre 1618, vornehmlich in Bezug auf die heliographische Breite derselben, und Hinweis auf eine erhebliche Störung dieser Periodizität während eines langen Zeitraumes, *Vjschr. Astron. Ges. Leipzig* 22:323-329.

Stuiver, M., and Braziunas, T.F., 1993, Sun, ocean, climate and atmospheric $^{14}CO_2$: An evaluation of causal and spectral relationships, *The Holocene* 3:289-305.

Tinsley, B.A., 1988, 'The solar cycle and the QBO influences on the latitude of storm track in the North Atlantic, *Geophys. Res. Lett.* 15:409-412.

Trudinger, C.M., Enting, I.G., Francey, R.J., and Etheridge, D.M., 1999, Long-term variability in the global carbon cycle inferred from a high-precision CO_2 and $\delta^{13}C$ ice-core record, *Tellus* 51B:233-248.

von Storch, H., and F.W. Zwiers, 1999, *Statistical Analysis in Climate Research*, Cambridge University Press, London.

Wanner, H., Pfister, C., Brázdil, R., Frich, P., Frydendahl, K., Jónsson, T., Kington, J., Rosenørn, S., and Wishman, E., 1995, Wintertime European circulation patterns during the Late Maunder Minimum cooling period (1675-1704), *Theor. Appl. Climatol.* 51:167-175.

Wanner, H., Holzhauser, HP., Pfister, C., and Zumbühl, H., 2000, Interannual to century scale climate variability in the European Alps. *Erdkunde (Earth Science)* 54:62-69.

Wuebbles, D.J., Wei, C-F., and Patten, K.O., 1998, Effects on stratospheric ozone and temperature during the maunder minimum, *Geophys. Res. Lett.* 25:523-526.

Xoplaki, E., Maheras, P., and Luterbacher, J., 2000, Variability of climate in meridional Balkans during the periods 1675-1715 and 1780-1830 and its impact on human life. *Clim. Change*, in press.

Xu, Z.T. et al., 2000, *East Asian Astronomical Observations (East Asian Archaeoastronomy: Astronomical Observations in East Asia Historical Records*, Gordon and Breach, in press.

Yarnal, B., 1993, *Synoptic Climatology in Environment Analysis. A Primer.* Belhaven Press, London, Florida.

Yarnal, B., and Frakes, B., 1997, A procedure for blending manual and correlation-based synoptic classification, *Int. J. Climatol.* 17:1381-1396.

Zielinski, G.A., 1995, Stratospheric loading and optical depth estimated of explosive volcanism over the last 2100 years derived from the Greenland ice sheet project 2 ice core, *J. Geophys. Res.* 100:20937-20955.

Zielinski, G.A., 2000, Use of paleo-records in determining variability within the volcanism-climate system, *Quat. Sci. Rev.* 19:417-438.

EARLY EUROPEAN INSTRUMENTAL RECORDS

Philip D. Jones

Climatic Research Unit
University of East Anglia
Norwich NR4 7TJ
UK

1. INTRODUCTION

The barometer and the first reliable thermometers were developed during the second half of the 17th century (Middleton, 1969). Only in a few cases, however, have series been developed for the entire length of available data. The earliest measurements are unlikely to be consistent with today's readings for a variety of reasons. Potential problems relate to uncertainties in the accuracy of the instrumentation, the units used, exposure, observation times, and the availability of the measurements in original or published form. Across Europe there has been no systematic attempt to exploit the potential wealth of the early instrumental data. Most national meteorological agencies (NMAs), set up in the mid-nineteenth century, are unaware of the early, pre-nineteenth century records. Indeed many only acknowledge data that is readily available in digital form, even ignoring long series in manuscript form from the late nineteenth or early twentieth centuries. In many parts of Europe, there is potential for the development of long series back to the mid 18th century. Their development would enable at least 150 years of record to be analysed before the 20th century and the possible human modification of the climate system.

Earlier work with European instrumental records has been concentrated in certain periods when this type of painstaking work was fashionable or been developed by several post-1950 exponents in some parts of Europe (Kington, 1997). Eighteenth century developments were the initial call for measurements to be taken (Jurin, 1723), the expansion of instruments with the Sociètas Royale de Médicine in France and the Societas Meteorologica Palatina for much of central Europe in Mannheim (Kington, 1997). In the mid-nineteenth century, Döve (1838+) began to collect thousands of monthly mean temperature records, from a few to over a hundred years in length, publishing the values together with, when not directly from the observer, source details. National yearbooks with summary data began to appear around 1860 in many countries (e.g. Sweden, see Moberg and Bergström, 1997). Summaries of many long series were published between 1880 and 1910 (e.g. Wahlen, 1886 for St. Petersburg and many other Russian locations).

History and Climate: Memories of the Future?
Edited by Jones et al., Kluwer Academic/Plenum Publishers, 2001

55

Apart from Döve, Wahlen and a few others, most considered the long series to be of little interest. The reasons for collecting the data were entirely related to weather forecasting. Publication of data in yearbooks was often considered more of a book-keeping exercise in Meteorological Services, and during the 1910-1940 period the study of climate change was scarcely considered. Indeed, many considered that if changes occurred, they were not that important and did not warrant much consideration (Lamb, 1997).

The modern phase of early instrumental climate analysis can be traced back to the most detailed study of temperature variations for England, that eventually led to the Central England temperature (CET) series (Manley, 1974). CET extends back to 1659 on a monthly basis and through recent developments on a daily basis also, to 1772 (Parker *et al.*, 1992). The extension to 1659 meant Manley had to incorporate early instrumental measurements with documentary evidence (frost counts, river freezing and harvest dates, etc.). The monthly series is both the longest 'instrumental' series in the world and the most analysed. Jones and Hulme (1997) in a recent analysis consider its representativeness to other parts of the British Isles and the wider Northern Hemisphere.

The success of CET owes much to a near lifetime's work by Manley, but also to his decision to construct a regional series, rather than one representative of a specific location. A regional series enables numerous fragmentary records to be incorporated, producing a final series for several centuries without any missing months. Concentration on a specific location is always likely to involve a few missing days. In series development it is always best to estimate values for these days, where possible from neighbouring locations. A number of series for European locations were published between 1945 and 1980 but none has been as analysed as CET. nor as widely known. The success relates partly to the length of the series and partly to the publication in a well-known journal. Many other series were published in NMA or observatory reports.

The development of CET came just before the digital age, and thus represents an even more remarkable achievement than more recent studies. Recent work has begun to see many of the available series digitised (Camuffo *et al.*, 2000), although there are still many waiting to be fully digitised and homogenized. Digitization enables all the necessary adjustments to the original data to be made in a reproducible way. Improvements in our understanding of the necessary corrections can easily be incorporated into a new set of adjustments, to produce a revised homogeneous series for its entire length. As part of the IMPROVE (EU) project, the complete daily (and sometimes sub daily) records will be made available on CD (both raw and corrected values) for 7 sites in Europe (Uppsala, Stockholm, St. Petersburg, Central Belgium, Milan, Padua and Cadiz).

The purpose of this chapter is to document early instrumental series (both those available and where considerable potential exists), separating them into temperature, precipitation and pressure. Examples of the sorts of analyses that can be made on the resulting series, including interrelationships between temperature and precipitation and atmospheric circulation, are also given.

2.   AVAILABILITY OF EARLY INSTRUMENTAL SERIES

2.1 Temperature

Temperature represents the most important climatic variable, as past climates are referred to by whether they were colder or warmer than today. For this reason temperature is the most studied climatic variable. The production of a long homogeneous series is beset by potentially more problems than precipitation and pressure. There is, however, a considerable literature on the assessment of modern (1850→) records (see review by Peterson *et al.*,

1998). The homogeneity of even earlier records has been recently discussed by Camuffo and Zardini (1997), Moberg and Bergström (1997) and Camuffo *et al.* (2000).

The following factors are necessary to develop long series:

- history of the location of the thermometers

- history of the exposure of the thermometers, e.g. screen design and introduction, north wall exposure

- changes to thermometer units and calibration of the new thermometers

- history of times of observation, together with a modern 30 year average of hourly readings to adjust all combinations of observation times to a modern standard

- neighbouring records with which to make comparisons

Table 1 lists 35 locations across Europe where homogeneous records back to before 1780 have been developed, or where, in a few cases, the potential exists. Spatially the records are available for a far greater area and start slightly earlier than the later precipitation and pressure records.

2.2 Precipitation

The first precipitation measurements in Europe that have survived were taken at Burnley, Lancashire, UK, by Richard Towneley in 1677. The recording of precipitation generally began a couple of decades earlier in the British Isles than elsewhere in mainland Europe. By the early 1720s at least ten sites in various parts of the country recorded rainfall amounts (Craddock, 1976). Records are available from Paris from the 1680s (Garnier, 1974) but have not been combined with modern data (see later in Table 2).

The greater number of records across the British Isles may be due to the development of the British Rainfall Organization in the 1860s by Symons (1865). Early records were located at the time and the measurements are now available in the Meteorological Archives in the UK. A comprehensive index of the archive has, however, never been published (Jones *et al.*, 1997). Early records for France were similarly collected and published (Raulin, 1881). A comprehensive list of most French data was republished in four volumes by Garnier (1974).

Precipitation recording is considerably easier than temperature and pressure measuring as readings only need to be taken every day or month. There are several serious impediments, though, to the development of long homogeneous records. Rain-gauge designs have varied considerably both within the British Isles and across Europe and have not been standardised even now. Today different countries favour different designs and since the late 19th century numerous intercomparisons have been undertaken, particularly with regard to snowfall. In comparisons, the gauge that catches the most snowfall is considered to be the best. Considerable improvements to wind-shielding designs have taken place over the 20th century, such that most long series from Russia, Canada, the USA and other countries, where snowfall is a major component of precipitation all indicate, probably erroneous, century-scale increases in winter precipitation totals (Groisman *et al.*, 1996 and references therein). Improvements in the design of gauges for collecting rainfall have also been made this century, although their effects are less dramatic than with snowfall. Gauges located near to ground level catch more rain than those sited above the ground. No gauge is perfect though and it is estimated that precipitation amounts are underestimated over land regions by about 10%, this figure rising to about 40% in polar regions in winter (Legates and Willmott, 1990).

Table 1. Locations of temperature sites with known/potential records before 1780.

Site	Location		Earliest Year	Source
	Lat(°N)	Long(°)		
Trondheim	63.4	10.5E	1761	B
St. Petersburg	60.0	30.3E	1743	Wahlen (1886)
Uppsala	59.9	17.6E	1722	Moberg & Bergstrom (1997)
Stockholm	59.4	18.1E	1756	Moberg & Bergstrom (1997)
Edinburgh	55.9	3.2W	1764	B
Moscow	55.8	37.6E	1779*	B
Copenhagen	55.7	12.6E	1768*	B
Lund	55.7	13.2E	1753*	B
Vilnius	54.6	25.3E	1777	B
Berlin	52.6	13.4E	1701	Schaak (1982), DWD
Central England	~52.5	~2.5W	1659	Manley (1974)
Warsaw	52.2	21.0E	1779	W
De Bilt	52.1	5.2E	1706	van Engelen & Nellestijn (1995)
Leipzig	51.4	12.4E	1759*	B
Central Belgium	~51.0	~4.0E	1767	Camuffo *et al.* (2000)
Jena	50.9	11.6E	1770*	DWD
Frankfurt	50.1	8.7E	1757*	B
Prague	50.1	14.3E	1771	Brázdil & Budíková (1999)
Cracow	50.0	20.0E	1792*	W
Karlsruhe	49.0	8.4E	1779*	B
Regensburg	49.0	12.1E	1773*	B
Paris	48.8	2.5E	1757	B
Vienna	48.2	16.4E	1775	Auer *et al.* (1999)
Kremsmünster	48.1	14.1E	1767	Auer *et al.* (1999)
Hohenpeissenberg	47.8	11.0E	1781	DWD
Basel	47.6	7.6E	1755	B
Budapest	47.5	19.0E	1780	B
Innsbruck	47.3	11.4E	1777	Auer *et al.* (1999)
Geneva	46.2	6.2E	1753	B
Milan	45.5	9.2E	1763	Camuffo *et al.* (2000)
Padua	45.4	12.0E	1725	Camuffo *et al.* (2000)
Turin	45.2	7.7E	1753*	B
Bologna	44.5	11.5E	1716*	Comani (1987)
Barcelona	41.4	2.2E	1780	Camuffo *et al.* (2000)
Cadiz	36.5	6.3W	1776	Camuffo *et al.* (2000)

B   Source details given in Bradley *et al.* (1985)
DWD   Deutscher Wetterdienst
W   From Joanna Wibig
*   Not Complete and/or not fully homogenised

Precipitation measurements are therefore less exact than temperature and pressure readings, so in the development of long series we consider homogeneity issues in the relative sense. Because of subtle changes in gauge design, measurement times, location and the environment around the gauge (with respect to buildings and trees, which disrupt wind flow patterns) it is almost impossible to derive long (>150 years) homogeneous daily records. Instead the longest of precipitation series are probably only homogeneous on timescales of a month and higher. On monthly timescales, errors related to gauge design and location will tend to be reduced.

Table 2. Locations of sites with precipitation records before 1780.

Site	Location		Earliest Year	Source
	Lat(°N)	Long(°)		
St. Petersburg	60.0	30.3E	1740	B
Uppsala	59.9	17.6E	1779	T
Lund	55.7	13.2E	1748	T
Pode Hole	52.8	1.3E	1726	T
Hoofdoorp	52.3	4.7E	1735	T
Oxford	51.7	1.2W	1767	T
Kew	51.5	0.3W	1697	T
Metz	49.1	6.2E	1779	T
Karlsruhe	49.0	8.4E	1779	T
Paris	48.8	2.5E	1770	T
Milan	45.5	9.2E	1764	T
Padua	45.4	12.0E	1725	T
Marseilles	43.3	5.4E	1749	T

B   See Bradley *et al.* (1985) for original source details.
T   See Tabony (1980) for original source details.

In most regions of central and western Europe it should be relatively easy to develop long homogeneous series back to the mid 19th century. Tabony (1980, 1981) has developed over 180 sites for Europe from southern Sweden to northern Italy and Ireland to Hungary. In the UK, which has the densest network of any European country, sufficient gauges exist to derive 4-5 homogeneous series on each river catchment of about 1000km^2. This density of gauges is sufficient to derive accurate runoff estimates on the monthly timescale with a simple hydrological catchment model (Jones and Lister, 1998). In the UK all monthly rain-gauge records are stored centrally, but only about one third of all data collected have been digitised (Jones *et al.*, 1997). Elsewhere in Europe, the density of raingauges may not be this great, but it is likely to be considerably greater than national meteorological service archives would imply, particularly for the years before 1940 (e.g. France, see Garnier, 1974). In many European countries many precipitation records are held by water authorities or other utilities and are not always known to the NMA.

From Tabony's collection of homogeneous records, 13 extend back beyond 1780 (see Table 2). Apart from St Petersburg all the records are from a core region of western and central Europe. In eastern and northern Europe recording generally began in the early 19th century, possibly because of problems with early gauges and snowfall. In southern Europe recording begins in the late 18th century, but few of the long potential series (e.g. Madrid, Barcelona, Lisbon, Rome) have been reliably homogenized earlier than the middle of the 19th century.

2.3 Atmospheric Pressure

The barometer was invented in the 17th century and measurements were regularly taken from that time at a number of European locations. The design of the barometer also varied more than the thermometer. As with early thermometers the height of the mercury was read on a graduated piece of wood behind the column. The exposure of the instrument (outside or in an unheated room) was not as important as with a thermometer, although it was realised later that knowledge of air temperature was vital to achieving consistent measurements. For later comparison with neighbouring sites it is vital to know the elevation of the instrument above sea level and the correction required for gravity at 45°N. Thus despite many records being as long and potentially as continuous as temperature, the length of homogeneous series is often slightly shorter, because more adjustments to the raw readings have to be applied. Another problem is that the graduated wood behind the glass column expands with humidity increases and in some cases this also must be allowed for (Camuffo and Zardini, 1997). Homogeneity assessments therefore are potentially more complex than for temperature, as most of the problems with this variable apply, although slight changes with location are less important, provided any elevation changes have been allowed for. Corrections to standard gravity were only applied after the mid-19th century, but are merely an offset from the true value that depends on latitude.

In some meteorological compilations of timeseries data such as Döve (1838+) and others from the 19th century, only temperature and precipitation data were considered. This probably reflects both the fewer barometric records and the greater complexities involved in correcting the readings. Table 3 lists 15 pressure series which are available for the pre-1780 period. Many of these come from the series put together by Jones *et al.* (1999) to reconstruct monthly mean-sea-level pressure (MSLP) data in gridded form for the 1780-1995 period. In this work, MSLP variations over the region from 35-70°N by 30°W-40°E were reconstructed from 51 locations over the region. The reconstructions were obviously better where data density was greater. They were also better in winter, when the circulation is considerably more organised, than in the summer season. This greater organization means that in winter, fewer locations are necessary to define the circulation with an acceptable accuracy. Reliable reconstructions in winter can be made with as few as 4-5 well located series (i.e. potentially back to the 1740s based on the list in Table 3).

In Table 3 the Paris record begins in 1764. Earlier data, taken by Morin for the period 1670-1713, have been corrected (LeGrand and LeGoff, 1992) as best as can be achieved using what is known about the instrument and its location. It is strange to reflect that this record was kept for 43 years, yet the only other known barometric readings at the time [made in Upminster (NE of London) by Derham from the late 1690s to the late 1700s] would have been unknown to the observer. That both sets of readings are essentially reliable is attested to by the day-to-day variability of the series, compared to modern day values and by comparisons between the Paris and London records (see Slonosky, 1999).

Table 3. Locations of sites with pressure records before 1780.

Site	Location		Earliest Year	Source
	Lat(°N)	Long(°)		
Trondheim	63.4	10.5E	1768	J
St. Petersburg	60.0	13.4E	1743	J*
Uppsala	59.9	17.6E	1723	Camuffo *et al.* (2000)
Stockholm	59.4	18.1E	1756	Camuffo *et al.* (2000)
Edinburgh	55.9	3.2W	1770	J
Lund	55.7	13.2E	1743	Bärring *et al.* (1999)
London	51.5	0.3W	1774	J
Paris	48.8	2.5E	1764	J
Vienna	48.2	16.4E	1775	J
Basel	47.6	7.6E	1755	J
Geneva	46.2	6.2E	1768	J
Milan	45.5	9.2E	1763	Camuffo *et al.* (2000)
Padua	45.4	12.0E	1725	Camuffo *et al.* (2000)
Barcelona	41.4	2.2E	1780	Camuffo *et al.* (2000)
Cadiz	36.5	6.3W	1776	Camuffo *et al.* (2000)

J See Jones *et al.* (1999) for original source details.
* Available in J from 1822 (earlier data known to exist, but not yet located).

3. ANALYSES OF THE LONG INSTRUMENTAL SERIES

3.1 Temperature - Individual Series

Figures 1-4 show annual temperature time series from some of the longest and most homogeneous European locations. Figure 1 shows five of the longest sites in western and northern Europe. Variability is lower over Central England partly because of its maritime location but also because it is the average of three locations rather than a single site. The course of temperature change over the last 250-350 years is similar in neighbouring locations, particularly in the zonal direction, with many extreme years standing out in most of the series (e.g. 1740, 1838, the early 1940s and 1989 and 1990). Variability in some series reduces at certain times (e.g. the 1880s in western Europe but not in the Swedish series). 1740 is the coldest year recorded for Central England, De Bilt and Berlin but is not that unusual at Uppsala. Anomalous warm and cool five to ten year periods are evident (e.g. warmth in the 1730s, 1820s, 1930s and the 1990s and the coldness in the 1740s, 1780s, 1810s, late 19th century and the 1960s). All five series show long-term warming from either the late 18th or early 19th centuries. Recent years are only marginally the warmest of the entire series because of the warmth of the 1730s (particularly in western Europe) and the 1820s (northern Europe).

Figure 2 shows four records from central Europe, Figure 3 five records from Alpine areas and Figure 4 three records from northeastern Europe. Again the more closely located the series the greater the agreement in their annual and decadal timescales values. The agreement gives confidence in the homogeneity checks and corrections because the individual series will have been assessed by different groups within each country. For central Europe (Figure 2) most extremes of more than 2°C from the underlying course of temperature are evident at all four locations. On longer timescales, the fall in temperatures

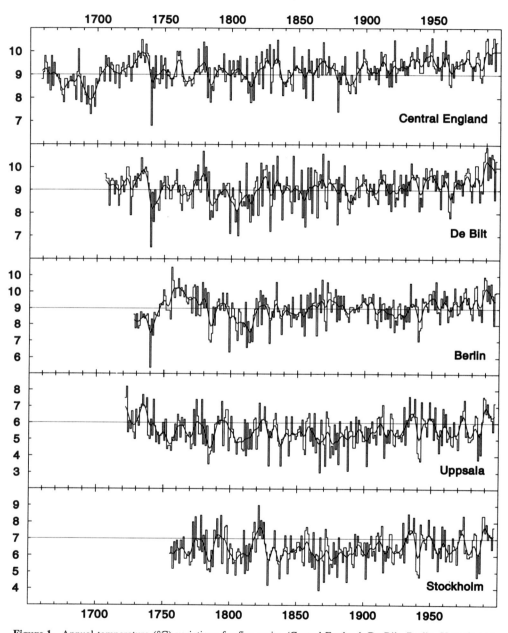

Figure 1. Annual temperature (°C) variations for five series (Central England, De Bilt, Berlin, Uppsala and Stockholm). The smooth line in this and subsequent plots highlights variations on decadal timescales (10 year Gaussian data adaptive filter). All Figures (including this one) plot station and regional time series data. The vertical scales used vary between stations. In absolute terms year-to-year variability for annual temperature, and to a lesser extent for precipitation and pressure, is greater in northern and eastern Europe.

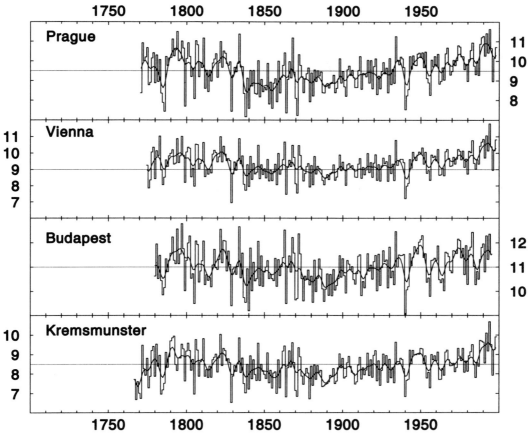

Figure 2. Annual temperature (°C) variations for four series in Central Europe (Prague, Vienna, Budapest and Kremsmünster).

from around 1800 to 1850 stands out, together with the cool years of the early 1940s and the rise in the last 15 years. Unlike western Europe, recent temperatures are not extreme and have been exceeded (except for 1994) by a number of years in the first hundred years of the records.

Alpine records (Figure 3) show many similarities to central Europe, although the level of recent temperatures is clearly higher than any earlier times. Year-to-year variability is greater in the pre-1880 period in all five series. This facet is also evident in the central European series and also at De Bilt, Berlin and Central England. Northeastern European (Figure 4) records show greater year-to-year variability throughout, as would be expected of a more continental climate. Recent years are generally the warmest although early warm decades are evident in the 1770s and 1820s. The coldest years in NE Europe were 1941 and 1942.

Even using the best records available for Europe reveals that some may not be wholly homogeneous all the time. Relative warming in some series (e.g. recently in Geneva, Milan and St. Petersburg) is suggestive of urbanization influences. Also the character of the Berlin series differs from De Bilt and Stockholm prior to about 1780. These features come to light from showing the long records together and may be less obvious when the records are compared with their nearer neighbours used in their construction.

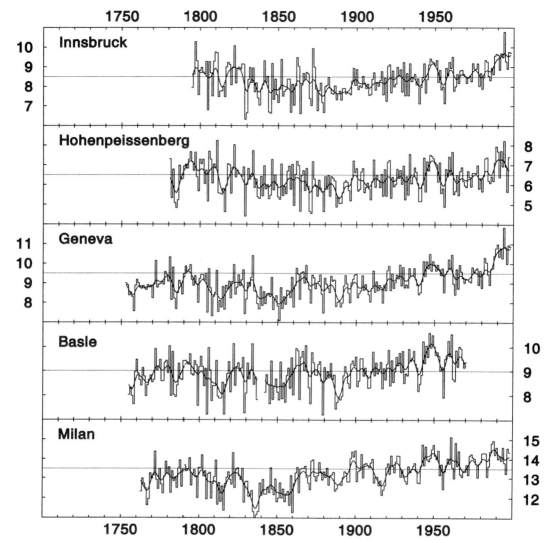

Figure 3. Annual temperature (°C) variations for five series in Alpine parts of Europe (Innsbruck, Hohenpeissenberg, Geneva, Basle and Milan).

3.2 Temperature - European Scale

Figure 5 shows an average European temperature series, on a seasonal basis, derived from five series (Central England, Uppsala, St. Petersburg, Hohenpeissenburg and Kremsmünster). The series has been produced by averaging anomalies (from 1961-90) from the five sites. It is shown in Figure 5 back to 1781. Combining series from all parts of Europe brings out the consistent features of Figures 1 to 4. On an annual basis the last ten years are the warmest of the last 220 years. Seasonally, the warmth is only evident during winters and springs. Summers and autumns were warmer, particularly during the 1780-1870 period in summer and the 1930-1980 period in autumn.

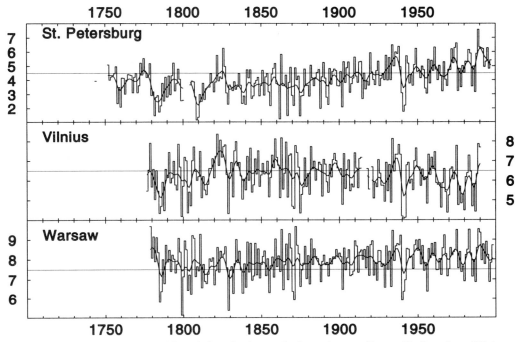

Figure 4. Annual temperature (°C) variations for three series in northeastern Europe (St. Petersburg, Vilnius and Warsaw).

3.3 Precipitation - Individual Series

Figures 6 and 7 show annual precipitation totals for stations in two regions (Scandinavia and the British Isles and western continental parts of Europe). The two figures show greater coherency across the region for precipitation variations on decadal and longer timescales. Extreme annual totals are less spatially coherent than temperature, although extremely dry years stand out in some series, particularly from southern England, the Netherlands and northern France. 1921 is the driest year of most of the records from Pode Hole and Hoofdoorp in the north to Milan and Padua in the south. All series show a couple of extremely large annual totals, sometimes with up to 50% more precipitation than the average for ten years before or after. Despite the large distances between the sites in southern Europe, the years 1772 and 1872 were excessively wet at at least two of the sites.

3.4 Precipitation - Regional

The longest regional precipitation series is that developed for England and Wales back to 1766 (Jones *et al.*, 1997 and earlier references). Figure 8 shows the series for seasons and the year. On an annual basis, the wettest year of the series is 1872 and the driest 1921. The series is shown to illustrate that annual series can mask subtle changes in the contribution of the seasons to the annual total. Winter precipitation totals rose by about 50mm to new levels around 1860 and both winter and spring totals have shown an increase over the last 30 years, although they have reduced during the 1990s. Summers in contrast have tended to show a decrease since the 1960s. The seasonal shifts are possibly much more widespread

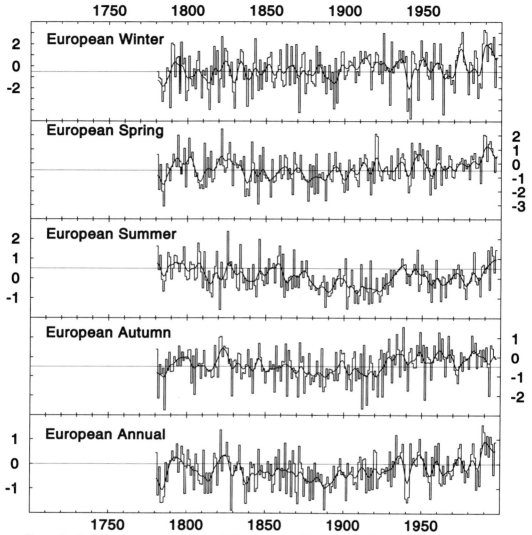

Figure 5. European average temperatures (°C) on a seasonal basis from 1781-1997. Seasons are standard (winter is December to February dated by the January, spring is March to May, etc.). The European average is composed of the unweighted average of five locations (Central England, Uppsala, St. Petersburg, Kremsmünster and Hohenpeissenberg).

across western Europe as they are evident in many of the individual series used here (not shown), and are evident over Scotland and Ireland (Jones *et al.*, 1997).

3.5 Pressure - Individual Series

Figures 9 and 10 show two sets of long annual European MSLP records. In Figure 9 the locations are mainly from the extremities of western Europe while in Figure 10 the sites are in central and southern Europe. The separation of the twelve series into two plots is a little arbitrary in southern Europe. As with the temperature and precipitation series, some

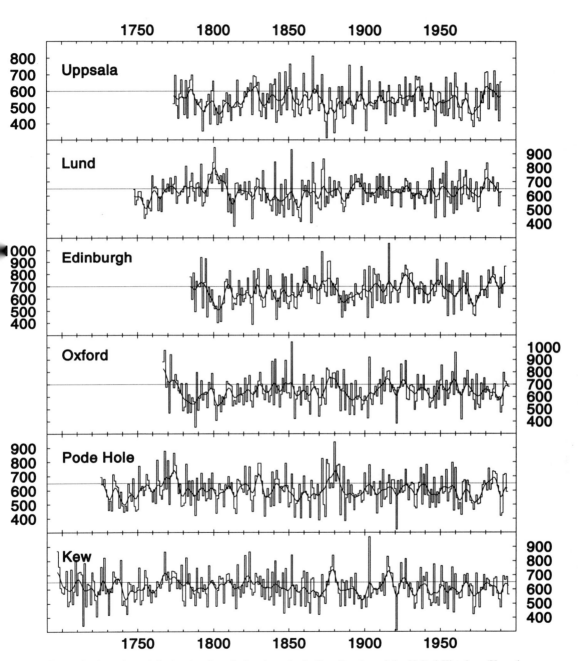

Figure 6. Annual precipitation (mm) totals for six series in Scandinavia and the United Kingdom (Uppsala, Lund, Edinburgh, Oxford, Pode Hole and Kew).

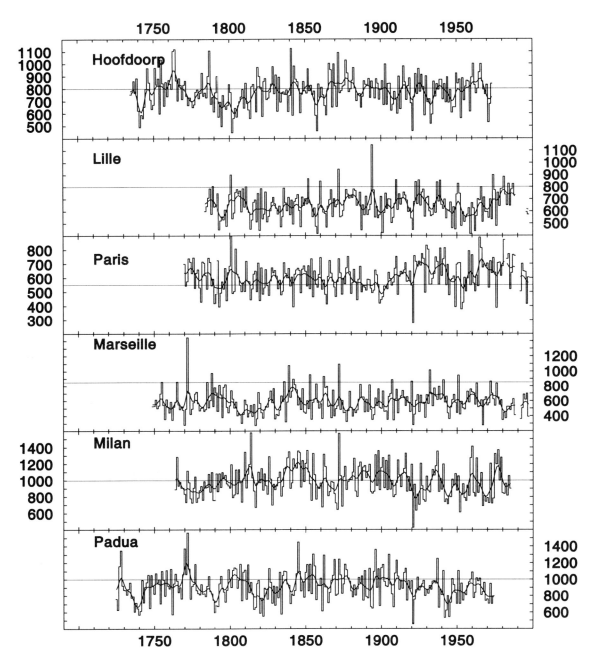

Figure 7. Annual precipitation (mm) totals for six locations in western and southern continental Europe (Hoofdoorp, Lille, Paris, Marseille, Milan and Padua).

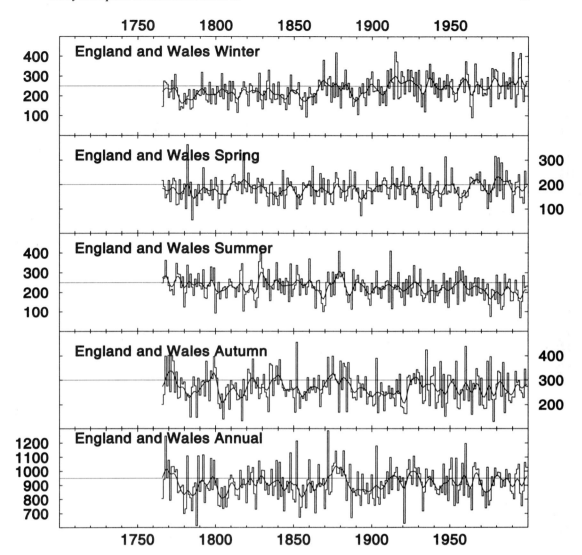

Figure 8. Seasonal and annual precipitation (mm) totals for England and Wales, 1766-1998. Seasons as defined in Figure 5.

extreme years stand out at neighbouring sites (e.g. 1834 for high pressure in central Europe) as do periods with higher and lower than average pressure.

3.6 Pressure - Circulation Indices

Two regional average pressure series have been produced for the 1781 to 1995 period. Figure 11 shows seasonal and annual MSLP values derived from an average of six Central European sites (Paris, Prague, Vienna, Basle, Geneva and Milan). As for temperature, this was produced by averaging MSLP anomaly values from the 1961-90 period. Variability of MSLP is markedly greater in winter than in summer. There are no longer-term trends in

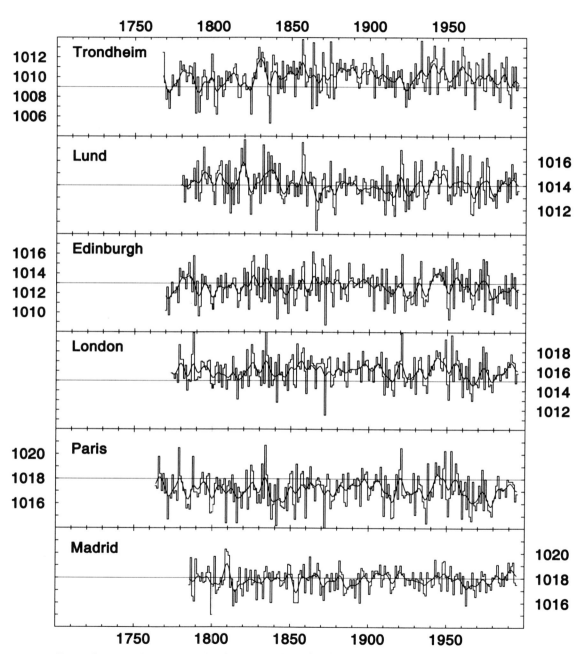

Figure 9.   Annual average sea-level pressures (hPa) for six locations in western and northern Europe
(Trondheim, Lund, Edinburgh, London, Paris and Madrid).

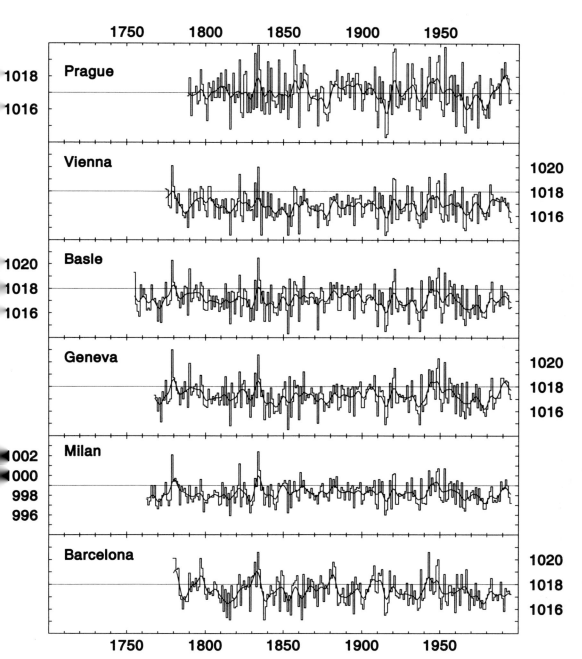

Figure 10. As Figure 11 but for six locations in central and southern Europe (Prague, Vienna, Basle, Geneva, Milan and Barcelona). The pressure for Milan is station level pressure.

pressure, the longest excursion being the higher pressures experienced during the 1940s and 1950s in spring.

Figure 12, in contrast, is a measure of the zonal flow strength across western and central Europe. It is produced from the difference between the average MSLP values at Madrid and Barcelona and those at Lund and Trondheim. Values are expressed as anomalies from the 1961-90 period. As with Figure 11, variability is greater in the winter half of the year compared to summer. On longer timescales, stronger zonal winds are evident in winter since the mid 1980s and weaker zonal winds during summers since the 1970s. Springs experienced weaker zonal winds before the mid 1850s.

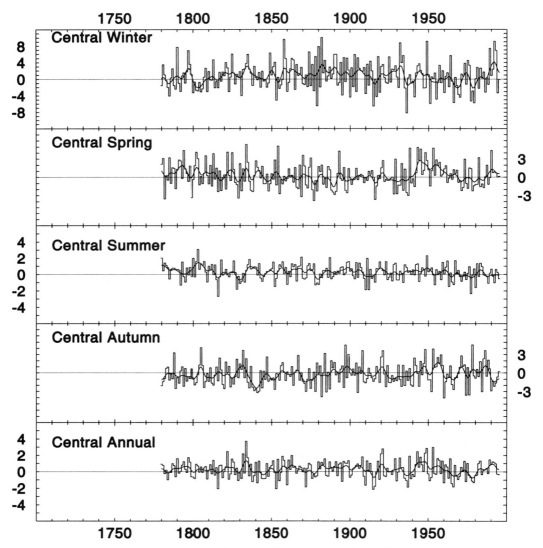

Figure 11. Seasonal and annual average of pressure (hPa) for western Central Europe. The average is made up of six sites (Paris, Prague, Vienna. Basle, Geneva and Milan). In calculating the average, the mean monthly pressure for the 1961-90 period was subtracted from the station data. The series plotted therefore represent anomalies from the 1961-90 period.

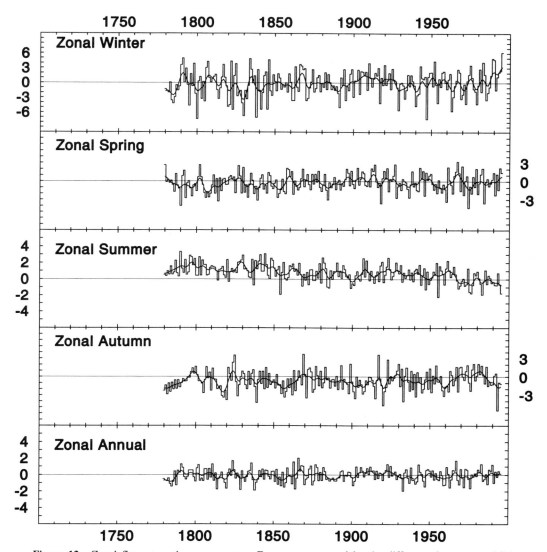

Figure 12. Zonal flow strength across western Europe as measured by the difference in pressure (hPa) between (Madrid and Barcelona) and (Trondheim and Lund). In calculating the index the mean monthly pressure for the 1961-90 period was subtracted from the station data. The series plotted therefore represents anomalies from the 1961-90 period.

3.7 Circulation/Surface Climate Relationships

The similarity of many of the neighbouring series shown in Figures 1-4 and 6-7 suggest that variations are the result of larger scale changes in circulation patterns. Table 4 gives correlations between some of the temperature (a) and precipitation (b) series and the two circulation indices illustrated in Figures 11 (central pressure) and 12 (zonal wind strength). The correlations are calculated on a seasonal basis over the years 1781-1995.

The correlations with temperature in Table 4a are generally as expected. Correlations are markedly higher for the zonal compared to the central pressure index, but vary strongly

Table 4a. Correlations between temperature series and the two circulation indices. Period of analysis is 1781-1995.*

	Years	Central Pressure (Figure 11)				Zonal Pressure (Figure 12)			
		DFJ	MAM	JJA	SON	DJF	MAM	JJA	SON
CET	215	**0.41**	**0.54**	**0.31**	**0.38**	**0.72**	**0.25**	**-0.52**	0.04
Berlin	215	**0.32**	**0.21**	0.08	0.00	**0.78**	**0.28**	**-0.44**	**0.36**
Uppsala	215	**0.54**	**0.43**	-0.06	0.11	**0.55**	**0.27**	**-0.26**	0.14
Stockholm	215	**0.52**	**0.42**	-0.02	0.09	**0.55**	**0.28**	**-0.27**	0.10
Basle	190	-0.04	**0.32**	**0.39**	-0.05	**0.62**	**0.21**	**-0.25**	**0.26**
Geneva	215	-0.09	**0.36**	**0.38**	0.02	**0.53**	0.20	**-0.28**	**0.21**
Hohenpeissenberg	215	0.10	**0.27**	**0.33**	-0.03	**0.60**	0.16	-0.19	**0.22**
Innsbruck	201	-0.10	0.06	-0.14	-0.19	**0.46**	0.11	-0.14	0.09
Kremuenster	215	0.07	0.15	-0.12	-0.16	**0.74**	**0.23**	-0.12	**0.32**
Vienna	215	0.05	0.08	-0.11	-0.19	**0.74**	**0.25**	-0.11	**0.32**
Budapest	215	-0.07	0.01	0.17	**-0.28**	**0.65**	0.19	0.06	**0.28**
Prague	215	0.15	0.16	0.16	-0.12	**0.75**	**0.24**	**-0.21**	**0.36**
Warsaw	215	**0.21**	0.01	**-0.26**	-0.13	**0.72**	**0.32**	**-0.26**	**0.41**
Vilnius	209	**0.31**	0.01	-0.09	-0.16	**0.70**	**0.33**	-0.01	**0.37**
St. Petersburg	215	**0.35**	0.10	-0.10	-0.05	**0.51**	**0.34**	0.00	**0.38**
European Average	215	**0.38**	**0.34**	0.17	0.06	**0.73**	**0.32**	**-0.28**	**0.31**

* Correlations in bold are significant at the 95% level, even after allowing for autocorrelation in the two series. Autocorrelation is low in the two pressure index series (see Figures 11 and 12).

Table 4b. Correlations between the precipitation series and the two circulation indices. Period of analysis is 1781-1995.*

	Years	Central Pressure (Figure 11)				Zonal Pressure (Figure 12)			
		DFJ	MAM	JJA	SON	DJF	MAM	JJA	SON
Uppsala	210	-0.13	-0.16	**-0.24**	**-0.33**	**0.37**	**0.41**	**-0.31**	**0.39**
Lund	210	-0.13	-0.12	**-0.21**	**-0.26**	**0.55**	**0.40**	-0.15	**0.64**
Edinburgh	210	-0.19	-0.19	**-0.37**	**-0.30**	**0.25**	**0.24**	-0.15	0.18
Oxford	215	**-0.40**	**-0.36**	**-0.45**	**-0.46**	**0.26**	0.14	-0.08	0.11
Pode Hole	214	**-0.54**	**-0.42**	**-0.48**	**-0.48**	0.02	0.13	-0.14	0.07
Kew	215	**-0.43**	**-0.43**	**-0.37**	**-0.50**	**0.29**	0.09	-0.15	0.15
Hoofdoorp	193	**-0.31**	**-0.37**	**-0.43**	**-0.51**	**0.44**	**0.33**	-0.20	**0.45**
Lille	212	**-0.42**	**-0.37**	**-0.36**	**-0.35**	**0.28**	0.13	**-0.22**	**0.34**
Paris	215	**-0.55**	**-0.52**	**-0.42**	**-0.50**	**0.26**	0.08	-0.04	**0.25**
Marseille	215	**-0.42**	**-0.32**	-0.15	**-0.23**	**-0.40**	**-0.23**	0.14	**-0.27**
Milan	215	**-0.48**	**-0.48**	-0.11	**-0.41**	**-0.35**	-0.16	0.03	**-0.22**
Paris	194	**-0.61**	**-0.58**	-0.09	**-0.44**	**-0.36**	-0.20	0.04	-0.15
England & Wales	215	**-0.39**	**-0.41**	**-0.48**	**-0.56**	**0.37**	**0.24**	**0.43**	**0.29**

* Correlations in bold are significant at the 95% level, even after allowing for autocorrelation in the two series. Autocorrelation is low in the two pressure index series (see Figures 11 and 12).

with season. The highest correlations are with winter, exceeding 0.7 in a belt over central Europe from Britain to Poland. Correlations are lower in Scandinavia and northern Russia and in Alpine regions. Correlations are lower in spring and autumn in the range of 0.15-0.35, still positive implying warmer temperatures with stronger westerly winds. In summer, in contrast, the relationship is inverse in the range -0.1 to -0.5. For the central pressure index, most correlations are weak and insignificant. The most significant values (0.2-0.5) occur over northern European stations in winter and to a lesser extent spring and over western Alpine stations in spring and summer. In all cases, the correlations indicate warmer temperatures accompany higher central pressures.

Table 4b gives similar correlation values between the two pressure indices and seasonal precipitation totals. For central pressure, precipitation is, as expected, inversely related. Correlations vary a little from season to season but are in the range -0.1 to -0.6. The highest values tend to be in western and southern Europe in winter, spring and autumn. Summer is the dry season in southern Europe. In western Europe, summer precipitation is still as strongly inversely related to the central pressure as in the other seasons. Even the best relationships only explain 25-35% of the variance of the seasonal precipitation totals (cf. the 50% explained for winter temperature by the zonal circulation strength). The zonal circulation explains some variance of the seasonal precipitation totals, but there are strong differences between regions. Positive correlations (0.2-0.6) with the zonal pressure index are found in northern Europe in winter, spring and autumn, but inverse relationships (-0.2 to -0.4) are evident in the same seasons in southern Europe. This pattern of correlation values between seasonal precipitation and the zonal wind strength has been recognised before (see, e.g., Hurrell, 1995).

CONCLUSIONS

Many locations in Europe have the potential for the development of instrumental climatic records for most of the 19th century enabling the variability of climate to be assessed for a period before any anthropogenic influence can have taken place. In several locations, records can be extended back to the first half of the 18th century. Only in a few of these has the potential extension been fully realised. Long instrumental records are capable of providing much more information than proxy climate indicators, particularly through the development of long daily series. Their construction needs to be undertaken with tremendous care to ensure that they are truly homogeneous for their entire records.

Long annual records for 16 temperature, 12 precipitation and 12 pressure sites, each extending back to 1780 have been shown. Neighbouring series show similar extreme years and similar trends on decadal and longer timescales, attesting to the independent and complex adjustment required by each series to ensure homogeneity. The relationships between the circulation and the surface climate were illustrated by correlating the temperature and precipitation series on a seasonal basis against a zonal and a central pressure index. Zonal pressure (defined by the difference in pressure between Madrid and Barcelona and Trondheim and Lund) explains up to 50% of the variance of winter temperature from Britain to Poland. Stronger zonal flow also implies warmer temperatures in spring and autumn, but the relationships are weaker. In summer, weaker westerlies lead to warmer temperatures, but only explain about 20% of the variance. Average pressure over central Europe (defined by the average of MSLP values at Paris, Geneva, Basle, Vienna, Prague and Milan) only explains about 10-20% of the variance of some temperature series in some seasons. In all cases, high pressure leads to warmer temperatures. For seasonal precipitation totals, central pressure explains between 10 and 35% of the variance in all seasons, except for summer (dry season) in southern Europe. Zonal pressure is positively correlated with precipitation totals in northern Europe and negatively correlated in southern Europe.

Acknowledgements

The author thanks Reinhard Böhm, Rudolf Brazdil, Maurizio Maugeri, Anders Moberg, Aryan van Engelen and Joanna Wibig for help in obtaining some of the long European temperature records. The German temperature series were made available by Deutscher Wetterdienst. This work has been supported by the European Union (IMPROVE, ENV4-CT97-0511) and the United States Dept. of Energy (Grant No. DE-FG02-98ER62601).

REFERENCES

Auer, I., Böhm, R., Schoener, W., and Hagen, M., 1999, ALOCLIM-Austrian-Central European Long-term Climate - The creation of a multiple homogenised long term Climate Data Set. *Proc. Second Seminar for Homogenisation of Surface Climatological Data, Budapest, November 1998*, 25pp.

Bärring, L., Jönsson, P., Achberger, C., Ekström, M., and Alexandersson, H., 1999, The Lund record of meteorological instrument observations: monthly pressure 1780-1997. *Int. J. Climatol.*, 19:1427-1443.

Bradley, R.S., Kelly, P.M., Jones, P.D., Goodess, C.M., and Diaz, H.F., 1985, A climatic data bank for Northern Hemisphere land areas, 1851-1980. *U.S. Dept. of Energy Technical Report TRO17*, U.S. Dept. of Energy, Washington, D.C.: 335pp.

Brázdil, R., and Budíková, M., 1999, An urban bias in air temperature fluctuations at the Klementinum, Prague, the Czech Republic, *Atmospheric Environment*, 33:4211-4217.

Camuffo, D., and Zardini, F., 1997, Controlling the homogeneity of a long meteorological series: the series of Padova (1725 today), in: *Applications of timeseries in meteorology and astronomy* (T. Subba Rao, M.B. Priestley and O. Lessi, Eds.) Chapman and Hall, London, 441-459.

Camuffo, D., Demarée, G., Davies, T., Jones, P., Moberg, A., Martin-Vide, J., Cocheo, C., Maugeri, M., Thoen, E., and Bergström, H., 2000, Improved understanding of past climatic variability from early daily European instrumental sources, in: *Proceedings of EU October 1998 meeting in Vienna*, A. Ghazi and I. Troen, eds.. Brussels (in press).

Comani, S., 1987, The historical temperature series of Bologna (Italy): 1716-1774. *Climatic Change* 11:375-390.

Craddock, J.M., 1976, Annual rainfall in England since 1725. *Quart. J. Roy. Met. Soc.* 102:823-40.

Döve, H.W., 1838+, Uber die geographische Verbreitung gleichartiger Witterungserscheinungen (Uber die nichtperiodischen Änderungen Temperaturve-teilung auf der Oberfache der Erde): *Abh. Akad. Wiss. Berlin*. I Teil (1838) 37,285; 38,286; 39,345. II Teil (1839) 40,30-307. III Teil (1844) 41,127. V Teil (1852) 42,3-4.

Garnier, M., 1974, Longues Séries de Mesures de Précipitations en France. *Mémorial de la Météorologie Nationale*, Fascicule No 1-4, Météorologie Nationale, Division de Climatologie, Paris.

Groisman, P.Ya., Easterling, D.R., Quayle, R.G., Golubev, V.S., Krenke, A.N., and Mikhailov, A.Yu, 1996, Reducing biases in estimates of precipitation over the United States: Phase 3 adjustments. *J. Geophys. Res.* 101:7185-7195.

Hurrell, J.W., 1995, Decadal trends in the North Atlantic Oscillation and relationships to regional temperature and precipitation. *Science* 269:676-679.

Jones, P.D., and Hulme, M., 1997, The changing temperature of 'Central England', in: *Climates of the British Isles: Present, past and future*, M. Hulme and E. Barrow, eds., Routledge, London:173-196.

Jones, P.D., Conway, D., and Briffa, K.R., 1997, Precipitation variability and drought, in: *Climates of the British Isles: Present, past and future*, M. Hulme and E. Barrow, eds., Routledge, London: 197-219.

Jones, P.D., and Lister, D.H., 1998, Riverflow reconstruction for 15 catchments over England and Wales and an assessment of hydrologic drought since 1865. *Int. J. Climatol.* 18:999-1013.

Jones, P.D., Davies, T.D., Lister, D.H., Slonosky, V., Jónsson, T., Bärring, L., Jönsson, P., Maheras, P., Kolyva-Machera, F., Barriendos, M., Martin-Vide, J., Rodriguez, R., Alcoforado, M.J., Wanner, H., Pfister, C., Luterbacher, J., Rickli, R., Schuepbach, E., Kaas, E., Schmith, T., Jacobeit, J., and Beck, C., 1999, Monthly mean pressure reconstructions for Europe for the 1780-1995 period. *Int. J. Climatol.* 19:347-364.

Jurin, J., 1723, Invitatio ad observationes meteorologicas communi consilio instituendes. *Phil. Trans. Roy. Soc.* 32:422-427.

Kington, J., 1997: Observing and measuring the weather. In. *Climates of the British Isles: Present, past and future*. (M. Hulme and E. Barrow, Eds.), Routledge, London, 137-152.

Lamb, H.H., 1997, *Through all the changing scenes of life: A meteorologist's tale*. Taverner Publications, East Harling, Norfolk:274pp.

Legates, D.R., and Willmott, C.J., 1990, Mean seasonal and spatial variability in gauge-corrected, global precipitation. *Int. J. Climatol.* 10:111-128.

LeGrand, J-P., and LeGoff, M., 1992, Les observations météorologiques de Louis Morin entre 1670 et 1713, in, Direction de la Météorologie Nationale, Monographic Nr. 6., Trappes.

Manley, G., 1974, Central England temperatures: monthly means 1659 to 1973. *Quart. J. R. Met. Soc.* 100: 389-405.

Middleton, W.E.K., 1969, *Invention of the Meteorological Instruments*, Johns Hopkins Press, Baltimore.

Moberg, A., and Bergström, H., 1997, Homogenization of Swedish temperature data. Part III: The long temperature records from Uppsala and Stockholm. *Int. J. Climatol.* 17:667-699.

Parker, D.E., Legg, T.P., and Folland, C.K., 1992, A new daily Central England temperature series, *Int. J. Climatol.* 12:317-342.

Peterson, T.C., Easterling, D.R., Karl, T.R., Groisman, P., Nicholls, N., Plummer, N., Torok, S., Auer, I., Böhm, R., Gullett, D., Vincent, L., Heino, R., Tuomenvirta, H., Mestre, O., Szentimrey, T., Salinger, J., Førland, E.J., Hanssen-Bauer, I., Alexandersson, H., Jones, P., and Parker, D., 1998, Homogeneity adjustments *in situ* atmospheric climate data: a review. *Int. J. Climatol.* 18:1493-1517.

Raulin, V., 1881, *Observations Pluviométriques Faites dans la France Septentrionale de 1688 à 1870.* Gauthier Villars, Paris, 820pp + Supplément.

Schaak, P., 1982, Ein Beitrag zum Preussenjahr 1981 en Berlin. *Beilage zur Berliner Wetterkarte* 38/82, S04/82. Institut für Meteorologie der Freien Universität, Berlin.

Slonosky, V., 1999, *Variability in surface atmospheric circulation over Europe from early instrumental records.* PhD Dissertation, University of East Anglia.

Symons, G.J., 1865, On the rainfall in the British Isles, *Report of the 35th meeting of the British Association for the Advancement of Science*:192-242.

Tabony, R.C., 1980, A set of homogeneous European rainfall series. Met. O.13 Branch Memorandum No. 104, Meteorological Office, Bracknell.

Tabony, R.C., 1981, A principal component and spectral analysis of European rainfall. *J. Climatol.* 1:283-294.

van Engelen, A.F.V., and Nellestijn, J.W., 1995, Monthly, seasonal and annual means of the air temperature in tenths of centigrades in De Bilt, Netherlands, 1706-1995. KNMI (Dutch Met. Service) Report, Climatological Services Division.

Wahlen, E., 1886, Wahre Tagesmittel und Tägliche Variation der Temperatur an 18 Stationen des Russischichen Reiches. Dritter Supplementband zum Repertorium für Meteorologie, St. Petersburg.

CIRCULATION CHANGES IN EUROPE SINCE THE 1780s

Jucundus Jacobeit[1], Phil Jones[2], Trevor Davies[2], and Christoph Beck[1]

[1]Institute of Geography, University of Würzburg, Germany
[2]Climatic Research Unit, University of East Anglia, UK

1. INTRODUCTION

Climatic variability and changes are inherently linked with variability and changes in the atmospheric circulation, the latter comprising causes as well as effects of the former (Hupfer, 1991). Therefore, climate research should always include investigations on the behaviour of the circulation. This is valid within an historical perspective. In view of possible changes in climate that might be expected in the near future due to anthropogenic greenhouse gas forcing, it is of crucial importance to learn about natural variabilities in climate and atmospheric circulation through the extension of empirical research into the historical past.

Since objective circulation analyses require homogenized pressure data, preferably with large-scale grid resolution, most studies have focussed on the last hundred years or even shorter periods (e.g. Barnston and Livezey, 1987; Rogers, 1990; Hupfer, 1991; Jacobeit, 1993; Wanner, 1994; Bartzokas and Metaxas, 1996; Klaus, 1997; Mächel et al., 1998; Kapala et al., 1998). For earlier periods, the classical reconstructions by Lamb and Johnson (1966) and Kington (1988) defined the state-of-the-art for a long time. Objective circulation analyses became possible when the first version of a reconstructed gridded SLP data set back to 1780 became available (Jones et al., 1987). Subsequently, enhanced efforts were made to establish an increased number of homogenized pressure time series extending further back into the historical past.

Recently, Jones et al. (1997) provided an extension of the NAO back to the 1820s. Schmutz and Wanner (1998) applied a classification technique to the sea-level pressure (SLP) fields from 1785. Case studies of anomalous conditions during the last 500 years were compiled by Pfister (1999), and investigations in the context of the European Community climate research project ADVICE (Annual to Decadal Variability in Climate in Europe) also focussed on the topic of circulation variability extended back to 1780 (Jacobeit et al., 1998; Beck, 1999). These investigations were based on recently improved reconstructions of gridded SLP data (see next section). Here, these reconstructions are used to present the major circulation changes that have occurred in Europe since the end of the 18th century, thus providing an extended frame for representing large-scale circulation variability.

History and Climate: Memories of the Future?
Edited by Jones et al., Kluwer Academic/Plenum Publishers, 2001

79

2.   DATA

The following analyses are based on gridded monthly mean sea-level pressure data that have been reconstructed back to 1780 for the North-Atlantic-European region by Jones *et al.* (1999) using homogenized long-term pressure time series from various European stations; starting in 1780 with 10 continuous series, increasing to 20 stations in the 1820s (including the important stations of Reykjavik and Gibraltar), and finally reaching a network of 51 stations by the 1860s.

The historical SLP data are available on a 5° latitude by 10° longitude grid over the area from 35 to 70°N and from 30°W to 40°E. For deriving gridded data from the station data, Jones *et al.* (1999) used EOF-regression models calibrated over the period 1936-1995 and verified for the period 1881-1935. Explained variances of above 90% are reached around the central grid points with decreasing values towards the periphery, especially for the earliest period with relatively few station time series. The more structured large-scale circulation during winter explains higher percentages of the variance than in summer. These data represent significantly improved reconstructions compared with earlier assessments (Jones *et al.*, 1999), and define a highly appropriate basis for considering circulation analyses for a longer period of time than previously available data sets would allow.

3.   METHODS

Analyses will focus on three different approaches: using major circulation indices; determining large-scale circulation types; and applying principal component analyses (PCA) to the whole grid in order to derive basic circulation patterns and their variations in time.

Calculating circulation indices from gridded pressure data would be most reliable if concentrated on those regions with the highest quality in the reconstructions, i.e. around the central grid points of the analysed area. This has been done by Jacobeit *et al.* (1998) who defined a Central European Zonal Index which was strongly related to regional temperatures in this area (Jacobeit *et al.*, 2000). On the other hand, the major mechanism governing the large-scale circulation variability across Europe is the North Atlantic Oscillation (NAO), and despite its centres of action being situated towards the periphery of the reconstructed pressure grid, an NAO index approximation has been calculated. It needs to be borne in mind that results are less reliable for the time before nearby station pressure series started during the 1820s.

There have been several different definitions of NAO indices: most commonly the difference between normalized station pressure time series is used. However, different stations have been used: Ponta Delgada - Akureyri (Rogers, 1984), Lisbon - Stykkisholmur (Hurrell, 1995), or Gibraltar - Reykjavik (Jones *et al.*, 1997), the latter going furthest back in time until the 1820s. Gridded SLP data are also used: e.g. for determining the moving pressure maxima and minima (Mächel *et al.*, 1998), or for deriving a grid-based principal component representing the NAO (Wanner *et al.*, 2000). A simple approach consists of using fixed grid point averages of normalized SLP (Jacobeit *et al.*, 1998), and here we take the grid points along 30 and 20°W to calculate the differences between normalized SLP averages within the latitudinal sections 35-40°N and 60-65°N (4 grid points for the Azores and the Icelandic centres, respectively). Additionally, cumulating the sum of successive index values leads to the determination of normalized cumulative anomalies. Predominantly rising/falling values in these cumulative anomalies indicate the prevalence of positive/ negative deviations, and major turning points in cumulative anomalies indicate transitions between periods dominated by opposite deviations. Therefore cumulative anomalies reveal decadal and interdecadal index variations particularly well (e.g. Mächel *et al.*, 1998).

The second approach of circulation analyses determines large-scale circulation types based on the degree of zonality, meridionality and vorticity of the monthly mean SLP isobars focussing on central grid points where reconstruction models performed well, the 25 grid points from 40 to 60°N and from 10°W to 30°E. Within this area, three SLP patterns have been defined representing typical westerly flow, southerly flow and central low pressure. Spatial correlations with these typical patterns have been calculated for all monthly mean SLP subgrids for the period 1780-1995. Based upon the resulting three correlation coefficients it is possible to classify each of the SLP subgrids into one of ten circulation types corresponding to the Central European Grosswettertypes (Hess and Brezowsky, 1977; Gerstengarbe and Werner, 1993). Low and High pressure types are defined by a maximum coefficient (positive or negative) with the central pressure type. Remaining cases are assigned to the eight main flow types (W, NW, N, NE, E, SE, S, SW) according to the minimum Euclidean distance of their zonality and meridionality coefficients from those defining the main flow types (i.e. the West type is defined by zonality/meridionality coefficients of 1.0/0.0, the Northwest type by 0.71/-0.71, the North type by 0.0/-1.0 etc., and zonality/meridionality coefficients of an actual SLP subgrid are compared with these defining values by means of calculating two-dimensional Euclidean distances, respectively). Seasonal frequencies of the resulting monthly Grosswettertypes have been calculated for 31-year periods (e.g. 1780-1810), iterated with time steps of one year.

Finally, pressure data for the whole grid has been submitted to varimax-rotated PCA, separate analyses for each month. Since we intend to derive basic circulation patterns, T-mode analysis has been applied with variables (SLP grids) differing in time and spatially varying cases (grid points). Since the number of grid points of the reconstructed historical SLP data is restricted to 60, it is not possible to enter all the 216 monthly mean SLP grids from the whole 1780-1995 period into one monthly T-mode analysis. Thus, PCAs have been performed for a series of 60-year periods starting with 1780-1839, continuing with periods each shifted by 12 years until the final period 1936-1995. All those SLP grids with a maximum loading on at least one of the resulting circulation patterns from these overlapping analyses were combined into a new set of variables for a concluding PCA. Thus, we may ensure that circulation patterns resulting from this concluding analysis reflect conditions from the whole 1780-1995 period. In order to generate continuous series of time coefficients, these circulation patterns (T-mode scores of the final extracted principal components) have been subsequently correlated with all the monthly SLP grids from the 1780-1995 period. The background for this procedure is that we have not applied an EOF, but a PCA technique - the former having eigenvectors of unit length, the latter eigenvectors weighted by the square root of their eigenvalues (see Richman, 1986). Thus, the T-mode loadings represent the correlation coefficients between variables (SLP grids entered into the analysis) and resulting PCA scores (circulation patterns). Correlation coefficients between these patterns and the SLP grids from the whole 1780-1995 period therefore represent a completed series of loadings, con-stituting the continuous series of time coefficients required for further investigations.

An essential aspect for pattern analysis is the number of extracted principal components, especially in view of varimax rotations, that have been applied to improve the reproduction of physical relations from the real world (Richman, 1986). According to principles discussed earlier (Jacobeit, 1993), we have only extracted those principal components having at least on one input variable the maximum loading among all the components. In this way, noisy components are rejected and it is ensured that extracted components represent dominant patterns that are realized at least for anomalous conditions.

Changes over time of these patterns are considered with regard to different aspects. The completed series of T-mode loadings (representing the patterns' time coefficients) are indicating more or less weight of a particular pattern within the original SLP grids.

Similarly, as for the NAO index, the normalized time coefficients have been used to calculate their normalized cumulative anomalies, thus allowing us to identify major periods with increased or decreased importance of a particular circulation pattern. Furthermore, the internal characteristics of these patterns, like pressure gradients or spatial dimensions of pressure centres, may have varied over time. Thus, SLP composites have been calculated for each circulation pattern and for each of the 186 31-year periods within the whole 1780-1995 period (starting from 1780-1810 until 1965-1995). Composites have been derived from those original SLP grids having their highest loading (correlation coefficient) on the respective circulation pattern. Composites have been determined as weighted means of these original SLP grids with the corresponding loadings as the statistical weights. Each pair of composites has been compared by means of its root-mean-square value, in order to identify periods with major differences in internal characteristics of a particular circulation pattern. Selected examples of significantly differing SLP composites will be given for the most important patterns.

4. RESULTS

4.1 Seasonal NAO Index

Fig. 1 shows the grid-based NAO index since 1780 for the winter (DJF) and summer (JJA) seasons. The different levels of values compared to those from Jones *et al.* (1997) are mainly due to different reference periods used for normalization. Instead of referring to the modern period 1951-1980, Fig. 1 is based on mean values and standard deviations from the whole period of investigation. Looking at the normalized cumulative anomalies some major periods of differing behaviour are identified. During winter, we get an early period, until the 1850s, with accumulating negative deviations, before an opposite tendency (prevailingly positive deviations) is dominating until the beginning 1930s. The well-known recent evolution is reflected in the cumulative anomalies by a decline from roughly 1950 until 1980, with an increase afterwards which still appears relatively normal (only for the very short period since 1989 are there significantly higher mean values than in the earlier periods, marked by positive deviations).

Contrasting developments are identified for the summer season. Cumulating positive anomalies occur during the early period until the 1870s, with an opposite tendency afterwards and during recent times. But winter and summer seasons do not always tend to inverse relationships, as might be indicated by roughly 40-year periods with changing trends in summer after the 1870s. The early period, until the middle of the 19th century, is clearly characterized by a less intense westerly circulation during winter and a more intense one during summer in relation to the long-term mean conditions. This should imply some changes in circulation pattern characteristics which cannot be revealed by simple indices - thus, types and patterns of circulation will be considered next.

4.2 Monthly Grosswettertypes

Frequencies of monthly mean Grosswettertypes determined as described in the preceding 'Methods' section are shown for winter (DJF) and summer (JJA) in Fig. 2. During winter, only minor contributions, on a monthly scale, of High and Low pressure types occur (not included within Fig. 2). The flow types described by large-scale wind directions have been integrated into groups of different circulation types: SW/W representing conditions with largely zonal advection of maritime air masses towards Europe; NW/N implying a subpolar source region of advected air masses from the North Atlantic (frequently linked with increased snow fall, for example, in the northern parts of the European Alps); NE/E comprising typical situations with cold air directed from the interior continent towards

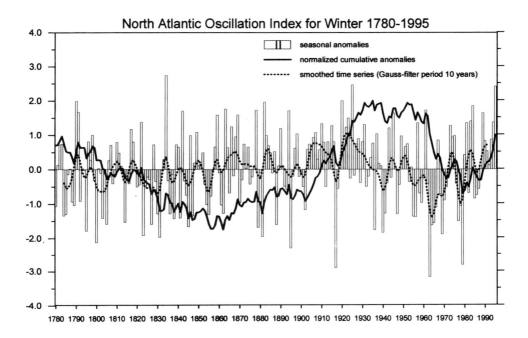

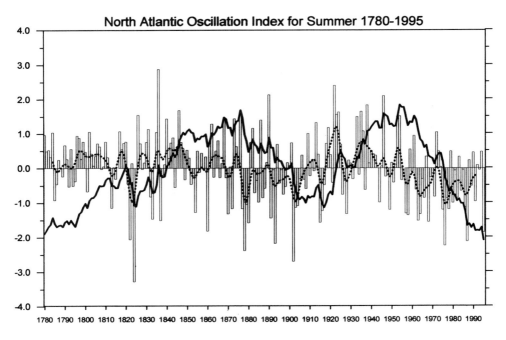

Figure 1. North Atlantic Oscillation Index derived from gridded monthly mean sea-level pressure for winter (DJF) and summer (JJA) in the period 1780-1995.

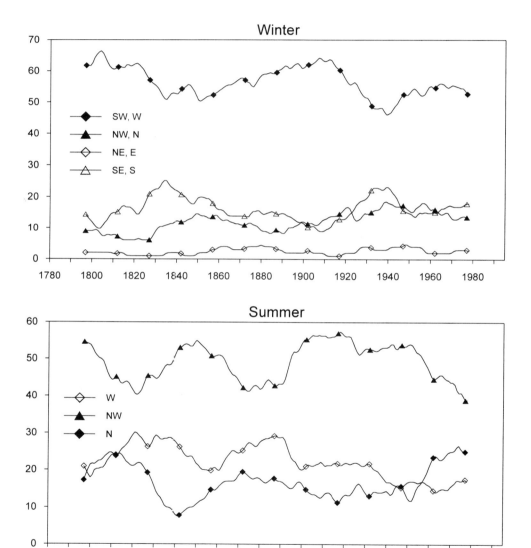

Figure 2. Absolute frequencies of selected monthly European Grosswettertypes for winter (DJF) and summer (JJA), calculated for 31-year periods with time steps of one year.

Central Europe; SE/S covering further meridional circulation types with a greater variety of air masses reaching large parts of Europe.

SW/W types decreased during the first half of the 19th century, and increased during the second half (Fig. 2), before another minimum was reached around 1940. SE/S types including considerable parts of SLP patterns with anticyclonic steering from the eastern continental area generally depict an inverse evolution, whereas NE/N types only slightly increased at the end of the 19th century and around the 1940s. Especially during the first half of the 19th century, SE/S types distinctly exceeded NW/N types. Together with the decrease in SW/W types this indicates a modified circulation regime for the last period of the Little Ice Age, with greater importance of anticyclonic influences on Central Europe.

During summer, NW and N Grosswettertypes reveal inverse frequency changes (Fig. 2), thus preventing integrations into groups of circulation types like those for winter. Since most of the Grosswettertypes account for only small percentages on a monthly scale during summer, Fig. 2 only reproduces particular frequencies of those Grosswettertypes of some importance for this season: W, NW, and N. For the westerly type there is an increased frequency during parts of the first half of the 19th century, in accordance with similar increases that have been identified on a daily scale for months with anomalous SLP distribution patterns during the same early instrumental period (Jacobeit *et al.*, 1998). After another peak at the end of the 19th century, the W type frequency in summer decreases during the 20th century with recent values clearly below the level of the earlier periods. The inverse frequency changes of NW and N types seem to be linked to the W type variation, such that higher values of the latter tend to follow increased N and decreased NW type frequencies. This link does not hold in the other direction, especially during the last decades when another convergence of NW and N types takes place with a low level of W type frequency. In general, the first half of the whole study period indicates a greater interdecadal variability among these main types, whereas the second half reflects more long-term trends (decreasing frequencies for W and NW, increasing tendency for N).

4.3 PCA Derived Circulation Patterns

Results from monthly T-mode PCAs are only shown for January and July representing high-season conditions for winter and summer. Figures 3 and 6 give the resulting basic circulation patterns represented by normalized SLP fields of those months with the absolute highest loading on the corresponding principal component. Figures 4 and 7 show the normalized time coefficients extended over the whole investigation period as described in the earlier 'Methods' section. Figures 5 and 8 show some examples of internal pattern changes given as SLP composites and their differences.

For both the January and July analyses, four principal components have been extracted, explaining some 96 and 97%, respectively, of the original variance during the whole study period (1780-1995) (see Table 1 for details concerning the particular components). According to the criteria for extraction mentioned in the earlier 'Methods' section, a fifth pattern in January would emerge. Since it is the dominant one only for two January months, it will not be considered for the following studies referring to the whole period of 216 years. Furthermore, since negative loadings only occur with minor amounts (> -0.5) never constituting a reflective pattern, the reproduced SLP fields of Figures 3 and 6 are, in fact, representing the basic circulation patterns of January and July, respectively.

Table 1. Variances (%) explained by principal components 1-4 of Figure 3 (January) and Figure 6 (July) during the period 1780-1995.

	PC1	PC2	PC3	PC4	Total
January	50.5	28.2	11.7	5.4	95.8
July	30.5	28.8	21.4	16.2	96.9

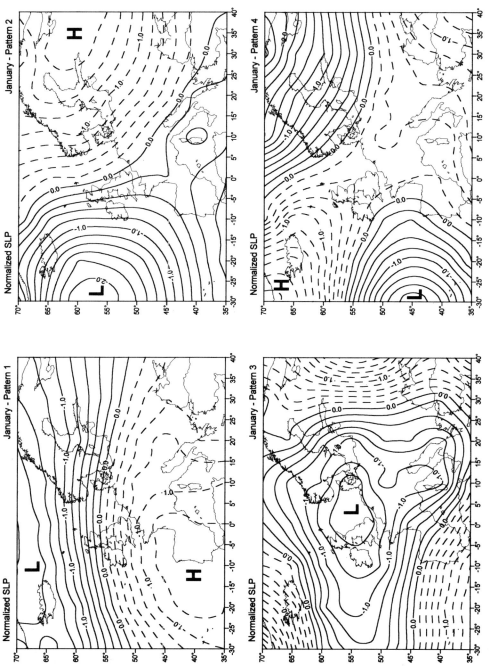

Figure 3. Circulation patterns for January derived from T-mode PCA of monthly mean North-Atlantic-European SLP grids of the period 1780-1995.

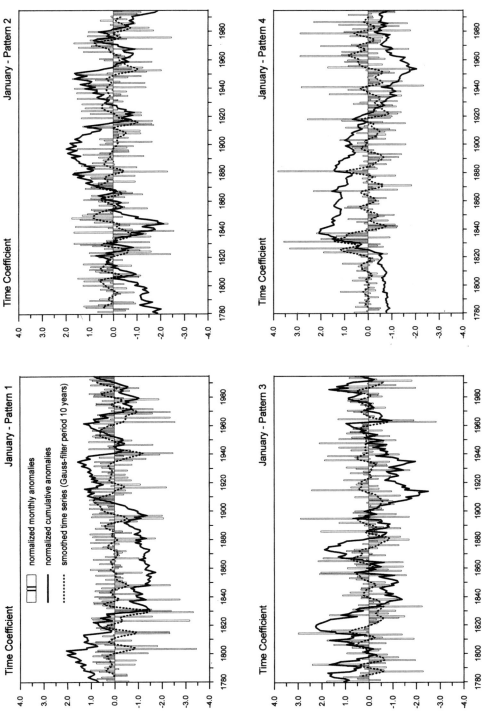

Figure 4. Normalized time coefficients (monthly, smoothed, and cumulative time series January 1780-1995) for the circulation patterns of Figure 3 (see next but one page).

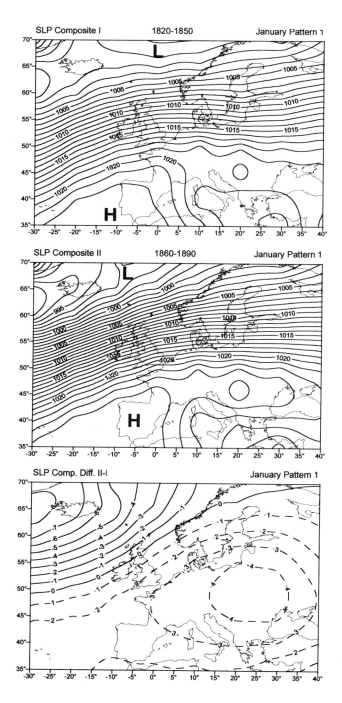

Figure 5. Selected SLP composites (hPa) and their differences derived from monthly mean SLP grids of those January months within the indicated 31-year periods with their highest loading on the indicated circulation pattern from Figure 3.

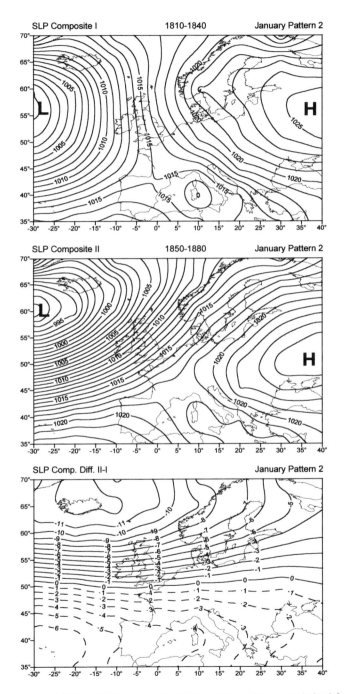

Figure 5. (continued) Selected SLP composites (hPa) and their differences derived from monthly mean SLP grids of those January months within the indicated 31-year periods with their highest loading on the indicated circulation pattern from Figure 3.

January

The high-winter situation is characterized by the following circulation patterns (Fig. 3):

- a westerly flow type with the Azores' high pressure extending towards the western Mediterranean (Pattern 1; representing month, January 1944);
- a Russian high extending as far as Central Europe, whereas the North Atlantic is covered by a strong low pressure system (Pattern 2; representing month, January 1848);
- a distinct low pressure system centred above the North Sea extending over much of Europe (Pattern 3; representing month, January 1814);
- a NAO reversal pattern with high pressure centred over the Icelandic region (Pattern 4; representing month, January 1881).

Most of the SLP variance (78.7%, see Table 1) is explained by the first two patterns; the other two only occur during rather anomalous situations like, for example, Pattern 4 in January 1963 with its NAO reversal (Moses *et al.*, 1987).

Recently, Corti *et al.* (1999) have analysed the northern hemispheric circulation for an extended winter time period (November to April) deriving just four different circulation regimes on the monthly time scales. These regimes do not agree in detail with the patterns described above, since there are differences in the analyses (e.g. concerning space and time domain, the atmospheric level, and particular methods that have been applied). But important regimes discussed by Corti *et al.* (1999) are still reflected in the above patterns; e.g. the positive phase of the NAO (Pattern 1) or the negative phase of the Arctic Oscillation (Pattern 4).

Information concerning circulation variability since 1780 may be drawn from the time coefficients of Fig. 4: the normalized cumulative anomalies show marked periods of dominating positive or negative deviations (i.e. periods of stronger or weaker representation of the corresponding circulation patterns within the original SLP fields). Thus, the westerly flow type (Pattern 1) had a major period of cumulating negative anomalies at the beginning of the 19th century, its Variance Explained (VE value henceforth) dropped to 41.2% between 1803 and 1831 (compared to 50.5% for the whole 1780-1995 period; see Table 1). On the other hand, the Russian high pattern sometimes reached above-average representations during the first hundred years, especially for the periods 1800-1821 (VE=33.8%) and 1845-1879 (VE=36.7%). The interval 1822-1844, during which VE dropped to 24.9%, approximately coincided with the maximum VE for Pattern 4 (18.4% between 1823 and 1837 compared to 5.4% for the whole period since 1780). Since Pattern 4 may be seen as a further westward extension of high pressure compared to its position over Russia in Pattern 2 (see Fig. 3), this sequence reflects a protracted period (roughly the first hundred years from 1780 onwards) with increased occurrence of high pressure influence on Europe culminating with the most westward extensions during the third and fourth decades of the 19th century. For the 1845-1854 decade, explained variances of the Russian high pattern (VE=49.9%) even exceeded those of the westerly flow pattern (VE=35.7%). In the second half of the 19th century, however, important changes occurred within these major circulation patterns (see below).

Conditions during the recent century are well-known (Klaus, 1997; Jacobeit, 1997) and will be mentioned only briefly. Fig. 4 shows that there were periods of well-developed westerlies at the beginning of the 20th century (VE=59.7% for Pattern 1 during 1898-1935), again during the 1950s and, especially, for the very recent time (VE=67.7% during 1981-1995). In contrast, there was one major period in between with increased importance of anticyclonic influences. This did not affect patterns with the most westward positions of high pressure (Pattern 4 with only occasional emergence e.g. in 1963 or 1987), but did affect those with extended Russian high pressure (VE=36.7% for Pattern 2 during 1924-

1950) combined with reduced importance (VE=41.2%) of the westerly flow pattern, implying a 18% reduction in VE differences between Patterns 1 and 2 compared to the long-term values since 1780 (Table 1).

The widespread low pressure pattern 3 accumulated negative deviations during the 1820s and 1830s, and for decades around the end of the 19th century (VE=7.8% for the 1874-1914 period), whereas its relative importance increased moderately during the 20th century (VE=12.3% for the 1915-1986 period).

Besides these variations in time coefficients and explained variances, circulation patterns also underwent internal changes illustrated by SLP composites differing from each other with a maximum RMS value within the 19th century (Fig. 5). Comparing westerly flow pattern composites for the periods 1820-1850 and 1860-1890 clearly indicates that, during the 19th century, this pattern shifted northward over Europe in association with significant increases in meridional pressure gradients. This occurred both over the continent, where pressure rose most strongly around Central Europe, and over the North Atlantic where pressure dropped most strongly in the Icelandic region.

The greatest RMS value for Pattern 2 composites included a period (1810-1840), going further back than available station records for Iceland (data close to the Azores had been available since the 1820s). This period was used for comparing with a later period (1850-1880), since important changes for this pattern were located near the central grid points and no other information emerged when using periods after 1820. In addition to another increase in meridional pressure gradients (see Fig. 5, pt. 2; composite differences map), the major change during the 19th century consisted in a southeastward retreat of the Russian high implying that western parts of Central Europe were no longer under its direct influence and became affected by southwesterly components in front of a deepened Atlantic low pressure system. These changes within Pattern 2 were an important stage in the transition from the Little Ice Age circulation to more recent conditions still before an increase in the westerly flow pattern set in at the end of the 19th century.

Since the other circulation patterns of January reach the highest loading within one month only in quite few cases, no further composites will be considered.

July

The high-summer situation is characterized by the following circulation patterns (Fig. 6):

- a zonal flow type with low pressure above the Nordic Sea and a zonal ridge of high pressure to the south (Pattern 1; representing month, July 1928);
- a widespread low pressure system with its centre over southern Scandinavia and the Azores high shifted to a southwesterly position (Pattern 2; representing month, July 1888);
- a diagonal ridge of high pressure extending from north of the Azores towards southern Scandinavia (Pattern 3; representing month, July 1825);
- a meridional ridge of high pressure centered above southern Scandinavia with blocking effects on the Icelandic low pressure system (Pattern 4; representing month, July 1994).

In contrast to the January situation, these patterns have a smaller range in percentages of explained variance (VE values for the 1780-1995 period ranging from 30.5 to 16.2%, see Table 1). This means that none of these patterns is restricted to only a few cases with rather anomalous conditions.

The normalized cumulative anomalies of the time coefficients (Fig. 7) again reveal some major periods of different importance in the circulation patterns. Thus, on a long-term

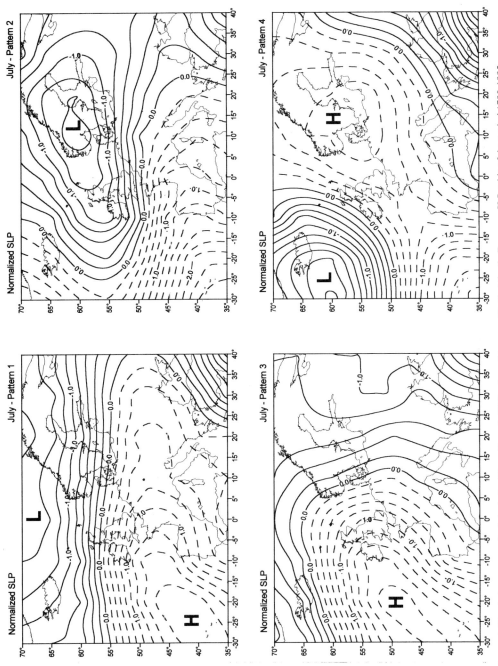

Figure 6. Circulation patterns for July derived from T-mode PCA of monthly mean North-Atlantic-European SLP grids of the period 1780-1995

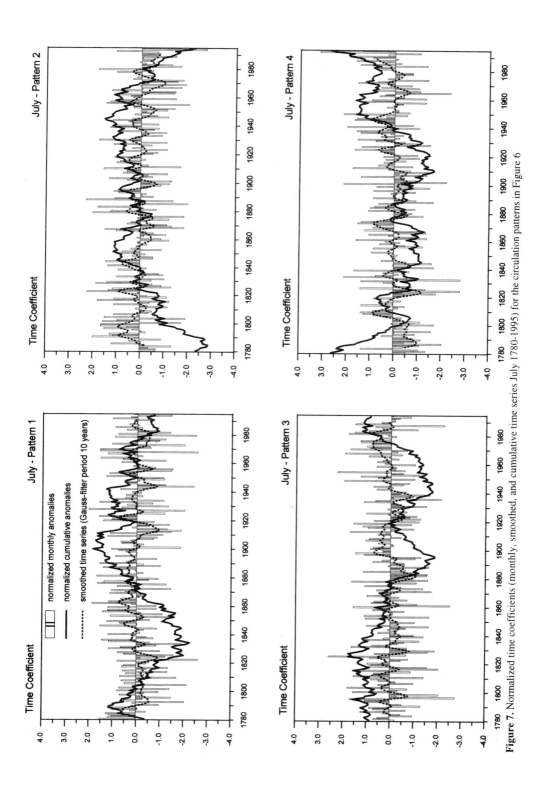

Figure 7. Normalized time coefficients (monthly, smoothed, and cumulative time series July 1780-1995) for the circulation patterns in Figure 6

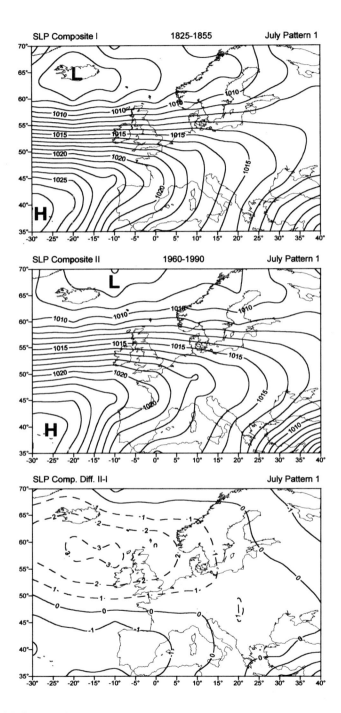

Figure 8. Selected SLP composites (hPa) and their differences derived from monthly mean SLP grids of those July months within the indicated 31-year periods with their highest loading on the indicated circulation pattern from Figure 6.

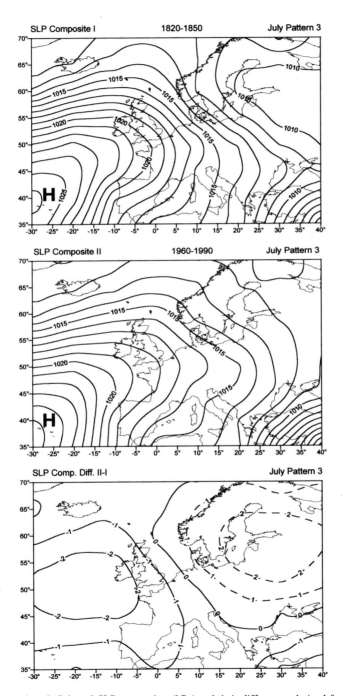

Figure 8. (continued) Selected SLP composites (hPa) and their differences derived from monthly mean SLP grids of those July months within the indicated 31-year periods with their highest loading on the indicated circulation pattern from Figure 6.

perspective, circulation patterns 1 and 3 show periods of inverse association. From the 1830s until the end of the 19th century the zonal pattern 1 accumulated positive deviations in its time coefficient; the Atlantic ridge pattern 3, in contrast, showed negative ones. Accordingly, explained variances differed more distinctly than in Table 1 (e.g. VE=33.6% for Pattern 1 and VE= 17.0% for Pattern 3 during 1828-1895). After a downward trend for the ridge pattern during the first four decades of the 20th century, the most recent period had positive deviations, in contrast to the zonal pattern which preferred negative deviations. This evolution amplified after 1970, with VE values for 1971-1984 of 23.0% (zonal pattern) and 34.7% (ridge pattern).

Another recent decrease in relative importance occurred for the cyclonic pattern 2, especially pronounced during the 1985-1995 period (VE=17.2%), when the meridional ridge pattern 4, in contrast, increased strikingly (VE=29.6%). Opposite conditions in July prevailed during the early period when the cyclonic pattern 2 reached its greatest long-term importance (VE=36.7% during 1790-1825), and the meridional ridge pattern 4 started at a distinctly low level (VE=8.3% during 1780-1802). Remarkably, the westerly pattern 1 (VE=24.0% during 1790-1825) also stayed behind the more pronounced cyclonic mode of Pattern 2 during this early period, before the above-mentioned period of cumulating positive deviations set in during the 1830s.

In respect of internal pattern changes, SLP composites with maximum RMS values between the first half of the 19th century and the second half of the 20th century will be compared for two contrasting circulation patterns; the westerly flow type of summer and the Atlantic ridging towards southern Scandinavia. According to Fig. 8 the zonal pattern 1 attained higher pressure above the central North Atlantic during the modern period (1960-1990) corresponding to a northward shift of the higher-value isobars in this region, and was downstream of the British Isles compared with the conditions some 130 years ago (1825-1855). A similar northward shift occurred within the cyclonic pattern 2 (not shown here), along with its decreased importance during summer in the modern period.

The diagonal ridge pattern 3, with its recently increased importance underwent internal changes at the same times: rising pressure around the Baltic region and falling pressure north of the Azores (Fig. 8) corresponding to a less developed anticyclonic ridge in its source region but extending - with a reduced pressure gradient - further northeastward towards southern Scandinavia, where cyclonic curvature existed during the first half of the 19th century.

5.   CONCLUSIONS

Based on gridded monthly mean SLP data reconstructed back to 1780 by Jones *et al.* (1999), calculations of seasonal NAO indices, frequency changes of European Gross-wettertypes on a monthly scale, and of PCA derived circulation patterns have been performed allowing an assessment of the major circulation changes that have occurred in Europe during the winter and summer seasons since the end of the 18th century. The first hundred years since 1780 have only been analyzed objectively in a few cases (see Introduction), considering some aspects of large-scale circulation variability. Key changes revealed by the analyses described in the preceding sections may be synthesized as follows.

During winter there was a period mainly in the first half of the 19th century - known as the concluding phase of the so-called Little Ice Age - with weakened westerlies in the North-Atlantic-European region. This is reflected in prevailingly negative deviations of the NAO index until the 1850s, in decreasing frequencies of SW and W Grosswettertypes accompanied by increases in SE and S types, in cumulating negative anomalies of the time coefficient of the zonal circulation pattern for January, and in a more southerly position of this pattern along with lowered meridional pressure gradients. At the same time, high pressure influence in Europe was distinctly greater, caused by more westward extensions of the Russian high (circulation pattern 2 for January) and, in an intervening period (third and fourth decades of the 19th century), by high pressure systems extending towards the Icelandic region (circulation pattern 4). During the second half of the 19th century an opposite regime was established: increasing frequencies of SW and W Grosswettertypes along with decreasing frequencies of SE and S types, and marked internal changes within the major circulation patterns indicating a northward shift of the zonal circulation pattern over Europe in association with significant increases in meridional pressure gradients, as well as a southeastward retreat of the Russian high within the second circulation pattern for January (Fig. 5). Around the turn from the 19th to the 20th centuries, for periods of well-developed westerlies set in for the first decades, for the 1950s, and for the very recent time interrupted by further periods of increased anticyclonic influence from the east.

For the summer season we get very different results. During the first half of the 19th century, zonal circulations prevailed indicated by cumulating positive anomalies in the NAO index, increased frequencies of the westerly Grosswettertype and dominating positive deviations in the time coefficient of the cyclonic circulation pattern 2 for July. These were replaced by similar positive deviations in the westerly pattern 1 during the second half of the 19th century, contrasting to the opposite development of the anticyclonic ridge pattern 3 (Fig. 7). During the 20th century, however, Grosswettertype W reduced in frequency, and roughly the last 50 years most clearly showed the reverse of the 19th century: predominantly negative deviations in the NAO index and in the time coefficient of the cyclonic circulation pattern, whereas the anticylonic ridge pattern accumulated positive anomalies and additionally extended further northeastward than during the first half of the 19th century (Fig. 8). This redistribution from cyclonically to anticyclonically controlled conditions is further emphasised by the long-term downward trend in the cumulative anomalies of the westerly circulation pattern 1 during the 20th century (Fig. 7), as well as by the fact that the meridional ridge pattern 4 which mostly had negative anomalies in the early stages of the whole period, recently has reached its maximum representation. Another indication for a modern increase in meridionality of the summer circulation might be the recent tendency for decreasing NW, but increasing N, Grosswettertypes on a monthly scale (Fig. 2).

In general, the extension of circulation analyses to a period of more than two centuries not only provides a perspective of immediate controls on past climate variations, but also offers an extended range of natural variability to be considered. This is important when judging the nature of recent changes, and of changes projected into the future. For example, looking at the time coefficient of the zonal circulation pattern 1 for January (Fig. 4) reveals that, with respect to the whole period since 1780, the modern changes from low values of the 1960s to high values of the 1990s are not as outstanding as might be judged by referring to shorter periods or to single indices (e.g. for the NAO) which are more restricted to particular areas. The time coefficient for the whole North Atlantic European zonal circulation pattern 1 (Fig. 4) shows even lower values during the first half of the 19th century than during the 1960s, as well as other important periods of positive deviations at the beginning of the 20th century and during the 1950s. Thus, in addition to attempts by analogue approaches to reconstruct past pressure patterns in the 16th century, at least for some outstanding anomalies (Jacobeit *et al.*, 1999), further objective recon-structions as recently provided by Luterbacher *et al.* (2000) for the Late Maunder Minimum period (1675-1715) should be encouraged in order to enable further extensions of circulation analyses into the historical past with additional improvements concerning possible ranges of natural variability.

Acknowledgements

This work has been supported by the European Commission under grant ENV4-CT95-0129.

REFERENCES

Barnston, A.G., and Livezey, R.E., 1987, Classification, seasonality and persistence of low-frequency atmospheric circulation patterns, *Mon. Wea. Rev.* 115: 1083-1126.

Bartzokas, A., and Metaxas, D.A., 1996, Northern hemisphere gross circulation types, *Meteorol. Zeitschrift N.F.* 5: 99-109.

Beck, C., 1999, *Zirkulationsdynamische Variabilität im Bereich Nordatlantik-Europa seit 1780,* Ph.D. Thesis, Faculty of Earth Sciences, University of Würzburg, 333 pp.

Corti, S., Molteni, F., and Palmer, T.N., 1999, Signature of recent climate change in frequencies of natural atmospheric circulation regimes, *Nature* 398: 799-802.

Gerstengarbe, F.W., and Werner, P.C., 1993, Katalog der Grosswetterlagen Europas nach P. Hess und H. Brezowsky 1881-1992, *Berichte des Deutschen Wetterdienstes* 113 (fourth edition).

Hess, P., and Brezowsky, H., 1977, Katalog der Grosswetterlagen Europas, *Berichte des Deutschen Wetterdienstes* 113 (second edition).

Hupfer, P. (Ed.), 1991, *Das Klimasystem der Erde,* Akademie Verlag, Berlin, 464 pp.

Hurrell, J.W., 1995, Decadal trends in the North Atlantic Oscillation: regional temperatures and precipitation, *Science* 269: 676-679.

Jacobeit, J., 1993, Regionale Unterschiede im atmosphärischen Zirkulationsgeschehen bei globalen Klimaveränderungen, *Die Erde* 124: 63-77.

Jacobeit, J., 1997, Atlantisch-europäische Bodenluftdruckfelder ombrothermisch anomaler Monate in Mitteleuropa als Hilfsmittel für die synoptische Interpretation analoger Anomalien im historischen Klima und in zukünftigen Klimaszenarien, *Petermanns Geographische Mitteilungen* 141: 139-144.

Jacobeit, J., Beck, C., and Philipp, A., 1998, Annual to decadal variability in climate in Europe - objectives and results of the German contribution to the European climate research project ADVICE, *Würzburger Geographische Manuskripte* 43: 163 pp.

Jacobeit, J., Wanner, H., Koslowski, G., and Gudd, M., 1999, European surface pressure patterns for months with outstanding climatic anomalies during the Sixteenth Century, *Clim. Change* 43: 201-221.

Jacobeit, J., Jönsson, P., Bärring, L., Beck, C., and Ekström, M., 2000, Zonal indices for Europe 1780-1995 and running correlations with temperature, *Clim. Change* (in press).

Jones, P. D., Wigley, T. M., and Briffa, K., 1987, Monthly mean pressure reconstructions for Europe (back to 1780) and for North America (to 1858), *United States Department of Energy* TR037: 99 pp.

Jones, P.D., Jonsson, T., and Wheeler, D., 1997, Extension of the North Atlantic Oscillation using early instrumental pressure observations from Gibraltar and southwest Iceland, *Int. J. Climatol.* 17:1433-1450.

Jones, P. D., Davies, T. D., Lister, D. H., Slonosky, V., Jonsson, T., Bärring, L., Jönsson, P., Maheras, P., Kolyva-Machera, F., Barriendos, M., Martin-Vide, J., Alcoforado, M.J., Wanner, H., Pfister, C., Schuepbach, E., Kaas, E., Schmith, T., Jacobeit, J., and Beck, C., 1999, Monthly mean pressure reconstructions for Europe for the 1780-1995 period, *Int. J. Climatol.* 19: 347-364.

Kapala, A., Mächel, H., and Flohn, H., 1998, Behaviour of the centres of action above the Atlantic since 1881. Part II: Associations with regional climate anomalies, *Int. J. Climat.* 18: 23-36.

Kington, J.A., 1988, *The Weather of the 1780s over Europe,* Cambridge University Press, Cambridge, 166 pp.

Klaus, D., 1997, Änderungen der Zirkulationsstruktur im europäisch-atlantischen Sektor, *Abhandlungen der Math.-Naturwiss. Klasse der Akademie der Wissenschaften und der Literatur Mainz,* Jahrgang 1997, Nr. 3.

Lamb, H.H., and Johnson, A.I., 1966, Secular variations of the atmospheric circulation since 1750, *Geophys. Mem.* 110: 125 pp.

Luterbacher, J. *et al.,* 2000, Reconstruction of monthly mean pressure over Europe for the Late Maunder Minimum period (1675-1715) based on canonical correlation analysis, *Int. J. Climat.* (in press).

Mächel, H., Kapala, A., and Flohn, H., 1998, Behaviour of the centres of action above the Atlantic since 1881. Part I: Characteristics of seasonal and interannual variability, *Int. J. Climat.* 18: 1-22.

Moses, T., Kiladis, G. N., Diaz, H. F., and Barry, R. G., 1987: Characteristics and frequency of reversals in mean sea level pressure in the North Atlantic sector and their relationship to long-term temperature trends, *J. Climatol.* 7: 12-30.

Pfister, C., 1999, *Wetternachhersage - 500 Jahre Klimavariationen und Naturkatastrophen,* Verlag Paul Haupt, Bern, Stuttgart, Wien, 304 pp.

Richman, M.B., 1986, Rotation of principal components, *J. Climatol.* 6:293-335.

Rogers, J.C., 1984, The association between the North Atlantic Oscillation and the Southern Oscillation in the Northern Hemisphere, *Mon. Wea. Rev.* 112:1999-2015.

Rogers, J.C., 1990, Patterns of low-frequency monthly sea level pressure variability (1899-1986) and associated wave cyclone frequencies, *J. Clim.* 3:1364-1379.

Schmutz, C., and Wanner, H., 1998, Low frequency variability of atmospheric circulation over Europe between 1785 and 1994, *Erdkunde* 52:81-94.

Wanner, H., 1994, The Atlantic-European Circulation Patterns and their Significance for Climate in the Alps, *Report 1/94 to the National Science Foundation*, Bern.

Wanner, H., Gyalistras, D., Luterbacher, J., Rickli, R., Salvisberg, E., and Schmutz, C., 2000, *Klimawandel im Schweizer Alpenraum*, Vdf Hochschulverlag AG, Zürich.

A MILLENNIUM OF WEATHER, WINDS AND WATER
IN THE LOW COUNTRIES

A.F.V. van Engelen, J. Buisman and F. IJnsen

Royal Netherlands Meteorological Institute
PO Box 201, 3730 AE De Bilt, The Netherlands

1. INTRODUCTION

In 1995 under the auspices of KNMI the first volume of a series of 6 books was published, where in some 4000 pages the weather in the Low Countries (present-day Benelux) for the period 1000 to present is depicted and classified. In the books[1], the course of the weather during both the winter and summer season as well is followed from year to year and placed in a historical context. In summer 2000, Vol. IV, covering the era 1575-1700, will appear, meaning that the non-instrumental period will be covered. Vol. V (to be published in 2001) and Vol. VI (2002) will deal with the instrumental period.

The reconstruction of the weather in the Low Countries is based on sources that are related to the area geographically covered by the present Netherlands and neighbouring areas of the southern part of the North Sea; Gt. Britain, Northern districts of France, the downstream basin of the Rhine, Westphalia and Northwest Germany. If relevant, remarkable or extreme weather in Middle Europe or even Northern Italy is also considered.

The major part of the text is devoted to detailed, well-documented, annotated descriptions and analyses of the weather in the past. Numerous compilations and classifications however offer a structured base for further interpretations. Without being complete the following are mentioned: sources per era and per area, climatology per 25 year period, harvest data, ice on major rivers, tree-ring data, classifications of winter and summer temperature, wet and dry seasons, storms and storm surges.

The books are written in the Dutch language and therefore not easily accessible to the international research world. It is for this reason that we present here in a brief overview some of the issues: the historical sources used, pitfalls that threaten dating due to the different types of calendars that were in use and the way the historical evidence could be classified into instrumental winter temperatures.

The most important publications from Volumes I-III are listed in the *References for the Period 763-1575*.

The *Appendix 1200-1250* contains an integral translation of the book text for 1248-1249, a table of the seasonal and annual temperatures of the first half of the 13th century, and a list of the most important historical documentary sources for this period.

History and Climate: Memories of the Future?
Edited by Jones *et al.*, Kluwer Academic/Plenum Publishers, 2001

2. SOURCES

Sources that provide useful information about the weather in the past can be subdivided into two main categories: the tangible sources and the non-tangible sources like toponyms, folk customs, traditions, legends and songs. The analyses of the past weather are predominantly based however on tangible sources. These sources yield essentially two types of data: direct, mostly written observational data of the weather and indirect or proxy data, within which the effects of weather and climate are recorded. In judging the reliability of the sources it is of importance to be aware of the number of levels of transmission, that can introduce noise, between observer and the historical fact. The evidence of a primary source, as provided by the observer himself or a contemporary, and only slightly transformed by transmission is potentially more reliable than that of a secondary source with a large distance in space or time between the historical fact.

The analyses of the weather in history of the Low Countries are based upon the following types of sources:

2.1 Instrumental observational series

Detailed descriptions of the course of the weather in the historical instrumental period (ca. 1700-1850) will be based on the digitised KNMI historical data-set with the more than 2,5 million three times a day readings, carried out at some 12 historical observational stations, of air temperature, pressure, windspeed and direction, precipitation, state of the sky and some astronomical observations. Station dictionaries with comprehensive descriptions of the instruments used, observers, conditions of observations and other relevant metadata have been composed for these and some 125 other observational locations in the Low Countries[2]. Descriptions from 1900 onwards will originate from the modern KNMI information systems. For the in-between period 1850-1900 the ''not yet digitised'' information from the so-called KNMI Yearbooks serves as the source.

2.2 Weather diaries and Weather journals

Weather diaries, offering daily qualitative descriptions of the weather, are popular from the beginning of the 17th century onwards. In the instrumental period they are often mixed with instrumental readings. The oldest weather diary, as far the author knows, was written by William Merle[3] from England, covering the period 1337-1344. Clergyman David Fabricius[4] (1564-1617) from East Friesland composed the oldest diary from the Low Countries. The Dutch barrister at fiscal law Anthonis Duyck[5], witnessing Prince Maurice and his army, wrote an important daily weather journal with notations for the weather from 1590-1602. Other rich sources are the diaries of Isaac Beeckman[6], giving the weather from day to day in Zeeland (1612-1615) and the weather observations carried out by probably the oldest weather observatory in the Netherlands at Dordrecht (1627-1637). Fabio Chigi (1599-1667) was papal nuncio at Cologne, Munster and Aachen and became well known as Pope Alexander VII. As an Italian he was fascinated by the cold northern climate and he noted down all frost, ice, snow and rain but paid no attention at all to sunny and warm days. We owe him an inheritance of descriptions of the weather for more than 2600 days[7].

2.3 Accounts

Useful proxies of the weather are the series of accounts of townships, ship cargoes, agriculture, water- and windmills and especially the accounts of river tolls.

In the middle ages and long after that period trade and traffic were hampered by tolls on both overland routes and rivers. Tolls were quite numerous, at least on the continent because they were abolished in England by order of the Magna Carta (1215). They were initiated by Landlords with the aim of benefiting river trade and were legalised by the so-called *royal stream prerogatives*. The oldest preserved toll accounts[8] date from the 14th century. They offer us very important information about situations hampering movement as heavy ice, low water levels and extreme weather conditions such as storms. Most important tolls were located along the Rhine (Cologne, Lobith, Arnhem), Meuse (Roermond), IJssel (Zutphen) and the Waal (Nijmegen, Tiel, Zaltbommel). Comparable weather information is hidden in the registers of the ship canal companies. Towboats were popular as public transport in the Low Countries from the 16th century onwards. The economist De Vries[9] compared the ship canal data with the De Bilt temperatures for the periods 1634-1757 and 1814-1839, yielding a reconstruction of winter-temperatures back to 1634.

In the past centuries a dense network of some 10,000 mills covered the Low Countries. Many of them were water windmills and watermills. A very complete and long series of accounts (1445-1540) with numerous annotations of the weather is found in the archives of the township of Zutphen[10], which exploited the watermills on the small river, the Berkel.

2.4 Letters

The majority of the writers of the thousands of letters that were examined considered the topic of weather as quite vulgar and generally did not spend many words on it. There are some notable exceptions however. The letters of Mr. Adriaen van der Goes[11] (1619-1686), lawyer at the Court of Holland, are one of the major sources of descriptions of the weather for the period 1659-1673. As a farmer his main interest was in harvests and working the soil. Weather related matters were important to him and he often described them in detail.

2.5 Diaries and Journals

Whilst diaries have a personal character predominantly, journals are to be considered as more or less official notes. These contemporaneous witnesses are dominated mostly by the personal or professional interest of the author. Though covering a relatively short period (1491-1498) the diary of Romboudt de Doppere (the Brave) registrar of the Minster of St. Donatius[12] (Brugges) gives us very important and quite detailed information about the weather conditions.

Ship journals are well known sources of weather (windspeed and direction, state of the sky) information in the past. As The Netherlands developed in the course of the 17th century as one of the world's most important trading country, thousands of journals were officially kept by their captains, serving for instance the *Verenigde Oostindische Compagnie* (VOC, 1602). Though they are mostly filled with information from far away (West and East Indies, African coasts), less relevant within the context of the climate in the Low Countries, many records reflect the sometimes long periods the ships lay at home, for instance the Dutch isle of Texel waiting for favourable winds or melting of obstructing floating ice in winter times.

Journals of war-vessels offer similar information. Many military journals are available from the sea wars between England and the Netherlands as those from 1652-1654, 1665-1667 and 1672-1674. It is quite probable that a complete and detailed reconstruction of wind and state of the sky, covering the areas of the coasts of Holland and Zeeland and the southern part of the North Sea could be based upon these ship-journals.

Diaries from farmers make mention of early and late frost, snow, dry and wet periods, storms, lightning, wind gusts and hail, all related to the crops, yields and prices. Crop data have a high potential as proxy data.

As many in the 16th and 17th century were inclined to explain the course of the weather by the constellation of sun, moon and stars, their astrometeorological journals can be considered as weather diaries, including the ephemerides. A well-known example is that from Isaac Beeckman[6].

2.6 Annals and chronicles

Annals and chronicles are quite simple forms of historiography. Annals originated from Easter tables, long standing tables of Easter dates, illustrated with annotations of various events during the years. Annals and chronicles are quite popular from the Middle Ages onwards. In contrast to the above summarised sources, they generally witness only partly the personal observations and experiences of the author. The distinction between both is that annals mostly are anonymous tabular lists of dates and events (earthquakes, storms, comets) and composed by various authors, for personal use mainly, whereby chronicles often (but not always) are written by one and the same author, more extensive and narrative and often popular as can be deducted from the many copies and prints. Town chronicles, predominantly from those nearby main rivers offer many weather related events as times of famine, storms, high and low waters and ice.

3. CALENDARS

After the conquest of Rome in the 6th century BC, the Romans introduced the *Roman Republican Calendar*, based, as was usual, on the phases of the moon with a length of 355 days. In order to prevent a shift of the seasons, the Romans introduced the *intercalans*; every two years the last five days of February were replaced by a period of 27 or 28 days. The beginning of the year was decreed on the first of March: the date on which the new consuls started their duties. Calendar and seasons thus got out of step. Advised by the Alexandrian astronomer Sosigenes, Julius Caesar settled affairs in 46 BC; he introduced a corrective intercalans of 67 days, in order to put the seasons in the right place and turned over to the solar calendar with an average length of the year of 365 days and 6 hours. Every four years the 24th of February was doubled as an intercalary day. This was the *Julian Calendar*.

In the Julian Calendar the vernal equinox and the winter solstice fell on the 25th of March and the 25th of December respectively. Every century is, according to the Julian Calendar, 18 hours too long. Gradually vernal equinox and solstice shifted forwards in time. As the vernal equinox was needed for the calculation of the date of Easter the *Council of Nicea* ordered this date to be fixed on the 21st of March. The astronomical vernal equinox did not obey this rule and in the 16th century the real astronomical vernal equinox fell on the 11th of March and the 'legal' equinox was still the 21st of the same month.

Table 1. From old style to new style.

Period	days
700 - 899	+ 4 days
900 - 999	+ 5 days
1000 -1099	+ 6 days
1100 -1299	+ 7 days
1300 -1399	+ 8 days
1400 -1499	+ 9 days
1500 -1582	+ 10 days

In 1563 the *Council of Trent* authorised the pope to carry out a reform. Pope Gregorius XIII announced on 24 February 1582 by the *Bull Inter Gravissimas* the *Gregorian Calendar*. The date October 4, 1582 was immediately followed by October 15, to compensate the lag of 10 days between astronomical and formal vernal equinox. Furthermore every four centennials the leap year was abandoned.

The catholic countries accepted this new calendar (*stilus novus*, new style) quite soon but the Protestant countries, on the contrary, kept the old calendar (*stilus vetus*, old style) sometimes for even more than a century. As a consequence in the Low Countries two calendars were in use that differed 10 days: an important fact when we have to judge our sources.

In the descriptions of the weather dating from before 1582, the old style is applied consistently, afterwards the new style. The differences (in days) of both styles are listed in Table 1.

4.   CLASSIFICATION OF WINTERS

Often historical quantitative evidence can be found of the coldness of the winter season as descriptions of the length of the period of frost or no evidence of frost at all, ice drift on the rivers, days that no toll could be charged.

In the KNMI books the method of IJnsen[13] is adopted for a quantitative classification of the winter temperatures, based upon the number of frost days (v, $T_{min} < 0°C$), ice days (y, $T_{max} < 0°C$) and very cold days (z, $T_{min} < -10 °C$). With these numbers, known for the winter season (November-March) at De Bilt (central Netherlands) from 1850 onwards the so-called *frost index*[14] (V) is calculated as:

$$V = 33\tfrac{1}{3} \bullet \left(\frac{v^2}{12100} + \frac{y}{50} + \frac{z}{30} \right)$$

For the instrumental period 1706-1850 only monthly temperatures for De Bilt were available[15] and so it was not possible to calculate directly the frost index. Therefore IJnsen developed a formula for what he called the *winter-number* (H) that characterises the coldness of the winter season, November-March. This formula is based upon T_f: average temperature November-March, T_h: average temperature December-February and T_k: average temperature of the coldest month where:

$$H = 74.88 - 4.61 \bullet T_f - 3.32 \bullet T_h - 2.3 \bullet T_k$$

H was calculated for the period 1707-1990. Over the overlapping period 1851-1990 H and V are functionally related by the equation:

$$V = \exp[51592 \bullet \tanh(0.01346 \bullet H)], \quad 0 \le H \le 100,$$

with a correlation coefficient of 0.97. This makes it possible to extend V back from 1850 to 1706.

V has no dimension and ranges from $V=0$: a winter without frost to $V=100$ for the most severe winter. The benefit is that, though calculated from figures for De Bilt, V has the expectance of the regional character of a winter, for an area with corresponding climatological characteristics: the Low Countries.

Based on the frequency distribution of the frost indices of the 284 winters covering the period 1707-1990, a symmetrical division into 9 categories (1–9), with 5 as centre, is made in which the frost indices are grouped and labelled with a definition.

In Table 2, this classification is given together with the expected frequency of occurrence (in %) of the various strengths of winter. To provide thermal information, the expected mean temperature of the climatological winter - the months December, January and February - and the one for the cold season November up to March inclusive for De Bilt are given with each category.

Table 2. Classification of winters according to IJnsen.

Category Cv	Frost Index V	Definition	Frequency in %	Temp. °C Winter (DJF)	Temp. °C Nov.-March
1	≤ 3.2	Extremely mild	1.0	6.2	6.8
2	3.3-5.7	Very mild	3.8	5.4	5.6
3	5.8-9.7	Mild	11.1	4.3	4.9
4	9.8-16.6	Fairly mild	21.0	3.3	4.2
5	16.7-28.4	Normal	26.2	2.3	3.5
6	28.5-44.3	Cold	21.0	1.2	2.7
7	44.4-73.0	Severe	11.1	-0.1	1.7
8	73.1-82.0	Very severe	3.8	-1.8	0.5
9	≥ 82.1	Extremely severe	1.0	-2.4	0.1

The median value of V is 21.51, so that the category 'normal' can be subdivided in the sub categories '5-' for: [16.7 ≤ V ≤ 21.5] and '5+' for [21.6 ≤ V ≤ 28.4], respectively 'on the mild side and 'on the cold side'

With the help of this classification into categories it is now possible, with the non-instrumental information, to characterise winters. In order to do this a procedure is followed, based on awarding marks to the following three aspects, which are significant for many winters:

- A_t: *thermal aspect*. The whole cold season (November-March), if possible, is examined, with emphasis on the climatological winter months December, January and February and special emphasis on the coldest month.
- A_d: *aspect of duration*. A measure for duration is the number of days with frost from November till March or a comparable parameter; in many cases it will only be possible to make a rough estimate.
- A_i: *aspect of intensity*. The intensity is low if there are only a few days with frost (or a comparable parameter), high if there are many very cold days. In the instrumental era, intensity is understood to mean the quotient of the frost index and the number of days with a minimum temperature below 0 °C.

Table 3. Estimate of the category of the winter-season by means of aspects

Thermal aspect (A_t)	Points	Frost Index (V)
Very mild	1	≤ 5.0
Mild	2	5.0 – 15.0
Normal	3	15.1 – 30.0
Cold	4	30.1 – 75.0
Very cold	5	≥ 75.0

Aspect of Duration (A_d)	Points	Number of Frost Days (v)
Short	0	≤ 47
Moderate	1	48-88
Long	2	≥ 89

Aspect of Intensity (A_i)	Points	Intensity (V/v)
Low	0	≤ 0.18
Moderate	1	0.19 – 0.55
Strong	2	≥ 0.56

With the many uncertainties and the often rough and concise character of the historical evidence in mind, the intervals of the subdivision of the three aspects should be wide. The scoring table (Table 3) is derived from the historical evidence of well-known winters from 1850 onwards. The outcomes of the scoring are tested for the period 1901-1987. The sum of the scores for the three aspects in the majority (84%) of the winters exactly equalled the number of the category (C_v) of the winter in concern.

Based on the qualitative historical descriptions of the weather, categories have been estimated independently by Buisman and IJnsen as well. The two series appeared to be very comparable.

These series were compared with several other series for overlapping periods: generally they showed excellent correlation (correlation coefficients range from 0.79 to 0.98, IJnsen[16]). For this reason, the following series contributed to the estimates of the frost indices:

1591-1613 Number of days with frost in East Friesland, according to David Fabricius
1613-1626 Records of the diary of Isaac Beeckman
1621-1650 Number of days with frost and extensive periods with frost per winter in the Kassel series from Hermann van Hessen[17]
1634-1706 Number of days with ice on the canals Harlem-Amsterdam and Harlem-Leiden
1660-1706 Central England monthly temperatures from Manley[18]

5. CLASSIFICATION OF SUMMERS

Historical evidence of the warmth of summer is less abundant than of the coldness of winter. In history, weather in summer had, in the Low Countries, less impact on society than the weather in the winter season. The witnesses were mostly more interested in summer drought than in summer warmth. Also an observable and distinct thermal level, as icing when temperature drops below °C, is absent. The classification that was developed for summer temperatures (IJnsen[19,20]) is based on a single parameter, the *summer-number S*, which is, analogous to the winter-number H, a function of the temperature sum (i.e. the monthly temperature times the number of days per month) S_w of the warm season May-September, the temperature sum S_z of the climatological summer (June-August) and the temperature sum of the warmest month S_m:

$$S = 0.0489 \bullet S_w + 0.0670 \bullet S_z + 0.1573 \bullet S_m - 246.2$$

On the basis of the frequency distribution of S, calculated from monthly temperatures at De Bilt for the period 1706-1990, the summers were classified into 9 categories Cs (1-9), with 5 as centre. Table 4 shows this classification, together with the expected frequency of occurrence (in %).

Table 4. Classification of summers according to IJnsen.

Category Cs	Summer Number S	Definition	Frequency in %	Temp. °C Summer (DJF)	Temp. °C May-Sept.
1	≤ 13.7	Extremely cool	1.0	14.0	13.1
2	13.8-24.1	Very cool	3.8	14.6	13.6
3	24.2-34.5	Cool	11.1	15.1	14.1
4	34.6-44.8	Fairly cool	21.0	15.6	14.5
5	44.9-55.2	Normal	26.2	16.2	15.0
6	55.3-65.5	Warm	21.0	16.7	15.4
7	65.6-75.9	Fairly warm	11.1	17.3	15.9
8	76.0-86.3	Very warm	3.8	17.8	16.4
9	≥ 86.4	Extremely warm	1.0	18.3	16.8

For the pre-instrumental period (i.e. before 1706), the category of a summer is estimated directly from documentary evidence. These values are supported by estimates based on the following series that were compared with the De Bilt series for the overlapping instrumental period (IJnsen[16]):

1354-1836 Data start of grape harvest Beaune and Dijon[21]
1659-1705 Central England monthly temperatures from Manley[18]

6. WINTER AND SUMMER SERIES AND CONCLUSIONS

All estimates have been combined to continuous series of categories for the pre-instrumental winters and summers that have been coupled to the De Bilt series for the instrumental period. Based on this series, a record of annual winter and summer temperatures for De Bilt was calculated for the period 800-present

In the period prior to approximately 1300, evidence is often too scarce to permit a subdivision into 9 categories. For the classification of winters and summers with insufficient data, the categories C_v of the winter season were grouped three by three into the three main categories I (*obviously very mild*), II (*about normal*) and III (*obviously very cold*) and the categories C_s of the summer season into the three main categories I (*obviously cool*), II (*about normal*) and III (*obviously warm*).

Figures 1 and 2 show the 25 year means of winter and summer temperature, respectively, in the Low Countries, expressed as temperatures for De Bilt for the period 800-present. Table 5 contains all annual estimates of the categories C_v and C_s for the same period. Recently the temperature reconstruction (as LCT: *Low Countries Temperature*) has been analysed and compared with relevant data. It can be concluded that over the period back to the 14th century the LCT reconstruction is consistent with a number of European temperature reconstructions[22].

Table 5. Categories 751-1000

Year	C_v	C_s	Year	C_v	C_s	Year	C_v	C_s	Year	C_v	C_s	Year	C_v	C_s
751			801	I		851			901			951		
752			802			852		III	902		III	952		
753			803			853			903			953		
754			804			854			904			954		
755			805			855			905			955		III
756			806			856	6		906			956		
757			807			857			907			957		
758			808	I		858			908			958		
759			809			859			909			959		
760			810			860	III		910			960		
761			811	III		861			911			961		
762			812			862	6		912			962		
763			813			863	I		913	III		963		
764	III		814			864			914			964	III	
765			815	III		865			915			965		
766			816			866			916			966		
767			817			867			917	III		967		
768			818			868			918			968		
769			819			869			919			969		
770			820		I	870		III	920			970		
771			821			871			921		III	971		
772		III	822	8		872	III	III	922			972		
773			823			873			923			973		
774			824	7		874	III	III	924			974		III
775			825			875			925			975		III
776			826			876			926			976		
777			827			877			927			977		III
778			828			878			928		III	978		
779			829	II		879			929	III		979		
780			830			880	III		930			980		
781			831			881	III		931			981		
782			832			882			932			982		
783		III	833			883			933			983		
784			834	II		884			934			984		
785			835			885			935			985		
786			836			886	II	I	936			986		
787			837			887	III		937			987		
788			838	II	III	888			938			988	II	III
789			839			889			939			989		
790			840			890			940	III		990		
791			841	II	I	891			941			991		
792			842	II		892			942			992		
793		III	843	III		893			943			993		III
794			844	II		894	III		944			994	III	
795			845	III		895			945			995		
796			846	II		896			946			996		
797			847			897			947			997		
798			848			898			948			998		
799			849	7		899			949			999		
800			850		III	900			950			1000		

Categories Winter (NDJFM): C_v and Summer (MJJAS): C_s Low Countries
C_v ranges from 1(extremely mild) to 9 (extremely severe). I=C_v 1,2,3 II=C_v 4,5,6 III=C_v 7,8,9
C_s ranges from 1(extremely cool) to 9 (extremely warm). I=C_s 1,2,3 II=C_s 4,5,6 III=C_s 7,8,9

Table 5. Categories 1001-1250.

Year	C_v	C_s	Year	C_v	C_s	Year	C_v	C_s	Year	C_v	C_s	Year	C_v	C_s
1001			1051		I	1101			1151	I	I	1201	6	I
1002			1052			1102	5		1152	5	I	1202		
1003	5	II	1053			1103			1153		II	1203		
1004			1054			1104			1154			1204		
1005			1055			1105	II		1155			1205	8	III
1006			1056	3	I	1106	5	III	1156	8	I	1206	3	6
1007			1057			1107			1157	4	III	1207	6	II
1008			1058			1108			1158	4	II	1208	5	III
1009			1059			1109		I	1159		II	1209		II
1010			1060	7		1110		I	1160	7		1210	7	
1011	7		1061			1111	7		1161			1211	6	II
1012			1062			1112			1162			1212	4	II
1013			1063			1113			1163	6		1213	6	II
1014			1064			1114			1164			1214	6	
1015			1065			1115	7		1165	7		1215	5	II
1016			1066	3		1116	8	I	1166			1216	7	
1017			1067			1117	2	I	1167	6		1217	7	III
1018			1068			1118		I	1168	8		1218	3	II
1019			1069	7	I	1119	3		1169		I	1219	7	I
1020	6		1070	5	III	1120			1170	4	III	1220	3	
1021			1071	3		1121			1171			1221	6	6
1022			1072			1122	3	II	1172	2		1222	4	III
1023			1073			1123	4		1173	7	III	1223	4	I
1024			1074	6		1124	8	I	1174		I	1224	4	II
1025			1075			1125	7	I	1175			1225	8	II
1026			1076			1126	7		1176	6	II	1226	5	II
1027			1077	9		1127			1177		III	1227	7	II
1028			1078		III	1128	5	III	1178	5	II	1228	5	III
1029			1079			1129	5	III	1179	7	I	1229		
1030			1080	5		1130	6	III	1180		II	1230	7	
1031			1081			1131			1181		II	1231		
1032			1082			1132	7		1182	3	II	1232	5	III
1033			1083		III	1133		II	1183		II	1233		I
1034			1084			1134	6	II	1184		III	1234	7	II
1035	3		1085			1135	II	I	1185		III	1235	5	III
1036			1086		I	1136			1186			1236	5	III
1037			1087			1137	5	III	1187	3		1237	3	II
1038			1088			1138		III	1188		III	1238	5	III
1039			1089	4	I	1139		II	1189		III	1239	4	II
1040			1090			1140			1190	4	II	1240		II
1041			1091	3		1141	3	I	1191		I	1241		III
1042		I	1092		I	1142			1192		II	1242	6	II
1043		I	1093	7		1143	8	II	1193			1243		
1044	7		1094	4	I	1144	4	I	1194		III	1244		III
1045			1095	5	III	1145			1195	5	II	1245		II
1046			1096			1146			1196		I	1246	5	I
1047	8		1097	4	II	1147			1197	6	I	1247		II
1048			1098	3	I	1148			1198	2	II	1248		III
1049			1099			1149		II	1199		II	1249	4	II
1050			1100	7		1150	8		1200			1250		II

Categories Winter (NDJFM): C_v and Summer (MJJAS): C_s Low Countries

C_v ranges from 1(extremely mild) to 9 (extremely severe). I=C_v 1,2,3 II=C_v 4,5,6 III=C_v 7,8,9

C_s ranges from 1(extremely cool) to 9 (extremely warm). I=C_s 1,2,3 II=C_s 4,5,6 III=C_s 7,8,9

Table 5. Categories 1251-1500.

Year	C_v	C_s	Year	C_v	C_s	Year	C_v	C_s	Year	C_v	C_s	Year	C_v	C_s
1251			1301	5		1351	7	6	1401		6	1451	5	3
1252	5	III	1302	2	2	1352	6	7	1402	4	5	1452	5	6
1253	6	II	1303	7	6	1353	4	5	1403		5	1453	6	3
1254	8	I	1304	3	7	1354	6	5	1404	4	4	1454	5	4
1255	5	4	1305		7	1355	7	5	1405	6	4	1455	5	4
1256	6	II	1306	8	5	1356		3	1406	5	2	1456	4	4
1257	5	I	1307	5	5	1357	4	5	1407	4	5	1457	6	7
1258	6	I	1308			1358	5	5	1408	9	4	1458	7	6
1259			1309			1359	7	1	1409	3	4	1459	6	5
1260	4	II	1310	7	4	1360	4	7	1410		6	1460	7	6
1261	5	II	1311	6	5	1361	6	8	1411	5	4	1461	5	7
1262	II	III	1312	5		1362	4	5	1412		6	1462	7	6
1263	6	6	1313		5	1363			1413		5	1463	6	3
1264		II	1314	6	3	1364	9	5	1414	6	5	1464	6	7
1265			1315		2	1365	3	3	1415	4	6	1465	7	3
1266	3	III	1316	5	3	1366		2	1416	4	4	1466	6	6
1267	6	III	1317	6	5	1367	7	5	1417	6	6	1467	5	4
1268		II	1318		7	1368	3	5	1418	6	6	1468	5	4
1269	II	II	1319			1369		4	1419	5	5	1469	5	6
1270	6	II	1320			1370	6	4	1420	7	9	1470	8	3
1271	5	I	1321		4	1371	3	7	1421	5	5	1471	5	8
1272		III	1322	7	3	1372	6	4	1422	5	9	1472	4	5
1273	5	II	1323	8		1373	5	5	1423	8	4	1473	5	9
1274		II	1324	5	7	1374	6	4	1424	6	8	1474	4	6
1275	5	I	1325	5	9	1375	6	7	1425	3	6	1475	5	4
1276	7	II	1326	7	9	1376	4	6	1426	6	7	1476	6	5
1277	6	III	1327			1377	5	5	1427	7	5	1477	7	3
1278	4	II	1328	6	7	1378	5	4	1428	5	2	1478	3	6
1279	5		1329		6	1379	6	5	1429	4	5	1479	4	7
1280	5	II	1330	6	1	1380	4	5	1430	4	6	1480	4	3
1281	6		1331	5	8	1381	6	5	1431	4	6	1481	8	3
1282	7	III	1332	4	6	1382		6	1432	8	6	1482	6	6
1283	3	I	1333		8	1383	4	8	1433	7	6	1483	2	5
1284	5	III	1334		6	1384	4	7	1434	6	8	1484	5	6
1285	3	III	1335	6	3	1385	4	8	1435	9	5	1485	5	1
1286	8	II	1336		5	1386	5	6	1436	5	1	1486	6	5
1287		III	1337	5	5	1387	5	4	1437	8	4	1487	5	5
1288	6	III	1338	5	7	1388	5	5	1438	7	4	1488	5	1
1289	5	II	1339	7	5	1389	6	5	1439	4	5	1489	5	4
1290	2	I	1340	6	6	1390		8	1440	6	4	1490	4	6
1291		II	1341	4	5	1391	4	5	1441	5	6	1491	8	1
1292	7		1342	6	5	1392		4	1442	5	8	1492	7	7
1293	4	III	1343	5	5	1393	5	8	1443	8	6	1493	6	4
1294	5	I	1344	4	6	1394	5	3	1444	5	5	1494	4	6
1295	6		1345		2	1395	5	5	1445	5	3	1495	5	8
1296	4	III	1346			1396	7		1446	4		1496	6	3
1297	6	III	1347		4	1397	4	6	1447	6	8	1497	5	4
1298			1348	5	5	1398	4	5	1448	6	3	1498	6	6
1299	7		1349	5	4	1399	7	5	1449	5	5	1499	6	5
1300	6	II	1350	3	4	1400	6	8	1450	4	5	1500	6	6

Categories Winter (NDJFM): C_v and Summer (MJJAS): C_s Low Countries

C_v ranges from 1(extremely mild) to 9 (extremely severe). I=C_v 1,2,3 II=C_v 4,5,6 III=C_v 7,8,9

C_s ranges from 1(extremely cool) to 9 (extremely warm). I=C_s 1,2,3 II=C_s4,5,6 III=C_s 7,8,9

Table 5. Categories 1501-1750.

Year	C_v	C_s	Year	C_v	C_s	Year	C_v	C_s	Year	C_v	C_s	Year	C_v	C_s
1501	6	5	1551	6	4	1601	7	4	1651	6	6	1701	5	7
1502	4	6	1552	4	5	1602	6	5	1652	6	6	1702	4	5
1503	6	7	1553	5	5	1603	6	6	1653	6	6	1703	5	5
1504	6	8	1554	7	7	1604	5	6	1654	4	4	1704	5	7
1505	5	3	1555	6	3	1605	4	5	1655	7	5	1705	5	5
1506	6	5	1556	4	9	1606	5	4	1656	6	6	1706	5	6
1507	2	6	1557	7	5	1607	4	5	1657	5	5	1707	5	6
1508	6	5	1558	6	5	1608	8	4	1658	7	3	1708	4	6
1509	5	7	1559	5	8	1609	4	5	1659	6	5	1709	8	4
1510	5	5	1560	4	3	1610	6	5	1660	7	5	1710	5	4
1511	8	2	1561	7	5	1611	6	5	1661	3	5	1711	4	4
1512	7	4	1562	4	4	1612	6	5	1662	3	5	1712	5	5
1513	5	6	1563	6	5	1613	4	5	1663	8	4	1713	5	3
1514	7	6	1564	6	3	1614	6	3	1664	4	6	1714	4	5
1515	3	3	1565	9	3	1615	6	6	1665	6	5	1715	5	3
1516	4	7	1566	5	4	1616	6	7	1666	5	8	1716	7	4
1517	7	4	1567	5	7	1617	4	3	1667	7	6	1717	5	4
1518	6	4	1568	5	5	1618	7	3	1668	5	5	1718	6	8
1519	4	3	1569	7	4	1619	5	5	1669	6	7	1719	5	8
1520	4	4	1570	5	4	1620	7	5	1670	7	5	1720	5	4
1521	3	7	1571	6	6	1621	8	2	1671	4	5	1721	5	6
1522	6	6	1572	6	5	1622	6	4	1672	7	5	1722	5	6
1523	5	6	1573	8	3	1623	5	6	1673	5	4	1723	5	5
1524	4	6	1574	5	5	1624	7	7	1674	6	3	1724	2	7
1525	6	6	1575	6	7	1625	5	4	1675	5	2	1725	5	1
1526	4	5	1576	2	4	1626	5	4	1676	3	8	1726	7	6
1527	5	3	1577	4	4	1627	6	3	1677	7	6	1727	5	6
1528	4	4	1578	5	7	1628	5	1	1678	6	6	1728	5	6
1529	3	2	1579	5	3	1629	4	7	1679	7	7	1729	7	6
1530	3	7	1580	5	6	1630	4	6	1680	5	5	1730	4	6
1531	4	6	1581	7	3	1631	5	7	1681	7	6	1731	5	6
1532	5	7	1582	6	4	1632	4	4	1682	5	4	1732	5	5
1533	6	3	1583	6	7	1633	4	4	1683	5	6	1733	4	7
1534	7	7	1584	5	5	1634	6	5	1684	9	6	1734	3	6
1535	6	3	1585	3	3	1635	7	6	1685	6	4	1735	4	6
1536	5	8	1586	7	4	1636	5	7	1686	2	7	1736	4	6
1537	7	4	1587	6	1	1637	5	7	1687	5	4	1737	2	4
1538	3	7	1588	5	3	1638	6	6	1688	6	4	1738	5	4
1539	5	5	1589	5	5	1639	5	5	1689	6	4	1739	3	5
1540	4	9	1590	7	7	1640	5	5	1690	5	4	1740	8	2
1541	7	3	1591	6	4	1641	6	5	1691	7	5	1741	5	4
1542	5	2	1592	5	4	1642	4	4	1692	7	3	1742	5	3
1543	6	4	1593	6	4	1643	5	5	1693	5	4	1743	6	5
1544	7	5	1594	5	4	1644	6	6	1694	6	3	1744	6	3
1545	6	4	1595	8	5	1645	6	6	1695	8	1	1745	5	3
1546	7	5	1596	5	4	1646	7	7	1696	5	4	1746	6	5
1547	6	5	1597	5	3	1647	6	5	1697	8	4	1747	5	5
1548	6	4	1598	6	5	1648	4	4	1698	7	3	1748	6	6
1549	6	4	1599	6	6	1649	7	3	1699	4	6	1749	3	5
1550	5	3	1600	7	3	1650	5	4	1700	4	4	1750	4	6

Categories Winter (NDJFM): C_v and Summer (MJJAS): C_s Low Countries

C_v ranges from 1(extremely mild) to 9 (extremely severe). I=C_v 1,2,3 II=C_v 4,5,6 III=C_v 7,8,9

C_s ranges from 1(extremely cool) to 9 (extremely warm). I=C_s 1,2,3 II=C_s 4,5,6 III=C_s 7,8,9

Table 5. Categories 1751-2000.

Year	C_v	C_s	Year	C_v	C_s	Year	C_v	C_s	Year	C_v	C_s	Year	C_v	C_s
1751	6	4	1801	5	4	1851	3	4	1901	6	6	1951	5	5
1752	4	5	1802	6	4	1852	4	7	1902	5	3	1952	3	5
1753	5	5	1803	7	5	1853	5	4	1903	5	3	1953	5	5
1754	5	4	1804	5	4	1854	6	4	1904	5	5	1954	6	3
1755	7	5	1805	7	2	1855	7	4	1905	4	6	1955	5	6
1756	3	6	1806	5	4	1856	5	4	1906	4	5	1956	7	3
1757	7	7	1807	4	6	1857	6	8	1907	6	2	1957	3	4
1758	5	6	1808	6	7	1858	5	7	1908	6	5	1958	4	5
1759	3	8	1809	6	4	1859	4	9	1909	6	3	1959	4	7
1760	7	4	1810	6	3	1860	6	2	1910	4	4	1960	4	4
1761	3	5	1811	6	6	1861	6	6	1911	3	8	1961	2	4
1762	6	4	1812	4	4	1862	5	4	1912	4	5	1962	5	2
1763	8	4	1813	6	4	1863	2	4	1913	3	3	1963	8	4
1764	3	5	1814	7	4	1864	5	3	1914	4	6	1964	6	4
1765	6	5	1815	6	4	1865	7	6	1915	3	4	1965	5	2
1766	6	5	1816	6	2	1866	3	4	1916	4	3	1966	5	4
1767	7	3	1817	4	4	1867	4	5	1917	7	7	1967	3	6
1768	6	5	1818	4	6	1868	5	9	1918	5	4	1968	5	5
1769	5	4	1819	4	7	1869	4	4	1919	5	3	1969	6	7
1770	4	5	1820	7	4	1870	6	5	1920	3	4	1970	6	5
1771	6	4	1821	6	3	1871	7	5	1921	5	6	1971	6	5
1772	5	5	1822	2	6	1872	5	7	1922	6	3	1972	4	4
1773	4	5	1823	8	4	1873	3	6	1923	3	5	1973	4	6
1774	4	5	1824	3	5	1874	4	6	1924	6	4	1974	4	4
1775	4	6	1825	3	5	1875	6	6	1925	4	6	1975	2	8
1776	7	6	1826	5	9	1876	6	6	1926	6	5	1976	4	8
1777	6	5	1827	6	5	1877	3	5	1927	3	4	1977	3	4
1778	6	8	1828	4	6	1878	3	5	1928	6	4	1978	3	3
1779	4	8	1829	7	4	1879	6	3	1929	7	5	1979	7	3
1780	6	7	1830	9	4	1880	7	6	1930	3	6	1980	4	5
1781	6	9	1831	6	6	1881	6	5	1931	5	4	1981	4	5
1782	5	6	1832	5	4	1882	3	3	1932	5	7	1982	6	7
1783	5	9	1833	6	4	1883	4	4	1933	5	6	1983	3	8
1784	8	5	1834	2	8	1884	2	7	1934	6	6	1984	4	5
1785	7	5	1835	4	6	1885	5	4	1935	3	6	1985	7	5
1786	6	4	1836	5	4	1886	6	5	1936	3	5	1986	6	5
1787	6	4	1837	4	5	1887	6	5	1937	4	6	1987	6	4
1788	5	7	1838	8	4	1888	6	3	1938	4	6	1988	2	5
1789	9	5	1839	5	5	1889	5	6	1939	5	6	1989	1	7
1790	3	3	1840	4	5	1890	5	3	1940	7	4	1990	1	6
1791	4	4	1841	7	5	1891	7	4	1941	6	6	1991	5	6
1792	5	5	1842	5	7	1892	5	4	1942	8	5	1992	3	8
1793	5	4	1843	5	5	1893	6	6	1943	4	5	1993	3	4
1794	5	6	1844	4	3	1894	4	4	1944	4	6	1994	4	9
1795	8	4	1845	8	3	1895	7	5	1945	4	6	1995	3	8
1796	3	5	1846	2	9	1896	4	6	1946	5	5	1996	7	5
1797	6	6	1847	7	5	1897	5	6	1947	8	9	1997	6	8
1798	4	6	1848	7	5	1898	2	5	1948	4	5	1998	2	5
1799	7	2	1849	5	5	1899	4	6	1949	4	5	1999	4	8
1800	7	4	1850	6	4	1900	5	6	1950	4	6	2000	2	

Categories Winter (NDJFM): C_v and Summer (MJJAS): C_s Low Countries

C_v ranges from 1(extremely mild) to 9 (extremely severe). I=C_v 1,2,3 II=C_v 4,5,6 III=C_v 7,8,9

C_s ranges from 1(extremely cool) to 9 (extremely warm). I=C_s 1,2,3 II=C_s 4,5,6 III=C_s 7,8,9

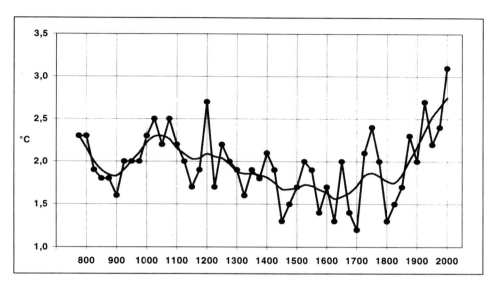

Figure 1. Winter (DJF) temperatures Low Countries (De Bilt) – 25 year means.

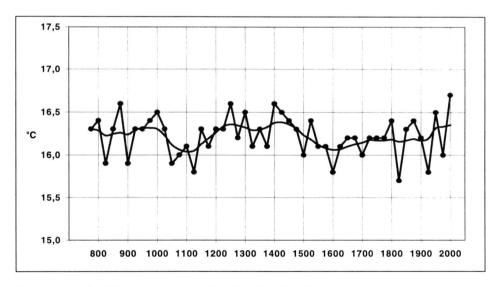

Figure 2. Summer (JJA) temperatures Low Countries (De Bilt) – 25 year means.

REFERENCES

1. J. Buisman, edt. A.F.V. van Engelen, *Duizend jaar weer wind en water in de Lage Landen*, Van Wijnen, Franeker (Netherlands). Vol. I, 763-1300, 1995, Vol. II, 1300-1450, 1996, Vol. III, 1450-1575, 1998.
2. A.F.V. van Engelen and H.A.M Geurts., *Historische Weerkundige Waarnemingen*, I-V, I: *Geschiedenis van weerkundige waarnemingen in het bijzonder in Nederland vóór de oprichting van het KNMI*, II *Vooruitstrevende ideeën over de meteorologie en klimatologie van Petrus van Musschenbroek – 1692-1761* IIa: *Inleidende tekst bij de meteorologische waarnemingsreeks Utrecht/Leiden 1729-1758, Petrus van Musschenbroek*, III: *Een rekenmodel dat het verloop van de temperatuur over een etmaal berekent uit drie termijnmetingen van de temperatuur*, IV: *Nicolaus Cruquius (1678-1754) and his meteorological obeservations*, V: *Beschrijving antieke meetreeksen*, KNMI publ. 165-I, II, IIa, III, IV, V, (De Bilt. 1983-1992)
3. Merle's MS, *Considerationes temperiei pro 7 annis..the earliest known Journal of Weather, kept by the Rev. William Merle, rector of Driby, Linc., repr...G.J. Symons* (London 1891).
4. David Fabricius, *Calendarium Historicum, Wettertagebuch, Resterhafe and Osteel, Ostfriesland*, in Statsarchiv Aurich
5. *Journael van tgene daegelykz gepasseert is in den oorloghe der Staeten Generael tegen de Spangiaerden ende andere vianden van de Vereenichde Nederlanden; Journaal van Anthonis Duyck, Advokaat-Fiscaal van den Raad van State, 1591-1602*, ed. op last van het Dept. Van Oorlog, inl. en aant. By Lodewijk Mulder, 3 volumes (Arnhem 1862-1866)
6. *Isaac Jacobus Beeckman, Journaal, hs. 6471*, Public Record Office Middelburg; transl. A.C. Meyer with corrections by P. Moors. Cf. *Journal tenu par Is. Beeckman de 1604 à 1634*, ed. C. de Waard, 4 volumes (The Hague 1939-1953)
7. Fabio Chigi, *Acta pacis Westphalicae*, Rheinisch Westfalischen Akad. Der Eissenschaften, Seies III, Abt. C. Dianien, Band 1: *Diarum Chigi, K. 39-1651*, 1 Teil, Text, Bearb. K. Repgen (Munster, 1984)
8. Publican accounts of tolls at Tiel, Zaltbommel, Nijmegen, IJsseloord etc. Ducal Archives and Audit Office of Gelderland, Public Record Office in the province of Gelderland at Arnhem
9. J. de Vries, *Histoire du climat et économie: des faits nouveaux, une interpretation differente*. In: Annales Economies Sociétés Civilisations, 32, 198-226 (1977).
10. Accounts of the steward in Old Archive Zutphen, Public Archive Zutphen.
11. *Briefwisseling tusschen de Gebroeders Van der Goes (1659-1673)* ed. G.J. Gonnet, 2 volumes (Amsterdam 1899, 1909)
12. *Romboudt de Doppere, [..] Chronique brugeoise de 1491-1498, découverte dans un ms de Jacques de Meyere*, publ. H. Dussart S.J. (Brussels, 1892)
13. F. IJnsen, *Onderzoek naar het optreden van winterweer in Nederland* (1981), KNMI publ. WR74-2
14. F. IJnsen, *Karaktergetallen van de winters vanaf 1706* in Zenit, 69-73 Febr. 1991
15. Recently at KNMI researches started to a reconstruction of daily De Bilt temperatures based upon a database with daily historical instrumental observations, carried out at some 12 locations, covering the period 1700-1900.
16. F. IJnsen, *Karaktergetallen en temperaturen voor winters en zomers, 1591-1705/1706*, priv. memo 1992,1994
17. *Observationes, Klimadaten von 1621-1650 nach Beobachtungen des Landgrafen Hermann IV von Hessen*, (Kassel) by Walter Lenke, in Berichte des Deutschen Wetterdienstes Nr. 63, Band 9, (Offenbach am M. 1960)
18. G. Manley, *Central England temperatures, monthly means 1659-1973* in Quart. Journ. Met. Soc. 100 389 (1974)
19. F. IJnsen, *De zomers in Nederland vanaf 1706, thermisch bekeken* KNMI publ. W.R. 76-15 (De Bilt, 1976)
20. F. IJnsen, *Karaktergetallen voor de zomers vanaf 1706* in Zenit, 313-315 Sept. 1991
21. H. Dubois, *Les foires de Chalon et le commerce dans la vallée de la Saône à la fin du Moyen-Age (vers 1280-vers 1430)*, Paris, 1976 and J. Lavalle, *Histoire et statistique de la vigne et de grands vins de la Côte d'Or*, Dijon, 1855
22. Shabalova, M.V., and A.F.V. van Engelen, *Evaluation of a reconstruction of temperature in the Low Countries AD 764-1998*, submitted to Climatic Change, March 2000. For the evaluation the following series/reconstructions were used: CET (1659-1996, Manley, 1974, updated), C. Europe (1550-1994, Pfister, 1985, updated), IWI (1501-1995, Koslowski & Glaser, 1999), EUR (1068-1979, Guiot, 1992), Fennoscandia (500-1980, Briffa *et al*, 1992), W.-C. Europe (750-1300, Pfister *et al.*, 1998), Tnh1 (1000-1991, Jones *et al*, 1998) and Tnh2 (1000-1998, Mann *et al.*, 1999).

SOURCES FOR THE PERIOD 763-1575

Alexandre, P., Le Climat en Europe au moyen age. Contribution à l'histoire des variations *climatiques de 1000 à 1425, d'après les sources narratives de l'Europe occidentale* (Paris 1987).

Alexandre, P., *Les seismes en Europe occidentale de 394 à 1259. Nouveau catalogue critique.* Observatoire Royal de Belgique, série Géophysique No. Hors-série (Bruxelles, 1990).

Anonymus, *An analysis of British Earthquakes*, in: British Earthquakes, an assessment by Principia Mechanica Limited for Central Electricity Generating Board, British Nuclear Fuels Limited, South of Scotland Electricity Board, vol.1, Report No: 115/82 (June 1982).

Augustyn, B., *Zeespiegelrijzing, transgressiefasen en stormvloeden in maritiem Vlaanderen tot het einde van de XVIde eeuw. Een landschappelijke, ecologische en klimatologische studie in historisch perspectief (with an English summary)*, 4 volumes in 2 bindings (ARA, Brussel 1992).

Bailey, M., *Per impetum maris: natural disaster and economic decline in eastern England, 1275-1350*, in: B.M.S. Campbell (ed.), *Before the black death, studies in the 'crisis' of the early fourteenth century* (Manchester, New York 1991).

Bertrijn. G., *Chronyck der Stadt Antwerpen, toegeschreven aan den notaris Geeraard Bertrijn*, G. van Havre, Maatschappij der Antwerpsche bibliophilen 5 (Antwerpen 1879).

Brandon, P.F., *Late medieval weather in Sussex and its agricultural significance*, in: Transactions of the Inst. of British Geographers 54 (1971).

Brandon, P.F., *Agriculture and the effects of floods and weather at Barnhorne, Sussex, during the late Middle Ages, Sussex,* in: Archaeol. Collections, CIX (1972).

Britton, C.E., *A Meteorological Chronology to A.D.1450,* Met. Office, Geophysical Memoirs No 70 (London 1937).

Camuffo, D., *Freezing of the Venetian Lagoon since the 9th century A.D. in comparison to the climate of Western Europe and England,* in: Climatic Change 10 (1987).

Dobras, W., *Wenn der ganze Bodensee zugefroren ist... Die Seegfrörnen von 875 bis 1963* (Konstanz 1983).

Dochnahl, F.J., *Chronik von Neustadt an der Haardt, nebst den umliegenden Orten und Burgen mit besonderer Berücksichtigung der Weinjahre unter Mitwirkung von Andreas Sieber bearbeitet von Friedrich Jac. Dochnahl, fortgeführt von Karl Tavenier bis Juni 1900* (Neustadt an der Haardt 1867, Nachdruck Pirmasens 1975).

Dufour, L., *Quelques événements météorologiques de la fin du XIVe siècle et du début du XVe*, in: Ciel et terre 93 (Bruxelles 1977).

Dufour, L., *Evénements météorologiques anciens d' après le Journal de Pierre Driart (1522-1535)*, in: *Inst. Royale Métérol. de Belgique, Publ.* Série B, N° 91) 1-23 (Bruxelles 1977).

Dufour, L., *Quelques événements météorologiques de la première moitié du XVIe siècle d' après le livre de raison de maître Nicolas Versoris*, in: La Météorologie VI-73 67-73 (1964).

Dufour, L., *Quelques événements météorologiques du XVIe siècle dans la principauté de Liège*, in: Inst. Royale Métérol. de Belgique, Publ. Série B, N° 102 1-26 (Bruxelles 1979).

Easton, G., *Les hivers dans l' Europe occidentale: étude statistique et historique* . (Leyden 1928).

Flohn, H., *Das Problem der Klimaänderungen in Vergangenheit und Zukunft* (Darmstadt 1985).

Gachard M, [=L.P.] and Ch. Piot ed., *Collection des voyages des souverains des Pays bas, I, Itinéraire de Philippe le hardi, Jean sans Peur, Philippe le Bon, Maximilien et Philippe le Beau*, Commission Royale d'Histoire de Belgique (Bruxelles 1876).

Gram-Jensen, I., *Sea floods: contributions to the climatic history of Denmark*, Klimatologiske meddelelser 13 (Kobenhavn 1985).

Gottschalk, M.K.E., *Stormvloeden en rivieroverstromingen in Nederland II* (Assen 1975).

Grieve, H., *The Great Tide, the story of the 1953 flood disaster in Essex*, County Council of Essex (Chelmsford 1959).

Heitzer, E., *Das Bild des Kometen in der Kunst* (Berlin 1995).

Hellmann, G., *Die Meteorologie in den deutschen Flugschriften und Flugblaettern des XVI. Jahrhunderts. Ein Beitrag zur Geschichte der Meteorologie.* Abhandlungen der Preussischen Akademie der Wissenschaften, Physikalisch-Mathematische Klasse 1921, Nr 1 (Berlin 1921).

Hennig, R., *Katalog bemerkenswerter Witterungsereignisse von den ältesten Zeiten bis zum Jahre 1800.* Abhandlungen des Königlich Preussischen Meteoroplogischen Instituts, Bd. II, No 4 (Berlin 1904).

Hetherington, B., *A chronicle of pre-telescopic astronomy* (Chichester 1996).

Houtgast, G., *Catalogus aardbevingen in Nederland. KNMI-publ.* 179 (De Bilt 1991).

Jones, P.D., Ogilvie, A.E.J. and T.M.L. Wigley, *Riverflow data for de United Kingdom. Reconstructed Data Back tot 1844 and Historical Data back to 1556.* Climatic Research Unit Univ. of East Anglia, Research Publication 8 (Norwich 1984).

Lamb, H.H., *Climate, History and the modern world* (London 1982).

Lamb, H.H., *Historic storms of the North Sea, British Isles and Northwest Europe* (Cambridge 1991).

Lowe, E.J., *Natural Phenomena and Chronology of the Seasons, being an account of remarkable frosts, droughts, thunderstorms, gales, floods, earthquakes, etc., which have occured in the British Isles since A.D. 220, chronologically arranged* (London 1870).

Lyons, M.C., *Weather, Famine, Pestilence and Plague in Ireland, 900-1500*, in: E.M. Crawford (ed.), Famine: The Irish experience, 900-1900, Subsistence Crises and Famines in Ireland (Edinburgh 1989).

Meaden, G.T., *Northwest Europe's Great Drought, the worst in Britain since 1251-1252*, in: Journ. of Met., Vol. 1, 12 (1976).

Müller, K., *Geschichte des Badischen Weinbaus*, 1953.

Nagel, W., *Die alte Dresdner Augustbrücke* (Dresden 1924, T.H. Diss. 1917).

Ogilvie, A.E.J., *Documentary evidence for changes in the climate of Iceland, A.D. 1500 to 1800*, in: Bradley R.S., and Ph.D. Jones eds., Climate since A.D. 1500 (London/New York 1992).

Pfister, C., *Klimageschichte der Schweiz 1525-1860. Das Klima in der Schweiz von 1525-1860 und seine Bedeutung in der Geschichte von Bevölkerung und Landwirtschaft*. Band I (Bern/Stuttgart 1984).

Pfister, C., *Variations in the spring-summer climate of Central Europe from the High Middle Ages to 1850*, in: H. Wanner, U. Siegenthaler eds. Long and short term variability of climate, Lecture notes in earth sciences (ed. by Somdev Bhattacharji *et al.*) 16 (Berlin 1988).

Pfister, C., *Wetternachhersage, 500 Jahre Klimavariationen und Naturkatastrophen (1496-1995)* (Bern/Stuttgart/Wien 1999).

Pfister, C., Luterbacher, J., Schwarz-Zanetti, G. and M. Wegmann, *Winter air temperature variations in western Europe during the Early and High Middle Ages (AD 750-1300)*, The Holocene, 8,5, (1998)

Pfister, C., Brázdil, R., Glaser, R., Barriendos, M., Camuffo, D., Deutsch, M., Dobrovony, P., Enzi, S., Guidoboni, E., Kotyza, O., Militzer, S., Rácz, L. and F.S. Rodrigo, *Documentary evidence on Climate in sixteenth-century Europe*, in Climatic change 43, 1999.

Le Roy Ladurie, E., *Histoire du climat depuis l'an mil II*, 2^e ed. (Paris 1983).

Rummelen, F.H. van, *Overzicht van de tussen 600 en 1940 in Zuid-Limburg en omgeving waargenomen aardbevingen en van aardbevingen welke mogelijk hier haren invloed kunnen hebben doen*, in: Med. Jaarversl. Geol. Bureau, No 15 (Maastricht 1942/43, 1945).

Sanson, J., *Les anomalies du climat Parisien du XIIIe au XVIIIe siècle d'après les sorties de la châsse de Sainte Geneviève*, in: La Météorologie 3e série 2 230-232 (1937).

Schmidt, M., *Der zugefrorene Bodensee, Beitrag zur Geschichte sehr strenger Winter in Südwest-Deutschland und der Schweiz* in: Meteorol. Rundschau, 20 Jg. Heft 1 16-25 (1967).

J.F. Schroeter, M., *Spezieller Kanon der Zentralen Sonnen- und Mondfinsternisse, welche innerhalb des Zeitraums von 600 bis 1800 n.Chr. in Europa sichtbar waren*. Mit 300 karten (Kristiania 1923).

Schubert E., and B. Herrmann (hrsg), *Von der Angst zur Ausbeutung, Umnwelterfahrung zwischen Mittelalter und Neuzeit* (Frankfurt am Main 1994).

Speerschneider, C.I.H., *Om isforholdene i Danske farvande i aeldre og nyere tid, aarene 690-1860*. Publikationer fra det Danske Meteorologiske Institut [...], Meddelelser no. 2 (Kjøbenhavn 1915).

Steensen T., (ed.), *Deichbau und Sturmfluten in den Frieslanden* (Bräist/Bredstedt 1992).

Tammann, G.A., and P. Véron, *Halleys Komet* (Basel 1985).

E. Vanderlinden, G.A., *Chronique des événements météorologiques en Belgique jusqu'en 1834*, Acad. royale de Belgique, Classe des Sciences, Mémoires, 2e série, VI (Bruxelles 1924).

Wee, H. van der, *The Growth of the Antwerp market and the European Economy (fourteenth-sixteenth centuries)*. III Graphs (Louvain 1963).

Weikinn, C., *Quellentexte zur Witterungsgeschichte Europas von der Zeitwende bis zum Jahre 1850*, in: Quellensammlung zur Hydrographie und Meteorologie, Band I, T. 1., Zeitwende -1500 (Berlin 1958).

Weikinn, C., *Quellentexte zur Witterungsgeschichte Europas von der Zeitwende bis zum Jahre 1850*, in: Quellensammlung zur Hydrographie und Meteorologie, Band I, T.2., 1501-1600 (Berlin 1960).

Witkamp, P.H., *Aardrijkskundig woordenboek van Nederland*, 4 vols. (Amsterdam 1877).

Yeomans, D.K., *Comets, A chronological history of observation, science, myth, and folklore* (New York 1991).

APPENDIX 1200-1250

Together with the last 25 year period of the earlier 12th century, the second quarter of the 13th century (1226-1250) experienced annual temperatures with a mean of 9.3°C, high compared with those of the last millennium with a mean annual temperature of some 9.0°C. The average summer (JJA) temperature amounts to 16.7° (normal 1961-1990 De Bilt: 16,2°C) and the winter (DJF) temperature to 2.0°C (normal 1961-1990 De Bilt: 2,6°C). With these values the Mediaeval Climate Optimum (MCO), as it is manifest in the Low Countries from ca. 1170 to ca. 1430, reaches its optimum in the first half of the 13th century.

During the period 1200-1225 winters are predominantly cold or normal, 6 are severe and three mild. The summers are generally normal or warm; only three summers are cold. In 1214 a storm surge swept the whole coastal area; in the years 1219 to 1221 the northern coasts are hit four times.

During the period 1226-1250 very few winters are cold, but three almost in a row (1227, 1230 and 1234) are severe. All summers except two are normal or warm. Storm surges occur from 1246 to 1249.

The course of the summer and winter temperatures and the averages for the year temperatures – expressed in terms for De Bilt – is shown in Figure A1. The values are tabulated in Table 6.

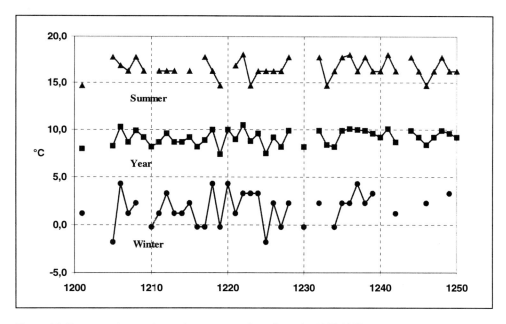

Figure A1. Summer, winter and annual temperatures Low Countries 1200-1250.

Table 6. Annual temperatures for the winter (DJF), summer (JJA) and year (Nov.-Oct.) Low Countries (De Bilt) 1200-1250.

°C	winter	summer	year	°C	winter	summer	year	°C	winter	summer	year
1200				1217	-0.2	17.7	8.9	1234	-0.2	16.2	8.2
1201	1.2	14.7	8.0	1218	4.3	16.2	10.0	1235	2.3	17.7	9.9
1202				1219	-0.2	14.7	7.4	1236	2.3	18.0	10.1
1203				1220	4.3		10.0	1237	4.3	16.2	10.0
1204				1221	1.2	16.8	9.0	1238	2.3	17.7	9.9
1205	-1.8	17.7	8.3	1222	3.3	18.0	10.5	1239	3.3	16.2	9.6
1206	4.3	16.8	10.3	1223	3.3	14.7	8.8	1240		16.2	9.2
1207	1.2	16.2	8.7	1224	3.3	16.2	9.6	1241		18.0	10.1
1208	2.3	17.7	9.9	1225	-1.8	16.2	7.5	1242	1.2	16.2	8.7
1209		16.2	9.2	1226	2.3	16.2	9.2	1243			
1210	-0.2		8.2	1227	-0.2	16.2	8.2	1244		17.7	9.9
1211	1.2	16.2	8.7	1228	2.3	17.7	9.9	1245		16.2	9.2
1212	3.3	16.2	9.6	1229				1246	2.3	14.7	8.4
1213	1.2	16.2	8.7	1230	-0.2		8.2	1247		16.2	9.2
1214	1.2		8.7	1231				1248		17.7	9.9
1215	2.3	16.2	9.2	1232	2.3	17.7	9.9	1249	3.3	16.2	9.6
1216	-0.2		8.2	1233		14.7	8.4	1250		16.2	9.2

HISTORICAL SOURCES PER REGION, 1200 -1250

The locations of the principal sources that were used are shown in the map of Figure A2. In the following paragraphs these sources are listed per country.

Low Countries

Luik/Liège [1194-1221]
> *Annales S. Jacobi Leodiensis*, ed. G.H. Pertz, M.G.H. SS 16 (1859) 635-680

Hoei/Huy [1230-1237]
> *Chronica Albrici monachi Trium Fontium, a monacho Novi Monasterii Hoiensis interpolata (Adnotationes Hoyenses)* ed. P. Scheffer-Boichorst, M.G.H. SS 23 (1874) 674-950

Wittewierum in Groningen [1219-1276]
> *Kroniek van het klooster Bloemhof te Wittewierum*, intr., ed. and translation H.P.H. Jansen, A. Janse (Hilversum 1991)

Egmond [1248-1250]
> *Annales Egmundenses, De jaarboeken van Egmond*, first Dutch translation by A. Uitterhoeve, Alkmaarse Cahiers, Vol. I (Alkmaar, 1990)

Britain

St.-Albans [1214-1236]
> *Rogeri de Wendover Chronica, liber qui dicitur Flores Historariarum (ab A.D. 1154)*, ed. H.G. Hewlett in R.B. SS, 84, I, II, III (London 1886-1889); Engl. Translation, 2 Vols, ed. Bohn (1849)

St.-Albans [1236-1259]
> *Matthaei Pariensis monachi Sancti Albani Chronica Majora*, ed. H.R. Luard, R.B. SS, 57, I-VII (London 1872-1884); and: The Weather, 1236-59, from *Chronica Majora*, trsl. J.A. Giles, *Matthew Paris' English History*, 3 Vol's. (1852-4), English historical documents, gen. Ed. David C. Douglas III 1189-1327, ed. Harry Rothwell (London 1875) 806-823

Coggeshall [1066-1223]

> *Radulphi de Coggeshall Chronicon Anglicanum* (1066-1223), ed. J. Stevenson, R.B. SS 66 (London 1875) 1-208

Winchester [1209-1350]

> J. Titow, *Evidence of weather in the Account Rolls of the bishopric of Winchester, 1209-1350*, in: The Ec. Hist. Review, ser. S, vol XII, 1-3 (Utrecht 1959-1960)

Waverley [-1291]

> *Annales monasterii de Waverleia (-1291),* ed. H.R. Luard, R.B. SS 36 II (London 1865) 127-411

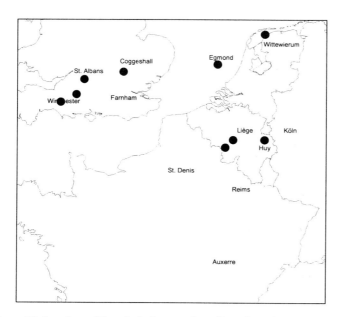

Figure A2. Locations of the principal sources Low Countries and surroundings 1200-1250

Germany

Köln [-.-]

> *Annales S. Pantaleonis Coloniensis maximi*, ed. G. Waitz, M.G.H., S.R.G. 18 (1880)

France

St.-Denis [1113-1368]

> *Chronique latine de Guillaume de nangis de 1113 à 1300 avec les continutations de cette chronique de 1300 à 1368,* ed. H. Géraud, S.H.F. (1843). Also: Continuatio S. Dionysii, ed. Fr. Delaborde, S.H.F. (1882) 300-333

Auxerre [-.-]

> *Roberti canonici S. Mariani Autissiodorensis Chronicon*, ed. O. Holder-Egger, M.G.H. SS 16 (1882) 226-276 and *Continuatio S. Mariano Autissiodorensis*, ibidem, 277-286

Reims [-.-]

> *Annales S. Nicasii Remensis*, ed. G. Waitz, M.G.H. SS 13 (1881) 84-87

[-.-] : scarce and scattered information

M.G.H., SS = Monumenta Germaniae historica, Deutsche Chroniken, Scriptores

M.G.H., S.R.G. = Idem, Scriptores rerum Germanicarum in usum scholarum

R.B., SS = Rerum Brittannicarum Medii Aevi, Scriptores

S.H.F. = Société de l'histoire de France

1248–1249 WINTER: FAIRLY MILD (4), SUMMER: ABOUT NORMAL (II)

Winter making history

The autumn of 1248, according to the monks at *Wittewierum* near *Ten Post* (*Low Countries*), was a dry one[1]. The following winter, so it is recorded in the Annals of the Abbey of *Egmond* (*Low Countries*), didn't amount to much till mid February. Only the wind direction created amazement.

> *In this same year, from the feast of the Evangelist Luke (18 October) till St Valentine's day (14 February) there was neither snow nor ice to be seen during the whole winter except for one or two nights. The wind was blowing so constantly from the west that it was only seldom that it shifted to the east for twenty-four hours and then - strangely enough - for two or three hours at night at the most.* [2]

Apparently it turned on 15 February, 1249 (Cf. 17 February in *England*). In *Cologne*, the winter of 1249 was very mild, with frost on only two days; furthermore this winter was wet[3]. In *England*, the people enjoyed springlike weather.

> *The temperature of winter was entirely changed to that of spring, so that neither snow nor frost covered the face of the earth for two days together; trees might be seen shooting in February, and the birds singing and sporting as if it were April, Matthew Paris, St Albans, reports.*

But, after mid February, the weather did change, according to our diarist[4].

Three severe storm surges

The most spectacular events in this extremely mild winter were provided by three storm surges, hitting with regular intervals of about 40 days large areas of the countries adjoining the *North Sea*. *England* was among the stricken countries and the coastal shores of *Flanders*, *Zeeland, Holland* and the northern part of the *Low Countries* and also parts of *Germany*. Menko, abbot of Bloemhof in *Wittewierum, Groningen* is again our crown witness. The first storm surge hit the north coast of the *Low Countries* on or immediately after 20 November.

> *In this year [1248], on Friday after the octave of St Martin [20 November], a heavy storm was rising, at first from the southwest, then changing to west and finally to northwest and north-northwest. A storm surge occurred and the sea dykes broke. The waters covered the surface of the earth, so that in Rozenkamp (Low Countries), where the waves leaped over the dykes at several places, they filled the whole courtyard. And so the flood caused great distress and there was a great deal of lament. While the sea was still seething, the wind changed, via north-northwest, north, and so, contrary to what usual was the case, most damage was done to the dykes facing north and south-southwest. Also winter corn and barley, sown abundantly because of the dry summer and autumn, suffered great damage. When, a short time later, the southern dyke along the sluice called the Delft [probably the sluice from Winneveer to Delfzijl, end of 12th c.] burst, the water flowed into all directions, especially into the southern Wolden [Duurswold]. A large part of the wheat was lost... This time, the western Wolden escaped from the disaster, as the wind did not blow from the east and its dyke remained undamaged.* [5]

Matthew Paris, too, records this surge tide, although his information is probably from hearsay.

> *On 24 November (perhaps November 19 is meant; see note 11), in this year, the sea overflowed its bounds to a great distance, and caused irreparable injury to those dwelling near the coast; for when the moon, according to the computation of the calendar, was in its fourth quarter, the tide flowed with swollen waters without any visible ebb or decrease.*

The flood must have been the result of the strong wind, blowing landwards, for the sea level was not particularly high. On the day of our Lord's Advent, the fourth day before Christmas [22 December ?], *England* was suffering from another earthquake, according to

Matthew Paris. He has this from the bishop of *Bath* in whose diocese it occurred. Gaps seemed to have appeared in the ruined walls, and what was most remarkable: while the tops of chimneys, parapets, and pillars were thrown from their places, their bases and foundations were not at all disturbed. At this side of the *Alps*, it was the third earthquake within three years: one in *Savoy*, and two in *England*[6]. The second storm surge happened on 28 December 1248.

> *When the year of our Lord 1249 had already begun [Menko is following the custom of beginning the year on Christmas Day], the 86th year after St Juliana Flood, the 55th year after St Nicolas Flood [1196], the 31st year after Marcellus Flood [1219] and the third year after the one that took place on the day of Luke the Evangelist [1246], on the nineteenth day of the lunar month, on the feast of John the Evangelist [27 December, 1248], Zephyr (west wind), quite unlike his accustomed sweetness raging like Boreas (north wind), tremendously increased in strength after evensong. At dead of night he had become so strong, that a roaring sound like this had not been heard in ages. Several houses were unroofed and some houses, even new and solidly built ones, collapsed... Around midnight a flickering light was observed in the sky; according to some it was lightning, but no sound of thunder followed... At the first cock-crow Zephyr made room for Circius [north-northwestern wind], the western neighbour of Boreas and the swirling water destroyed the dykes and overflowed its bounds, covering the country...*

The only fortunate thing was that the changing of wind direction had not happened a bit earlier, when the tide was at its highest, for then hardly anything would have remained of *Friesland*[7]. Probably, the low-lying country alongside the *Elbe* with the city of *Hamburg* was struck, too, on 28 December[8]. It is remarkable that, in *England*, on the same day, the sound of thunder was heard[9]). On 4 February, the third storm surge of this winter followed.

> *On the day after the feast of Blasius, the virgil of the virgin Agatha, a thursday [4 February], the wind enormously increased in strength, first from the west and then from north-northwest. Again there was a surge of the sea and again the water overflowed the dykes and covered the country. And since Cirrus kept blowing uninterrupted for three days, the sea was thrown onto the land on each consecutive day. But the tide was not so extraordinarily high that the houses adjoining the shore were destroyed. Therefore not so many people were drowned at that time as were before, during the St Marcellus Flood, when the sea, in one enormous surge, all but emptied itself on the land and surprised many of those dwelling near the coast... As soon as the first flood had spread out over several parts of the country, a second one came, just as is the sea's nature ... , passing the broken dykes easily and not only extending over the coastal region, but penetrating into regions lying further inland; the fact was that there were nine floods without interruption. And then the west dyke, called the "new one", burst [This could have been the Woldijk, but also the Zandsterdijk, the second one east of the Fivel] (Menko, abbot of Wittewierum in Groningen)10*

According to the annalist of the abbey of *Egmond*, *Holland* and *Flanders* were also hit.

> *In the year of our Lord 1248 around Martinmas [11 November, but perhaps 18 November was meant; see note 11], there were very severe storms and floods - the worst ever experienced within living memory - sweeping all the coasts, but especially those of the counties of Holland, Flanders and East- and West-Friesland [East-Friesland is present-day Friesland]. All along the coast, the sea dykes burst or were swept away. It was dreadful to watch how the beasts of burden and the cattle suffered death by suffocation, while men and women and children were hardly able to escape to higher spots or were sitting on the beams of their houses, till they were taken down, almost lifeless, by compassionate neighbours who put them in boats. In Delft, sea fish (called bullik and rivisk in the vernacular) was caught then with nets in the river11.*

We find Melis Stoke, still young in this year, referring to a heavy storm tide in Flanders and Zeeland and Holland.

	Translated freely:
Ghesciede ene sware plaghe.	*A great calamity happened*
In dien selven jare mede	*In this same year*
Dorghinc de vloet meneghe stede,	*Many towns were flooded,*
Vlaendren, Zeelant alte male,	*Flanders, Zeeland again...*
Ende oec Hollant also wale.....	*And also Holland completely...*
(Melis Stoke12, geb. ca 1235; cf. 1288).	*(Melis Stoke12, born ca. 1235; cf.128*

In the *Westerschelde*, the only recently (1244) dyked (and later to be engulfed again) island *Koezand* was flooded; it had to be dyked again. The village *Vroondijke*, northwest of *Axel* in the shire *Assenede* was destroyed by the flood of 1249; it had to be rebuilt. In later years, the *Braakman* creek would come into being here[13] (See 1375-1376).

As we have already seen (November) *England*, too, was hit by these storm surges, but information is vague and even unreliable, and especially our main source, Matthew Paris, usually so conscientious, is dubious here for he was staying abroad during these months. Apart from the eastern coast, *Holland*, the southern coast of England must have been hit. In the little seaport *Winchelsea*, in particular, the destruction must have been enormous[14]. And in just less than forty years it would again be the victim of the sea's violence and would fare even worse: the village had to be abandoned (cf. 1287-1288).

After the third storm surge, it got wintry again. A general tournament was appointed to be held at *Northampton*, on Ash Wednesday, that is on 17 February, to which many people had been looking forward because of the fine weather. But it all went wrong: on the day in question heavy snow fell and it continued snowing for two days. The tournament was cancelled. On 19 February, the snow covered the earth to the depth of a foot, breaking down the branches of the trees. When the thaw came, there was such an abundance of water that it caused the furrows in the fields, now dilated like caverns, to fill with the rivulets which ran down them[15].

The water, having risen higher and higher, only slowly got away. In the lower lying parts of the country this made ploughing, around St Liudger (?) (26 March) almost impossible, though the effect of salt and night frost had made the tilling of the fields quite easy. (*Wittewierum, Gr.*)

Summer (1249)

Early July, after a deluge of rain, floods in the vicinity of *Oxford*. But there was also a period of drought in the *South of England*.

Disappointing corn and grape harvests (*Cologne*). In *Groningen* again a year of hardship, the fourth since 1246.

ENDNOTES

1. *Kroniek van het klooster Bloemhof*, op.cit. 373
2. *Annales Egmundenses*, Egmond, op.cit. 166
3. Alexandre, 391.
4. Matthew Paris, op.cit. 812, 813. See also 1250-1251
5. *Kroniek van het klooster Bloemhof*, op.cit. 371 ff.
6. Matthew Paris, op.cit. 821. For the dating, see Gottschalk, note 13, below
7. *Kroniek van het klooster Bloemhof*, op.cit. 371 ff
8. Weikinn, I, 110, cites some sources reporting a storm surge on 28 December, 1248, in the *Elbe estuary*, with *Hamburg*. In a *Sächsische Weltchronik* (13th c.), there is mention of a violent southwest storm, even bringing down the trees in the woods. It seemed like an earthquake. Following this there was the storm surge that burst the dykes in the "Niderlanden". Cf. 1396.
9. Britton, 98, quoting contemporary annals of *Tewkesbury* Abbey in *Gloucestershire*
10. *Kroniek van het klooster Bloemhof*, op.cit. 371 ff.
11. *Annales Egmundenses*, op.cit. 166. For the dating see Gottschalk, note 13, below.
12. Melis Stoke, *Rijmkroniek*, ed. W.G. Brill, in: Werken uitgegeven door het Historisch Genootschap, nieuwe reeks 40 (1885) 3e boek, 820-824.
13. Gottschalk, I (1971) 174 ff naturally discusses at length the several storm surges. She draws attention to the fact that, possibly, the date in the English source shouldn't read VIII kalends Dec. (= November 24) but XIII.kalends Dec. (= November 19): an error quite easily made by a clerk copying Roman numerals. The *Annales Egmundenses* have "around St Martin" = November 11. If we assume that the word "octave" has been mistakingly left out, then the date becomes ca. November 18. In both cases we arrive approximately at the date mentioned by Menko: November 20. For the account of what happened with Vroondijke see: Gottschalk (1984) 101 ff.

14. Matthew Paris, op.cit. 813-814, 816. Cf. also Gottschalk I (1971) 188-190. In his report of (probably) the years 1250 and 1251 (apparent from the context), Matthew Paris speaks a few times about storm surges. His descriptions are not exactly conspicuous for their accuracy and their clarity and even contain some untruths. Two storm surges can be distinguished by him: one on 1 October (1250(?) and one in the spring of 1251(?). Of the first many details are given: in the harbour of *Hertbourne* several ships perished, in *Winchelsea*, salt-houses, fisherman's cottages, bridges and mills, more than 300 houses and some churches were swallowed by the sea. *Holland* (Linc.) as well as *Holland* on the other side of the sea and furthermore *Flanders* were hit. The second storm surge supposedly struck Friesland. On the fair at *Boston* (Linc.), Frisian merchants were (later) selling silver and golden jewelry found on drowned bodies. Obviously, Matthew Paris's record was based on garbled reports. Not surprisingly, knowing that, sent by the Pope, he spent these particular years in an abbey on the Norwegian island of *Nidarholm* (Antonia Gransden, *Historical writing in England, c. 550-1307 I*, London 1974, 356) and completely missed the infamous floods. After his return, he patched up his historical record with hearsay. His absence also shows in the limited number of details he supplies about the weather. Remarkably enough, the storm surge of 1248 he dates accurately, although without any significant details and without giving any precise locations. He never realized that, firstly, a number of details mentioned later on by him belonged to one and the same storm surge and that, secondly, the second storm surge still occurred in the same winter. Nothing at all is known about a possible storm surge in *Holland* and *Flanders* in 1250 or 1251. Conclusion: the information given by Matthew Paris for the years 1250 and 1251 (?) must refer to the storm surges of 1248-1249.
15. Matthew Paris, 813. Thomas Wykes (latter part 13th c.) talks about cold weather end of March - begining of April, continuing till mid May 1250 (=1249?).

LONG CLIMATIC SERIES FROM AUSTRIA

Ingeborg Auer, Reinhard Böhm, and Wolfgang Schöner

Central Institute for Meteorology and Geodynamics
Hohe Warte 38, A-1190 Vienna, Austria
E-Mail: ingeborg.auer@zamg.ac.at
reinhard.boehm@zamg.ac.at
wolfgang.schoener@zamg.ac.at

1. INTRODUCTION

Long term meteorological observations offer a large research potential for the study of climatic variability and change. Within the national funded Project ALOCLIM (Austrian Long Term Climate) a homogenised data set, consisting of 274 single series (see Table 2) has been developed. In addition, results from previous investigations increase the number of homogenised time series to 350. However, it is not possible to describe the complicated climatic system with time series using one single element, e.g. temperature or precipitation only. For that reason the ALOCLIM data set was created as a multiple one consisting of as many as possible climate elements (including air pressure, vapour pressure, sunshine duration, etc.). The longest Austrian meteorological time series of Kremsmünster began in 1767 which means that the time series analyses cover a period longer than 200 years. Most of the project's working capacity has been spent on homogenising the series but scientific analysis has recently begun. Homogenisation was based on two steps. In step one those inhomogeneities documented by metadata were removed. In step two statistical tests were applied to detect and remove the remaining inhomogeneities. The paper will describe the methods of homogenisation and give examples of the first analyses based on the data set.

2. THE ALOCLIM METADATA BANK

Metadata increases the quality of the homogenisation process and, therefore, metadata should be an integrative part of the homogenising procedures, if at all possible. Homogenisation, without using the available metadata information, must be characterised as a narrow road between the two abysses of subjectivity and statistical estimates.

The general information part of available metadata includes information about changes concerning the whole network (e.g. meteorological units, observation hours, introduction of thermometer screens, heights of thermometers and/or rain gauge orifices above ground, etc.), pointing at systematic inhomogeneities affecting larger regions. Such general break

History and Climate: Memories of the Future?
Edited by Jones *et al.*, Kluwer Academic/Plenum Publishers, 2001

125

points are not easily detectable by relative homogeneity tests, as all series will be similarly affected.

2.1 ALOCLIM General Metadata Information

During the instrumental period there have been increasing efforts to unify and standardise the climatological network, causing several serious inhomogeneities due to the lack of undisturbed comparative series. Until 1870, units like deg. Reaumur, Pariser Linien (1 Pariser Linie = 2.25583 mm), Wiener Klafter (1 Wiener Klafter = 1.896484 m), Wiener Fuß = Wiener Schuh (1 Wiener Fuß = 0.31603 m) had been in use, and only since 1.1.1871 were metric units common. There were also several changes of the observation hours, fortunately all well documented. At the beginning of the instrumental period all meteorological observations should have been done according to "True Solar Time" due to Kreil (1848), however the old astronomic observatories (which incorporated the principal climate measurements before 1850) used "Mean Local Time" (MLT). After 1850 Mean Local Time used to be the basis of measurements for most of the climate elements. It should be mentioned that before 1873 the following astronomic manner of writing was used: noon = 0 hours, midnight = 12 hours; since 1873: noon = 12 hours, midnight = 0 or 24 hours.

In 1873 the Vienna Congress formulated the need to standardize the observation hours (7, 14, 21). Daylight Saving Time (DST) was in public use from 1938 to 1948 and has been in use since 1980. Recently observations were carried out at MLT; however, in the years 1938-48 inclusive, DST was effectively used for the observations at most of the Austrian stations. For all ALOCLIM stations it was possible to find the used observing hours. The process of standardising took quite a long time; an 80% level of standardised observing hours was reached in the 1880s, 100% not earlier than in the 1930s (Figure 1).

Another process, the introduction of Stevenson screens started in the 1870s and lasted longer than 50 years. Similar findings can be seen in other countries (e.g. Nordli *et al.*, 1997). Heights of thermometers and rain gauge orifices were not constant. In the ALOCLIM network there has been a systematic development from unshielded thermometers in the shadow of buildings (very rare and hardly usable) to smaller or larger metal screens usually fixed to north facing-walls of buildings, to bigger double louvered screens similar to the

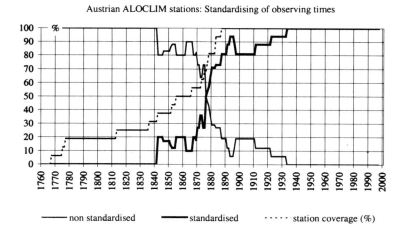

Figure 1. Time Series for the development of standardising the observation hours at the ALOCLIM stations.

Table 1. Austrian standards for climatological observation hours
MLT = mean local time, DST = daylight saving time

	1873-1937	7, 14, 21	MLT	
	1938-1948		DST	except Vienna and Kremsmünster
period	1949-1970	7, 14, 21	MLT	
	since 1971	7, 14, 19	MLT	
	since 1980		DST	but not used for climate observations

international standard known as "Stevenson screen" or "English screen" (Figure 2). They were usually located in open surroundings remote from buildings.

Modern standards require a standard thermometer height of 2 m above ground, but during the instrumental period there has been a trend through time from higher (above ground) thermometer sites to lower ones (Figure 3a), thus producing a systematic tendency of increasing maximum temperatures and decreasing minimum temperatures in the original data. Similarly, in the case of rain gauge orifices, a free and open position was thought to be more effective than the near to ground installations of today. This resulted in a similar development, as in the case of thermometer heights, from higher installations of the rain gauges to lower ones, thus causing a systematic bias of increasing precipitation in the original data (Figure 3b).

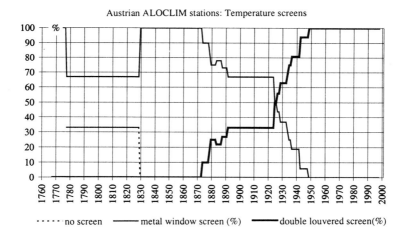

Figure 2. Time series of thermometer screens in use within the Austrian ALOCLIM network.

2.2 ALOCLIM Station History Information

Station history information is of fundamental importance, mainly for determination of the break points and as a support to statistical tests. Statistically insignificant breaks may well produce a significant bias in the whole series if they have the same sign. On the other hand, one has to be very careful with statistically significant breaks that could not be confirmed in the station history files.

a) Austrian ALOCLIM stations: Heights of thermometers above ground

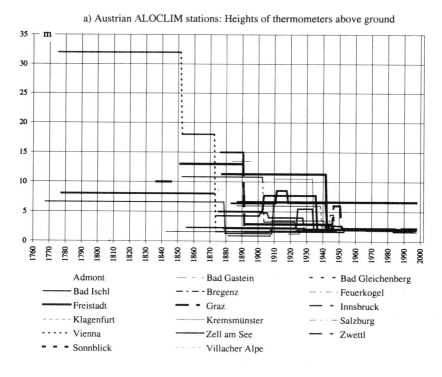

b) Austrian ALOCLIM stations: Heights of ombro-orifices above ground

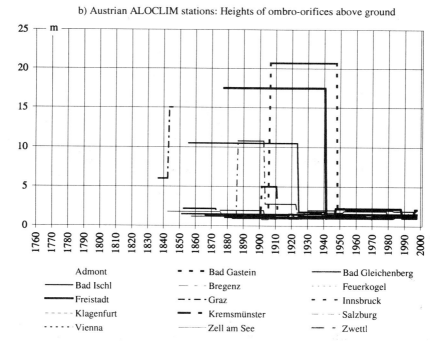

Figure 3. Development of heights of thermometers and rain gauge orifices above ground within the ALOCLIM network.

The ALOCLIM station history files include as much information as possible structured into four groups: 1) short descriptions of the surroundings of the station (topographic and land-use maps and the degree of urbanisation including recent population numbers); 2) the general quality of the series; 3) the main historical features; and 4) a detailed description of each sub-section of the series concerning observers, observation hours, instrument sites and instruments. However, to enable an easier and quicker use of the metadata during the procedure of applying statistical tests, a large part of the information of the station history files has been compressed into so-called "meta quick looks". A single graph gives a quick overview of the station history; i.e. general information as well as many items of the detailed description (Figure 4). Over the horizontal time axis the bars indicate the observers, instruments and so on. The interruptions stand for breaks, if the bars are replaced by lines there is no specific information about the item in those years.

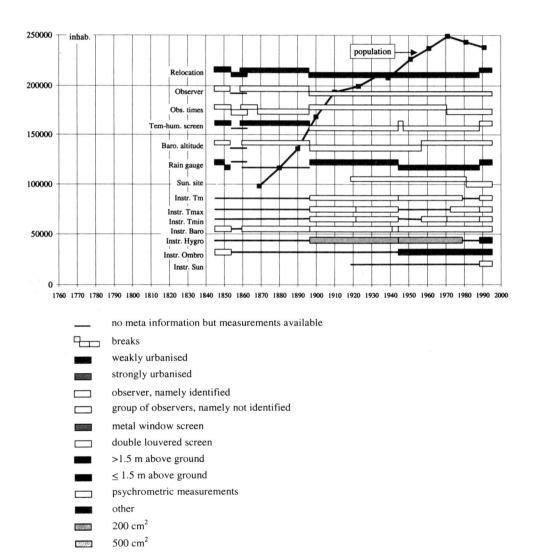

Figure 4. 'Meta quick look' of the ALOCLIM station Graz.

3.   HOMOGENISATION OF THE ALOCLIM DATA SET

As already mentioned, homogenisation was carried out in two steps; first, the elimination of documented breakpoints with the help of metadata information, followed by the application of relative homogeneity tests to remove the remaining inhomogeneities.

3.1   Documented Breakpoints

Breakpoints with documented adjustment values are most commonly related to:

*   station relocations with parallel measurements
*   instrumental changes with parallel measurements
*   changes in observation hours
*   different rules for the computation of mean values

The homogenisation of such breakpoints is based on well documented metadata as found in meteorological yearbooks, observation forms, inspection protocols, station correspondence and instrumental descriptions. For ALOCLIM we could rely on the ALOCLIM metadata bank, demonstrated here by the following examples.

3.2   Example 1: Changes of observation hours and/or changes in the formulae for the calculation of mean values

To overcome these inhomogeneities, mean monthly values of air pressure, air temperature, relative humidity and vapour pressure series were adjusted to 24-hour mean monthly values, utilising a ten year hourly data set of 50 automatic weather stations (Böhm, 1999). Based on this information (horizontally and vertically regionalised, and also for urban and rural regions in single cases), relative daily courses of these elements were calculated. These regionalised daily courses enable the computation of adjustment values for all necessary combinations of observing hours or average formulae to derive 24-hour mean values. The regionalisation showed, for all the four elements, five different zones, one extra alpine, and four vertically distinguished inner-alpine zones. For the elements of air temperature, relative humidity and vapour pressure an additional group of urban stations had to be selected. Figure 5 shows an example for air pressure.

For all elements, the daily amplitude decreases with increasing altitude and with it the systematic deviations of the estimated means from the true mean. Daily amplitudes for temperature and humidity are much higher than for air pressure. For the annual course the daily amplitudes of the summer months exceed those of the winter months. In each case it turned out that the original standard of observations at 7 am, 2 pm and 9 pm defined at the Vienna Congress in 1873 is nearest to the true mean. The worst in each case is the modern tendency to replace the mean by the very rough approximation of (max+min)/2.

3.3   Example 2: Elimination of inhomogeneities due to instrument changes

The automation of the measurement of sunshine duration during the 1980s and 1990s caused systematic inhomogeneities in the Austrian series (and most probably also elsewhere) of hours of bright sunshine. Fortunately the differences between the conventional Campbell-Stokes and the electronic Haenni-Solar sensor are well known (Dobesch and Mohnl, 1998). Figure 6 shows the systematic annual variations of differences split into high elevation and low elevation stations. The changes of instruments require

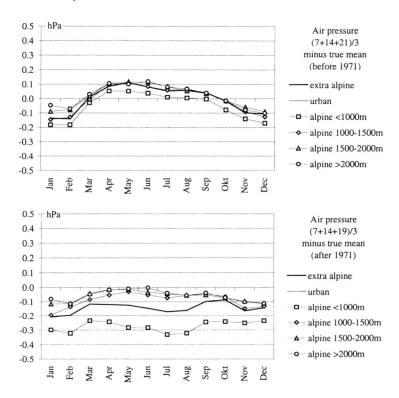

Figure 5. Errors of two different methods for the calculation of monthly means of air pressure (deviations from true mean) within the ALOCLIM network before and after 1971.

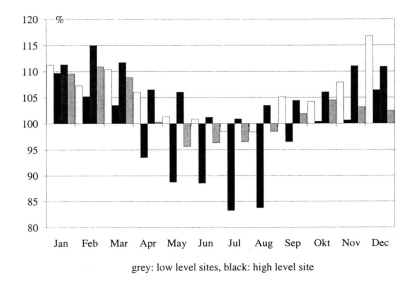

grey: low level sites, black: high level site

Figure 6. Mean monthly ratio (percent) of the hours of bright sunshine of Haenni Solar instrument in relation to Campbell Stokes.

corrections of up to +15% during winter months at low elevation stations and up to -15% during summer months at the high alpine stations.

The reason for the strong differences at high and low elevation sites is the different kinds of clouds. Two examples are; more intermittent convective clouds during summer afternoons at high elevation; and more stratiform foggy clouds at low elevation during winter mornings.

3.4 Non documented breakpoints – relative homogeneity testing (external)

This step of homogenisation was applied just after all quantitatively well known breakpoints had been removed. The MASH Test (Szentimrey, 1997) and the procedure of HOCLIS (Auer *et al.*, 1999) were the main tools for the detection and homogenisation of non-documented breakpoints within the ALOCLIM project (breakpoint detection seasonal, adjustments monthly). External testing stands for comparisons of time series at different climate stations for one climate element.

Each series tested within ALOCLIM turned out to be inhomogeneous. This confirmed our intention to avoid the use of homogeneous reference series. In fact such series never exist in reality and the homogeneity testing concept has to be altered towards a matrix-like inter-comparison of each series with each other series. The candidate series and the reference series change place from sub-period to sub-period and only shorter periods are used to calculate adjustment values. This way of proceeding avoids the import of trends and homogenisation notably increased the data quality, including for climate elements for which there has not been any practical experience (neither within the ALOCLIM group or within the international literature (e.g. Peterson *et al.*, 1998)). Elements like cloudiness, relative humidity, precipitation days, and others, reached a satisfactory state of homogeneity. The following example in Figure 7 shows ratios of homogenised and original annual series of sunshine duration.

3.5 Validation: Final homogeneity testing by inter-comparisons of different elements within one site (internal testing).

This method, only possible if a multiple data set is available, was applied to give support to the results of the first two steps of homogenisation. However, for some cloudiness series the connection between relative sunshine duration and cloudiness was used to complete the homogenisation procedure. As long as turbidity is constant, the sum of cloudiness and relative sunshine duration should also stay at a constant level (Lauscher, 1957). Since sudden turbidity changes should not be expected to happen, no sudden breaks in the sum of homogeneous relative sunshine and cloudiness series should occur. This kind of homogeneity testing was of special interest when referring to the question about the underestimation of high cloudiness at the turn of the century (for which no clear information was found in the metadata). For the stations Kremsmünster and Vienna, for example, no hints of such an underestimation could be detected, as is documented in Figure 8. The long-term sum of relative sunshine duration and cloudiness (both in %) shows no breaks. The slight low frequency long-term variability in the curves should not be associated with inhomogeneities but with real changes of turbidity, perhaps.

3.6 Homogenisation of Daily Values

For some elements, such as the number of frost days, ice days, summer days, hot days, clear days and overcast days, the results revealed that the procedures described could produce problematic results – mainly due to the non linear relations among them because of their upper and lower limits. So for six ALOCLIM sites daily data were digitised back into

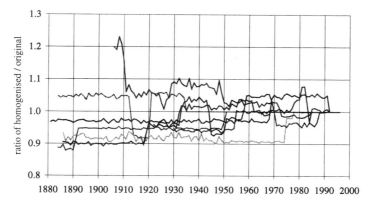

Figure 7. Comparison of homogenised and original ALOCLIM annual sunshine duration series. Ratio of homogenised and original data. Each line indicates one ALOCLIM station.

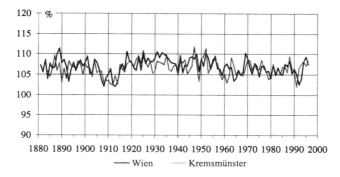

Figure 8. Sum of cloudiness (percent) plus relative sunshine duration (percent of possible).

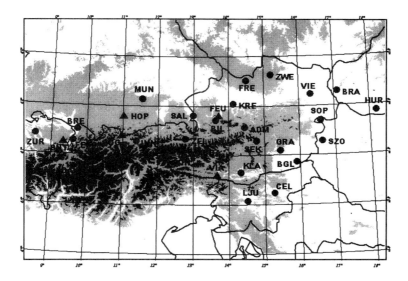

Figure 9. Network of ALOCLIM. Dots: locations below 1500 m asl, triangles: locations above 1500 m asl. Topography: white:<500 m, grey: 500-1500 m, dark grey: >1500 m.

Table 2. The ALOCLIM data set

full name	abbr.	long.	lat.	alt. (m)	air pressure mean	temp. mean	temp. mean daily max.	temp. mean daily min.	temp. abs. max.	temp. abs. min.	frost days	ice days	summer days	hot days	precip. sum	precip. max. daily sum	precip. days >1 mm	sunshine duration	cloud. mean	cloud. clear days	cloud. overcast days	relat. humidity mean	vapour pressure mean
Admont	ADM	14°27'	47°34'	646		1883																1913	
Bad Bastein	BGA	13°07'	47°06'	1100		1854	1855	1855	1855	1855					1858	1883	1888		1864			1879	1879
Bad Gleichenberg	BGL	15°54'	46°52'	303		1881	1882	1882	1882	1882					1879	1879	1888		1879			1860	1860
Bad Ischl	BIL	13°38'	47°43'	469	1855	1855	1882	1882	1882	1882					1858	1864	1888		1864				
Bratislava	BRA	17°06'	48°17'	280	1852	1850	1891	1891	1891	1891					1857	1857	1857	1930	1872			1871	1871
Bregenz	BRE	09°44'	47°30'	424	1875	1871	1880	1880	1880	1880					1874	1869	1888	1934	1872			1874	1874
Celje	CEL	15°16'	46°14'	234		1851																	
Davos	DAV	09°51'	46°47'	1590		1901											1901	1886					
Feldkirch	FEL	09°37'	47°16'	440		1875									1876								
Feuerkogel	FEU	13°43'	47°49'	1618		1930	1930	1930															
Freistadt	FRE	14°30'	48°30'	548		1876									1878								
Graz	GRA	15°27'	47°05'	366	1837	1837	1882	1881	1882	1881	1894	1894	1894	1894	1864	1874	1888	1922	1864	1894	1894	1862	1856
Hohenpeissenberg	HOP	11°01'	47°48'	986	1781	1781					1894						1879					1879	1880
Hurbanovo	HUR	18°12'	47°52'	124	1872	1872	1877	1877	1878	1878	1894	1894	1894	1894	1871	1871	1871	1934	1872			1872	1872
Innsbruck	INN	11°24'	47°16'	577	1830	1795	1891	1891	1891	1891	1894	1894	1894	1894	1866	1875	1877	1906	1866			1883	1893
Klagenfurt	KLA	14°20'	46°39'	447	1844	1813	1860	1860	1860	1860	1876	1876	1876	1876	1851	1864	1888	1884	1844	1877	1877	1860	1844
Kremsmünster	KRE	14°08'	48°03'	383	1822	1767	1836	1836	1836	1836	1876	1876	1876	1876	1851	1864	1874	1884	1842	1877	1877	1862	1840
Ljubljana	LJU	14°31'	46°04'	316		1851											1891						
München	MUN	11°33'	48°08'	535	1825	1825											1879		1843			1842	1842
Salzburg	SAL	13°00'	47°48'	430	1842	1842	1876	1876	1876	1876	1877	1877	1877	1877		1864		1888	1843	1877	1877	1862	1853
Säntis	SNT	09°21'	47°15'	2500	1883	1864					1877	1877	1877	1877			1901	1888	1883			1901	1901
Seckau	SEK	14°47'	47°17'	874		1891	1891	1891	1891	1891					1891	1891	1891		1891			1891	1891
Sonnblick	SON	12°57'	47°03'	3105	1887	1887	1887	1887	1887	1887	1887	1887			1927	1891	1891	1887	1887	1887	1887	1887	1886
Sopron	SOP	16°36'	47°41'	234		1871																	
Szombathely	SZO	16°38'	47°16'	221		1874																	
Villacher Alpe	VIA	13°40'	46°36'	2140	1880	1851	1883	1882	1883	1882					1845	1853	1888	1884	1879			1881	1881
Wien	VIE	16°21'	48°14'	203	1781	1775	1836	1836	1836	1836	1872	1872	1872	1872	1875		1873	1881	1842	1872	1872	1837	1837
Zell am See	ZEL	12°47'	47°20'	766		1875																	
Zwettl	ZWE	15°12'	48°37'	505		1883	1883	1883	1883	1883					1883	1883	1888		1883			1883	1883
Zugspitze	ZUG	10°59'	47°25'	2962	1901	1901											1901	1901	1901			1901	1901
Zürich	ZUR	08°34'	47°23'	569	1864	1864	1883	1883	1883	1883							1901	1886	1864			1901	1901

the 1870s and quality checked with the ZAMG routine quality programme. Next, the adjustment values of homogenisation of monthly values were applied to the daily values, thus producing six multiple homogenised daily data sets. From these the threshold derived elements could be calculated. In addition, the homogenised daily data sets are valued in that they enable the calculation of mathematically and physically consistent inter-comparisons and combinations; for example, in the case of non linear connections for which the linear monthly means can be problematic.

They can also be used to construct new monthly time series of elements that were not part of the climatological routine in the past (very cold days and other threshold values, heating degree days and others).

With all these methods of breakpoint detection and homogenising, the ALOCLIM project has produced a data set of 274 homogenised single series. The locations of the ALOCLIM stations are presented in Figure 9; Table 2 includes additional information about stations, climate elements and starting years of the series.

4. CLIMATE VARIABILITY DURING THE PAST 230 YEARS IN AUSTRIA

The ALOCLIM data set, consisting of homogenised long term series of various climate elements, at altitudes between 200 and 3100 m, allows investigation of the variability of climate in Austria during the past 230 years.

Starting with mean temperature (Figure 10), the records from mountain stations and those from valley stations enable us to demonstrate the high degree of agreement in the long-term fluctuations. Horizontal low elevation groups (West, East, North, South) and two vertical subgroups (200-1000 m and 2100-3000 m asl.) show similar features in timing,

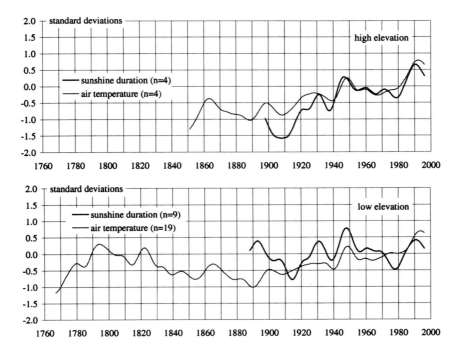

Figure 10. Annual sunshine and temperature series in high and low elevation altitudes, smoothed by 20 year Gaussian low-pass filter, relative to 1961-1990.

dominated by a bicentennial wave with a first maximum in the early 19th century, a minimum around 1890 and the main maximum in the 1990s. The coldest periods were approximately 1 K below the average of 1961-1990, the warmest 0.7 K (in the 1990s), 0.3 K (early 19th century and the 1890s) above normal (for seasonal evolution proceed to the next section, Figure 16).

In contrast, the eastern alpine sunshine series do not show this consistency versus altitude (Figure 10). Although the features of the low elevation stations show only minor differences amongst themselves, the high elevation curves do differ significantly from the low elevation sites. With their centennial increasing trend of bright sunshine hours, high elevation stations show a strong similarity to the temperature curves. On the other hand, this similarity is not true for low elevation records. High alpine sunshine duration has increased significantly since 1900. In the relatively undisturbed atmosphere, the incoming short wave radiation seems to play a non-negligible role in the 20th century alpine warming. However, the recent de-coupling with the stronger increase of temperature, in respect to sunshine, should be carefully observed in the future. The second half of the recent century is especially characterised by a reduction of sunshine at low elevation sites, so a significant trend of 1.6 hours per year of the difference series of high and low elevation stations becomes obvious. Cloudiness series cannot explain this result. The difference between the series of high elevation and of low elevation cloudiness is characterised by variations at a shorter time scale than by the long term trend seen in the sunshine series (Figure 11). At least for the last 50 years we find a possible explanation in the time series of incoming short wave radiation. The time series from Vienna suggests that this may be due to the strong economic development after World War II, when increasing turbidity modified solar radiation. The ratio of direct solar radiation to scattered radiation has clearly and significantly decreased from approximately 1.2 in 1950 to 0.8 for recent years. At the same time cloudiness decreased slightly, but not significantly (Auer *et al.*, 1998).

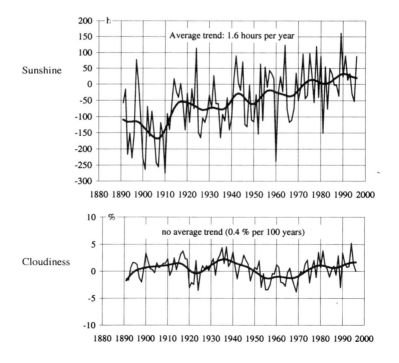

Figure 11. Difference series high minus low elevation of sunshine duration and of cloudiness.

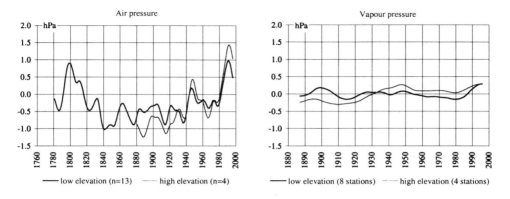

Figure 12. 20 yrs. smoothed annual air pressure and vapour pressure series, relative to 1961-1990.

Air pressure (Figure 12) has been monitored in Austria since 1780. The fluctuations of the low elevation series are similar throughout the whole country, but some interesting contrasts become evident at higher altitudes. Air pressure series of stations below 700 m asl. show two maxima with positive deviations of 0.9 hPa around 1800 and 1.0 hPa around 1990. The mean air pressure increase of the high elevation series since 1880 exceeds that of the low elevation series by far, with 1.48 hPa per 100 years for the high alpine region and only 0.7 hPa for the lowlands. This asymmetric increase of low- and high elevation air pressure may be used as a measure for the increasing temperature of the air column in between (Böhm *et al.*, 1998b).

The dependence of mean (virtual) temperature of an air column on the logarithmic ratio of air pressure at the upper and lower ends of the column is described by the barometric height equation. The transformation of temperature into virtual temperature needs a vapour pressure series as well. A relative increase of high versus low elevation air pressure may be interpreted as a measure of an increasing temperature, not only in the thin layer of thermometer height, but also for considerable parts of the lower troposphere which is remote from a number of possible biasing effects.

A precondition for the existence of an "East Alpine Standard Air Column" (EASAC) (with the barometrically averaged altitudes of the three high elevation sites as the upper and the barometrically averaged altitudes of eight low elevation pressure sites as the lower boundary of the column) is the non-existence of horizontal significantly different climate variations in the area. For air temperature and air pressure this was calculated before, and it is also valid for vapour pressure and virtual temperature.

Virtual temperature (Tv) can be calculated as

$$Tv = T + 0.378*T*e/p$$

with: T = temperature
$\;$ p = air pressure
$\;$ e = vapour pressure

The transition of temperature to virtual temperature does not cause any biasing of the original temperature series. The long-term trend and variations of the virtual temperature excess keep within a range less than 0.1 K. As any horizontal regional differences are far from significance it is possible to define an EASAC which can be assumed representative

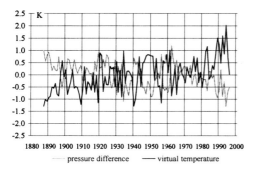 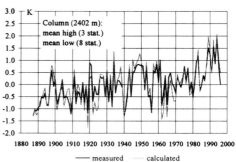

Figure 13. a) Time series of air pressure difference (low minus high) and mean virtual air temperature of EASAC (both relative to 1887 to 1996)

b) Time series of calculated and measured virtual air column temperature of EASAC (both relative to 1887 to 1996)

for the ALOCLIM territory. The lower and upper boundaries of EASAC are: $z_0 = 452$ m, $z_1 = 2854$ m, vertical extension $z_1 - z_0 = 2402$ m.

The comparison of time series of measured mean column temperature and the difference time series of lower boundary minus higher boundary air pressure (Figure 13a) demonstrates that the temperature increase is accompanied by a decreasing air pressure extension of the air column. The warming air causes a reduction of air density and thereby a relative decrease of lower boundary in relation to higher boundary air pressure. From the barometric height equation, the following formula for mean (virtual) column temperature can be derived.

$$Tv,mean = \Delta z * g / R * \ln(p1/p0)$$

with: $\Delta z = 2402$ m
  $g = 9.811$ m/s^2
  $R = 287$ J kg^{-1}K^{-1}

Applied to EASAC, this formula allows the calculation of the virtual mean column temperature series, obtained without any thermometric measurements, being valid not only for the surface level but also for a 2400 m thick layer of the lower troposphere. A comparison of the calculated mean column temperature with the measured temperature (obtained as a mean of the high and low elevation group) is generally good. Although derived by two independent procedures, and based on independently homogenised climate elements the two time series show a high degree of correlation. The method has successfully reproduced the 100 years temperature increase of 1.8 K (Figure 13b) and also the high frequency correlation is very high ($R^2=0.86$).

Comprehensively we can state that the records of the element group temperature, humidity, air pressure, sunshine and cloudiness from mountain observatories enable us to compare the climatic fluctuations in mountains and in the lowlands. The long term variability of air temperature, as well as of virtual temperature, has been similar in the whole territory of Austria. However sunshine duration, air pressure and vapour pressure varied differently at high and low elevation stations. As the high elevation sunshine series are strongly coupled to the temperature series, incoming radiation has to be considered as a forcing factor. This coupling of temperature and bright sunshine has decreased with decreasing elevation during the last 50 years. Nevertheless, changes or increases of cloudiness cannot be the reason for the dampening of the low level sunshine duration. Strong economic development, together with an increasing turbidity of the boundary layer supported by the direct and scattered radiation series of Vienna could, more likely, serve as an explanation. Applying the principle of relative air pressure, the barometric height equation allows calculation of mean air column (virtual) temperature time series which are highly correlated to directly measured air temperature series. Thereby the directly measured warming of +1.8 K since the 1880s in the eastern Alps could be confirmed by an independent and non thermometric measure; i.e. air pressure.

In contrast to all the time series discussed before, the ALOCLIM precipitation series (Figure 14) show no spatial consistency at all. Subgroups of low elevation stations behave quite differently in the long-term sense, and only a few common features are evident which are also obvious in the smoothed low elevation mean time series. In this way, the warming since 1890, which is the spatially uniform characteristic of the temperature change, has been accompanied, in some regions and in some periods, by trends towards drier conditions and, in others, to towards humid conditions (Auer and Böhm, 1994). Up to now, we do not have information about the long term precipitation evolution in the high alpine region. It is doubtful that this information will be collated in the near future. We know that high elevation precipitation series are rare, spatial correlation of precipitation series is low, small changes of location etc. cause quite strong inhomogeneities, due to high wind speeds and the frequency of snowfall. High alpine precipitation data are, and will continue to be, far from reliable.

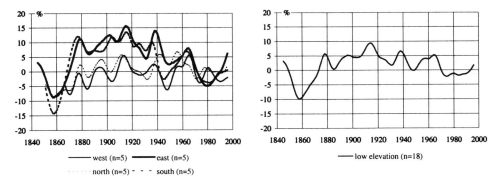

Figure 14. Annual precipitation series of low elevation stations.

4.1 Austrian Temperature Series and the NAO Index

The North Atlantic Oscillation (NAO) can be assumed to be an important advective factor also for locations in Central Europe which is linked to the North Atlantic by dominant westerly to southwesterly winds. The typical feature to be expected for the thermal interactions between an oceanic and a continental region should be a strong advective potential in summer (ocean cooler, continent warmer) and in winter (vice versa) with weak effects in spring and autumn (Figure 17b). In fact, sea surface temperatures of the eastern North Atlantic source region (in respect to Central Europe) are 15 °C higher in winter, whereas in summer they are more or less the same. Thus any NAO signal in Central European temperatures should most likely be detected in winter.

A considerable amount of work has already been spent on the study of NAO influence on European climate (Jones *et al.*, 1997, Hurrel, 1995 and 1996 and others). In most cases the study period has covered single seasons up to some few decades, especially in the upper

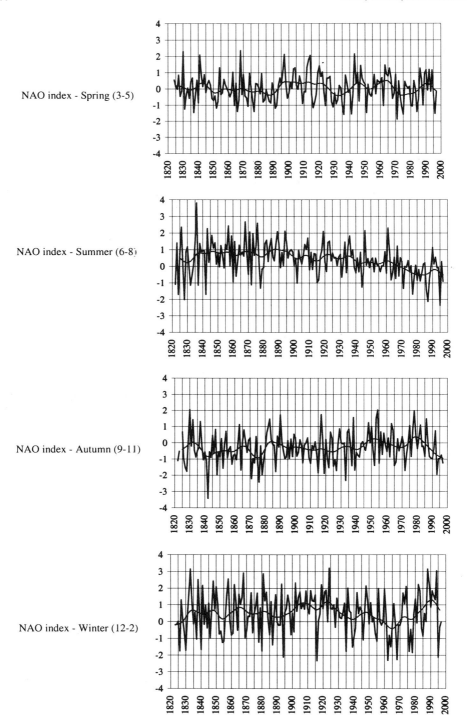

Figure 15. Seasonal NAO-index 1825-1997 – single seasons and 20-year. Gaussian low-pass smoothing (source: Jones *et al.*, 1997).

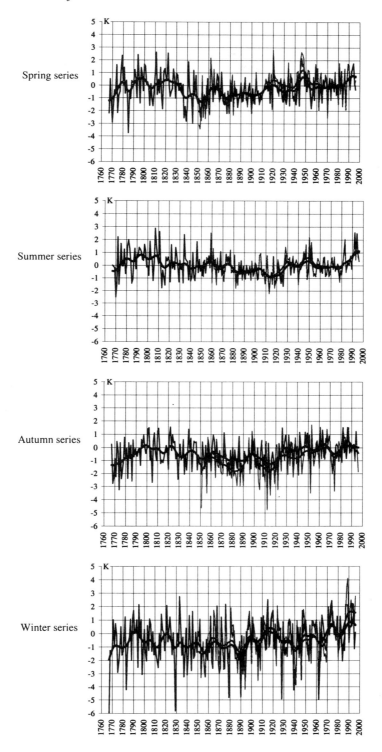

Figure 16. Austrian mean seasonal temperature series (deviations from 1961-90 average) grey: high level mean, black: low level mean, single seasons and 20 year Gaussian low-pass filtered.

air data period for which appropriate three dimensional data sets are available or existing. Our aim was an extension of the period of investigation as far back as possible. This could only be achieved by a reduction of the spatial NAO patterns to a single index - based on (normalised) air pressure differences between locations near the Icelandic low and the Azores' high pressure systems. Jones *et al.* (1997) have recently extended the NAO-index series (based on Gibraltar and Reykjavik) back to the 1820s (Figure 15). Concerning the temperature of Central Europe, the ALOCLIM project has produced a number of carefully tested and homogenised time series of equal length. Both data sets are based on monthly means and allow the study of the influence of NAO on Central European temperatures and the long-term stability of this relationship.

The most dominant feature of NAO is that, since the late 19th century, the summer NAO has decreased. This can expected to be of minor influence on summer temperatures due to the low advective potential of the Atlantic source region in summer. In contrast, the winter series obviously show similarities between NAO and Central European temperature. The strong NAO increase during the most recent decades corresponds to the warming of the winter atmosphere – it was mainly this fact that has drawn the attention of climatologists to the possible links between NAO and European temperatures.

As has been mentioned already, there are quite different temperature variations in the different seasons, but one common feature has to be stressed: low and high elevation variability is very similar, and also differences between locations horizontally are negligible in the study area (Figure 16).

For each month the mean correlation between NAO and Austrian temperature was calculated for low and high elevation temperature means over the period 1851-1997 and additionally for the low elevation stations over the full length of the NAO-period 1825-1997 (Figure 17a). Temperatures have been transformed into normalised values, relative to 1951-1980, corresponding to the NAO-normalising reference period (according to Jones *et al.*, 1997). From the annual course of correlation, it becomes clear that there is, in fact, an annual variation, with maximum in winter, a minimum in summer and the transition seasons in between. There is no significant difference between low and high elevation sites. The correlation between NAO and temperature corresponds quite well with the temperature differences of SSTs in the NAO source region (40 to 10 deg W and 40 to 45 deg N) minus Austrian temperatures (reduced to sea level). This temperature difference can be assumed to be the advective potential of the NAO on Central European temperatures (Figure 17).

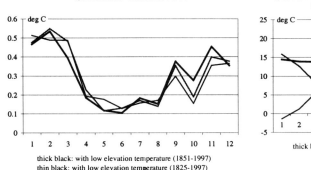
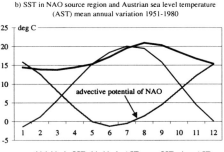

a) Mean annual variation 1851-1997

b) SST in NAO source region and Austrian sea level temperature (AST) mean annual variation 1951-1980

thick black: with low elevation temperature (1851-1997)
thin black: with low elevation temperature (1825-1997)
grey: high level temperature (1851-1997)

thick black: SST, thin black: AST, grey: SST minus AST

Figure 17. a) Annual variation of correlation between NAO and Austrian temperature
b) The advective potential of NAO on Austrian temperature (mean SST in source region 10-40W, 40-45N, data source: Bottomley *et al.*, 1990)

Long-term NAO-correlation with Austrian temperature is about 0.5 in the months of January, February and March, drops to 0.1-0.2 from April to August, and is between 0.2 and 0.4 from September to December. The higher Autumn than Spring values correspond to the time lag of ocean temperatures which produces a higher advective NAO potential in Autumn compared with Spring and Summer.

The annual cycle of the correlation does not correspond to the usual seasons in mid-latitude climate. Therefore, for further proceedings we used only 3 seasons: January to March with the highest NAO-correlation with Austrian temperatures; April to August with the lowest; and September to December with intermediate correlation.

Consideration needed to be given also to the question: what if shorter sub-periods show stronger (or weaker) connections of NAO with Central European temperature? Thus the correlation coefficients were calculated for 21-year sub-intervals and these sub-intervals moved in one year steps over the entire NAO-period. This "moving windows" technique was performed for each of the three NAO-seasons, and for each single station record, as well as for the high level and low level means presented in Figure 18. Two aspects become clear at first view. There is no temporal stability in either of the correlation time series. There are oscillations, sudden jumps, short or long-term collapses and, in some cases, even changes in sign. High Alpine sites show less spatial differences in the temporal course of the correlation.

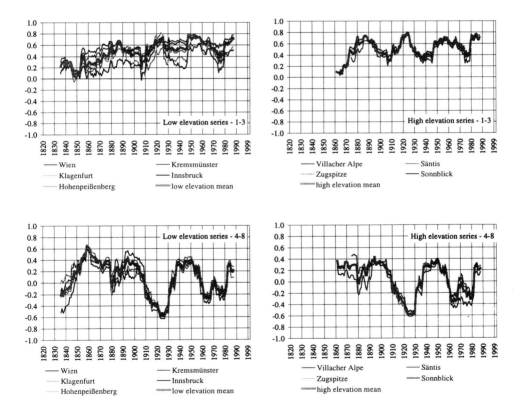

Figure 18. Correlation of NAO with Austrian low and high elevation temperature in moving windows of 21 years for single sites and low and high elevation means for winter and summer season. (Single sites: thin, means: bold).

The correlation time series confirm the analysis based on the entire long-term period, in some aspects. Jan-Mar correlation is the highest of all seasons, Apr-Aug is the lowest. Wintertime (1-3) correlation is always positive throughout the whole period of investigation. In some sub-periods it exceeds 0.6, with maximum values of 0.8. On the other hand, it also drops towards values near zero. NAO correlation with Austrian low elevation sites shows a long-term slightly increasing trend, high elevation sites are more characterised by oscillations without a long-term trend. Summertime (4-8) correlation oscillates more or less symmetrically around zero. The amplitudes of oscillation are high. NAO is positively correlated to ALOCLIM-temperatures with values up to 0.6 in some sub-periods of 30 to 40 years length; in others negatively at the same absolute level. Sep-Dec correlation (not shown) is also (as in Winter) nearly always positive but at a lower level, and with lower peak values than the Jan-Mar correlation.

The features of the correlation series are not easy to explain but one thing at least is certain: studies of European climate sensitivity to the North Atlantic Oscillation should rely on long-term data sets. Results obtained for only some decades are obviously not representative for the long-term influence of the NAO. There may even be opposite results due to the oscillatory behaviour of the NAO forcing (comparisons to be made, for example, of the 1910-1930 summer NAO-correlations with those 1930-1960). The summer influence of NAO on European climate is not necessarily insignificant, as shown by the data of the last 4 to 5 decades only.

Correlation analysis concentrates on high frequency variability; long term effects are often hidden behind the stronger short term variability. So we compare the smoothed NAO and the Austrian temperature series for the three "NAO-seasons". Both are normalised to the same reference period (1951-1980) which allows also quantitative comparisons (Figure 19).

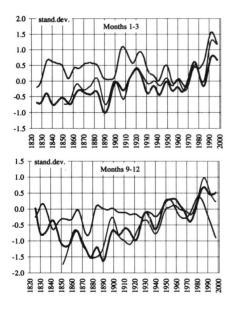

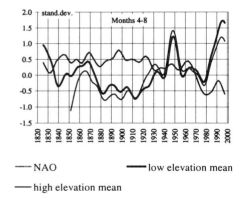

Figure 19. Smoothed time series of NAO-index and ALOCLIM-temperatures for the three NAO-seasons, (low and high elevation mean normalised to 1951-80 reference period, smoothing by 20 year Gaussian low-pass filter).

The Jan-Mar NAO corresponds well with ALOCLIM-temperatures at time scales of 20 to 40 years. With only few exceptions (around 1840, 1900-1920 and in the 1980s), each NAO maximum and minimum corresponds to a temperature maximum and minimum. In the second part of 20th century the trends are also similar. Only the long term trend of the entire period 1825-1997 is different. The long-term temperature increase cannot be seen in the NAO series. The NAO is at a higher level than Austrian winter temperatures during the first 100 year period of investigation.

The Apr-Aug NAO does not show any similarities to Austrian temperatures on all time scales less than 100 years. Only the long term trends for the entire 20th century show a certain inverse correspondence of a centennial linear decrease of the NAO of roughly one standard deviation compared with an opposite temperature increase of the same amount. For the Sep-Dec NAO, the situation is somewhat in between. Some of the short term features are similar (e.g. the period 1930-1980), some dissimilar, and others are contradictory. Similar to the winter series, the NAO cannot explain the long-term increasing temperature trend evident since 1880.

4.2 Maximum and Minimum Daily Temperatures and the Diurnal Temperature Range

Figure 20 shows the time series of mean daily maximum and minimum temperature for two separated groups of the ALOCLIM stations (low and high elevation samples). Again,

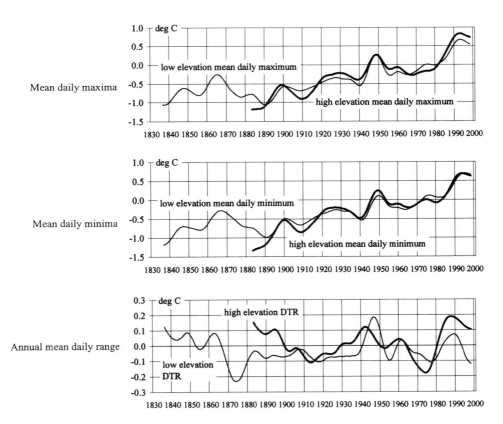

Figure 20. Smoothed time series of annual mean daily maximum, minimum, and diurnal temperature range of low and high elevation stations relative to the average 1961-1990.

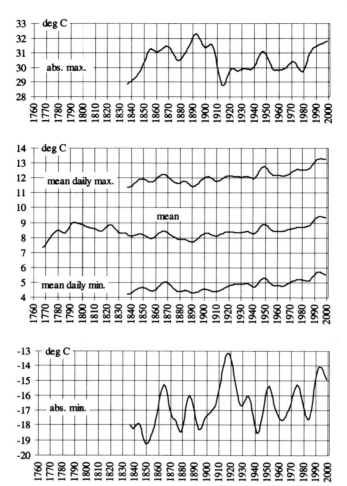

Figure 21. Mean, mean maximum and minimum and absolute maximum and minimum temperature in Kremsmünster.

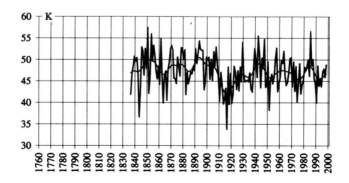

Figure 22. Annual temperature range, single values and 20-years Gaussian smoothed values in Kremsmünster.

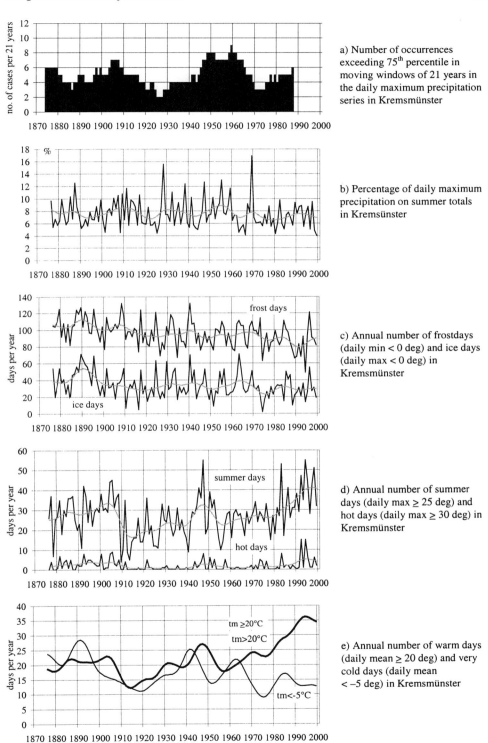

a) Number of occurrences exceeding 75th percentile in moving windows of 21 years in the daily maximum precipitation series in Kremsmünster

b) Percentage of daily maximum precipitation on summer totals in Kremsünster

c) Annual number of frostdays (daily min < 0 deg) and ice days (daily max < 0 deg) in Kremsmünster

d) Annual number of summer days (daily max ≥ 25 deg) and hot days (daily max ≥ 30 deg) in Kremsmünster

e) Annual number of warm days (daily mean ≥ 20 deg) and very cold days (daily mean < –5 deg) in Kremsmünster

Figure 23. Examples for long-term series of elements derived from daily data sets.

there has been a parallel evolution of the two different temperature groups as well as an indication of a slightly greater temperature increase in the mountains.

The question concerning a decreasing diurnal temperature range (DTR) in Austria and Central Europe has been treated in several papers (e.g. Böhm and Auer, 1994, Brazdil *et al.*, 1995, Weber *et al.*, 1997), and is now re-analysed for the ALOCLIM temperature series. The greater number of series, and the length of the series, confirmed the former results for the territory of Austria. In fact there is no systematic long-term trend towards decreasing DTR. The long-term variability is characterised more by oscillations similar to the sunshine duration series (compare Figure 20c with Figure 10); DTR is higher in sunny and warm years and lower in cloudy and cool years.

4.3 Absolute Maximum and Minimum Temperature and the Annual Range of Temperature

In contrast to homogenised time series of mean temperature the coverage of long-term series of extreme values is very poor for most parts of the world. For Austria ALOCLIM offers a sample of six homogenised time series of daily values (five for low elevation sites, one for the high alpine region), but still our investigations are at the level of single station results; they are not able to describe the fluctuations of climatic extremes for the country as a whole. Figure 21 shows, as an example, the five temperature series (mean, mean daily extremes, and absolute extremes) gained from the homogenised daily data set of Kremsmünster. Figure 22 illustrates the absolute annual temperature range of that site which shows strong similarity to the mean daily range curves of Figure 20c.

4.4 Examples for Time Series derived from Homogenised Daily Data Sets

The final series in Figure 23 underlines the potential of homogenised daily data sets to open a new dimension in climate variability studies. Not all questions concerning the homogenisation of long-term daily data sets are yet solved; but are certainly worth pursuing.

5. CONCLUSIONS

In the archives of the national meteorological services there are many hidden treasures of climate data waiting to be discovered and processed into an appropriate state for analysis. They have to be digitised, homogenised and placed in the public domain. The homogenised series of the described project ALOCLIM (Table 2) can already be accessed by contacting the authors. Access is free of charge for scientific application for the Austrian and the Slovakian series, for the other countries the respective data-holders have to be contacted.

Homogenisation depends on metadata information and statistical tests. Therefore the work of homogenisation should be done by the meteorological services, for their knowledge of metadata is not easy to transfer. ALOCLIM is an example of such a national effort.

We wanted to demonstrate that:

1) homogenisation is possible for climate elements which are not normally considered to be in homogeneous form and,

2) a multiple homogeneous data set can contribute to a better understanding of the mechanisms of climate variability.

Some examples of analyses using the ALOCLIM dataset have been shown and discussed in the paper in order to underline point 2.

Long-term temperature variability has been very similar across the study region, even for low and high elevation sites from 124 to 3105 m asl., but there are seasonal differences. Summer temperature series are characterised by a bi-centennial wave structure with high temperatures at the beginning (near 1800) and at the end of the series. Winter temperatures were at a more or less lower level throughout the first part of the series (1770–1900) and increased during the 20th century to a recent maximum, which is considerably higher than the 19th century level. DTR series do not follow the prevailing decreasing global trend. They seem to react more to sunshine and temperature in the sense of a positive correlation (warmer years having higher DTR but not vice versa). The long-term air pressure series of low-elevation sites show a striking similarity to the air temperature series, a fact that suggests air pressure (via a northward shift of the subtropical high) to be one of the main forcing factors of temperature in the region. This supposition is supported by the undisturbed high-elevation sunshine series which have the same long-term features as temperature and air pressure. Low-elevation sunshine series on the other hand behave differently; they do not show the increase of the high-elevation ones, thus causing a noticeable increase of the difference between high and low elevation sunshine in the Alps. Cloudiness series do not appear to show a relevant feature, therefore we suppose that an increase of turbidity in the lower levels of the troposphere may be the cause. Also, high Alpine air pressure series increased relative to low-elevation ones. This effect could be used to derive an independent indication for the warming from late 19th to late 20th century of a 2400 m thick "mean East Alpine Standard Air Column" (EASAC) from barometric time series. The model (calculating air temperature of the EASAC from high- and low elevation air pressure and vapour pressure series) could explain 86% of the measured temperature variability and successfully reproduced the centennial East Alpine temperature increase of 1.8K.

Contrary to temperature, air pressure, humidity, sunshine and cloudiness series, that showed no significant regional horizontal differences in the study area, precipitation series behaved differently. Precipitation increased in the west, fluctuated without a long-term trend in the central parts, and increased before the 1940s followed by a strong decrease afterwards in the east and southeast; thus showing clearly that there is no unambiguous response of precipitation to climate warming at regional scales. A first analysis of NAO forcing on east Alpine temperatures showed the well known mean annual structure with higher correlation in winter and lower in summer. An additional correlation analysis, using moving time-windows, resulted in a fluctuation of the strength of the NAO-forcing during the instrumental period, with considerably lower winter forcing during several periods, with even sub-periods having a rather high NAO-influence on summer temperatures.

Our results are valid only for a very small part of the Earth. Global coverage of homogenised multiple datasets is not yet available, but we hope that there will be ongoing national and international efforts to close the existing gaps on the way to a homogeneous multiple climatic data set of global dimensions. Another goal for the future should be the creation of homogenised series of long-term daily data. Analyses based on such data sets would strongly increase our knowledge about climate variability and change.

Acknowledgements

Project ALOCLIM (Austrian Long Term Climate) worked in collaboration with the Slovak Hydrometeorological Institute, the Meteorological Service of Hungary, the Hydrometeorological Institute of Slovenia and the Czech Hydrometeorological Institute. Special thanks are addressed to the Swiss Meteorological Institute and the German Weather Service for providing data from Switzerland and Germany. We also wish to thank Stella Canavan and the two reviewers for their linguistic revision, and in addition to the two reviewers for their remarks that strengthened the content of our paper.

REFERENCES

Auer, I., 1993, Trends and variations in precipitation in Austria since 1845, *Early Meteorological Instrumental Records in Europe*, Methods and results, Barbara Obrebska-Starklowa, ed., Uniwersytet Jagiellonski, Krakow, 173–182.

Auer, I., Boehm, R., and H. Mohnl, 1993, Climatic change on Sonnblick- a multi-elemental approach to describe climatic change using a centennial data set, *8th Conf. Applied Climatology*:249-252, Anaheim, California.

Auer, I., and Böhm, R., 1994, Combined temperature-precipitation variations in Austria during the instrumental period. *Theor. Appl. Climatol.* 49:161-174.

Auer, I., 1995, Extreme precipitation events - time series analyses of Viennese data, *Proc. Int. Conf. on Past, Present and Future Climate*, Helsinki, 22-25 August 1995:211-217.

Auer, I., and Böhm, R., 1997, Data, metadata and the question of homogeneity for the Austrian climatological network, *Proceedings of the 1st Seminar for homogenization of surface climatological data*, Budapest, HU, 6-12. Oct. 1996:83-95.

Auer, I., Böhm, R., Hagen, M., and Schöner, W., 1998, 20th Century increase of boundary layer turbidity derived from Alpine sunshine and cloudiness series, *Proc. 8th Conf. on Mountain Meteorology*, 3–7 August 1998, Flagstaff, Arizona:77-80, AMS Boston.

Auer, I., Böhm, R., Hagen, M., and Schöner, W., 1999, ALOCLIM – Austrian – Central European long-term climate – creation of a multiple homogenised long term climate data set, *Proc. 2nd Seminar for Homogenisation of surface climatological data*, (Budapest, Hungary, 9-13 November 1998), WCDMP-No.41, WMO-TD No. 962:47-71, WMO Geneva.

Böhm, R., 1993, Air temperature fluctuations in Austria 1775–1991, a contribution to greenhouse warming discussion, *Prepr. 8th Conf. on Applied Climatology*, Jan 17-11, Anaheim, CA:J 26-J30.

Böhm, R., 1999, Dangers and advantages of automation concerning the homogeneity of long-term time series, *Proc. of ICEAWS 1999* (2nd International Conference on Experiences with Automatic Weather Stations, Sept.1999, Vienna) CD-ROM:17pp.

Böhm, R., and Auer, I., 1994, A search for greenhouse signal using daytime and nighttime temperature series, *Proc. European Workshop on Climate Variations* held in Kirkkonummi (Majvik), Finland, 15-18 May 1994, Publ. Acad. of Finland 3/94:141-151.

Böhm, R., Auer, I., Hagen, M., and Schöner, W., 1998a, Possibilities of multi-elemental climate time series research: First results of project ALOCLIM (Austrian Long-term Climate), *Proceedings of ECAC 98*, Vienna (CD-ROM):8pp.

Böhm R., Auer, I., Hagen, M., and Schöner, W., 1998b, Long alpine barometric pressure series in different altitudes as a measure for the 19th/20th Century Warming, *Proc. 8th Conf. on Mountain Meteorology*, 3-7 August 1998, Flagstaff, Arizona:72-76, AMS Boston.

Brazdil, R., Budikova, M., Auer, I., Boehm, R., Cegnar, T., Fasko, P., Lapin, M., Gajic-Capka, M., Zaninovic, K., Koleva, E., Niedzwiedz, T., Ustrnul, Z., Szalai, S., and Weber, R.O, 1996, Trends of maximum and minimum daily temperature in Central and south eastern Europe, *Int. Journal of Climatology* 16:765-782.

Bottomley, M, Folland, C. K., Hsiung, J., Newell, R.E., and Parker, D.E., 1990, *Global Ocean Temperature Atlas*, Met. Office and Dept. of Earth, Atm. and Planet. Sciences, MIT.

Dobesch, H., and Mohnl, H., 1998, Meßwertdifferenzen zwischen den Sonnenscheingebern Campbell-Stokes und Haenni Solar 111B - Konsequenzen in den klimatologischen Aussagen, in: *Annalen der Meteorologie*, Nr. 37:125-126, Selbstverlag des Deutschen Wetterdienstes, Offenbach am Main.

Gutmann, J., 1948, *Beobachtungs- und Meßmethoden des Wetterdienstes* (Anleitung zur Ausführung und Verwertung meteorologischer Beobachtungen), Zentralanstalt für Meteororologie und Geodynamik, Publ. No. 158, 143 Seiten, Wien.

Hann, J., 1884, *Jelinek's Anleitung zur Ausführung meteorologischer Beobachtungen nebst einer Sammlung von Hilfstabellen*. Neu herausgegeben und umgearbeitet von Dr. J. Hann, Druck der kaiserlich-königlichen Hof- und Staatsdruckerei, 185 Seiten, Wien.

Hurrell, J.W., 1995, Decadal trends in the north Atlantic oscillation and relationships to regional temperature and precipitation, *Science* 269:676-679.

Hurrell, J.W., 1996, Influence of variations in extratropical wintertime teleconnections on northern hemisphere temperatures, *Geophys. Res. Letters* 23:665-668.

Jelinek, C., 1869, *Anleitung zur Anstellung meteorologischer Beobachtungen und Sammlung von Hilfstabellen*. Erste Ausgabe, Druck der kaiserlich-königlichen Hof- und Staatsdruckerei, Wien.

Jelinek, C., 1876, *Anleitung zur Anstellung meteorologischer Beobachtungen und Sammlung von Hilfstabellen*. Zweite umgearbeitete und vermehrte Ausgabe, Druck der kaiserlich-königlichen Hof- und Staatsdruckerei, Wien.

Jones, P.D., Jónsson, T., and Wheeler, D., 1997, Extension to the North Atlantic Oscillation using early instrumental pressure observations from Gibraltar and south-west Iceland, *Int. J. Climatol.* 17:1433-1450.

K.k. Central-Anstalt für Meteorologie und Erdmagnetismus, 1893, *Jelinek's Anleitung zur Ausführung meteorologischer Beobachtungen nebst einer Sammlung von Hilfstafeln.* In zwei Theilen. Erster Theil: Anleitung zur Ausführung meteorologischer Beobachtungen an Stationen II. und II. Ordnung. Vierte umgearbeitete Auflage, Druck der kaiserlich - königlichen Hof- und Staatsdruckerei, 71 Seiten, Wien.

K.k. Central-Anstalt für Meteorologie und Erdmagnetismus, 1895, *Jelinek's Anleitung zur Ausführung meteorologischer Beobachtungen nebst einer Sammlung von Hilfstafeln.* In zwei Theilen. Zweiter Theil: Beschreibung einiger meteorologischer Instrumente und Sammlung von Hilfstabellen. Vierte umgearbeitete Auflage, Druck der kaiserlich - königlichen Hof- und Staatsdruckerei, 101 Seiten, Wien.

K.k. Central-Anstalt für Meteorologie und Geodynamik, 1905, *Jelinek's Anleitung zur Ausführung meteorologischer Beobachtungen nebst einer Sammlung von Hilfstafeln.* In zwei Teilen. Erster Teil: Anleitung zur Ausführung meteorologischer Beobachtungen an Stationen I. bis IV. Ordnung. Fünfte umgearbeitete Auflage, Druck der kaiserlich - königlichen Hof- und Staatsdruckerei, 127 Seiten, Wien.

K.k. Zentralanstalt für Meteorologie und Geodynamik, 1906, Bericht über die internationale meteorologische Direktorenkonferenz in Innsbruck, September 1905. *Anhang zum Jahrbuch 1905*, K.k. Hof- und Staatsdruckerei, 154 Seiten, Wien.

K.k. Central-Anstalt für Meteorologie und Geodynamik 1910, *Jelinek's Anleitung zur Ausführung meteorologischer Beobachtungen nebst einer Sammlung von Hilfstafeln.* In zwei Teilen. Zweiter Teil Sammlung von Hilfstabellen. Fünfte umgearbeitete Auflage, Druck der kaiserlich - königlichen Hof- und Staatsdruckerei, 94 Seiten, Wien.

Klinger, E., 1986, *Die Wetterbeobachtungen an Klimastationen (Anleitung zur Durchführung meteorologischer Beobachtungen und Messungen).* 107 Seiten, Herausgeber, Verleger, Druck: Zentralanstalt für Meteorologie und Geodynamik, Wien.

Kreil, K., 1848, Entwurf eines meteorologischen Beobachtungssystems für die österreichische Monarchie. Abdruck aus dem *III. Hefte der Sitzungsberichte vom Jahre 1848*.

Lauscher, F., 1957, Zur Frage: Sonnenschein + Bewölkung = 100%? *Wetter und Leben* 9:143-146.

Nordli, P.Oe., Alexandersson, H., Frich, P., Foerland, R.J., Heino, R., Jonsson, T., Tuomenvirta, H., and Tveito, O.E., 1997, The effect of radiation screens on nordic time series of mean temperature *Int. J. Climatology* 17:1667-1681.

Peterson, T.C., Easterling, D.R., Karl, T.R., Groisman, P., Auer, I., Böhm, R., Plummer, N., Nicholis, N., Torok, S., Vincent, L., Tuomenvirta, H., Salinger, J., Förland, E.J., Hanssen-Bauer, I., Alexandersson, H., Jones, P., and Parker, D., 1998, Homogeneity adjustments of in situ climate data: a review, *International Journal of Climatology* 18:1493-1517.

Pozdena, R., 1913, Das neue Normalbarometer "Marek" der k.k. Zentralanstalt für Meteorologie und Geodynamik. *Jahrbücher der Zentral-Anstalt für Meteorologie und Goedynamik*, Jahrgang 1911, N. F. XLVIII. Band, S XIII-XXIII, Wien.

Quayle, R.G., Easterling, D.R., Karl, T.R., and Hughes, P.Y., 1991, Effects of recent thermometer changes in the cooperative station network, *Bulletin AMS* 72(11):1718-1723.

Schlein, A., 1915, *Anleitung zur Ausführung und Verwertung meteorologischer Beobachtungen*. Sechste, vollständig umgearbeitete und vermehrte Auflage von Jelinek's Anleitung zur Anstellung meteorologischer Beobachtungen und Sammlung von Hilfstafeln, 1. Teil. Herausgegeben von der k.k. Zentralanstalt für Meteorologie und Geodynamik in Wien. 48 Figuren im Text, 17 Figuren auf 17 Tafeln, 180 Seiten. Druck der k.k. Hof- und Staatsdruckerei, Wien und Leipzig Franz Deuticke.

Szinell, C., 1997, Methods for homogenization of data series, *Proc. 1st Seminar for Homogenization of Surface Climatological Data*, Budapest, Hungary 6-12 October 1996:9-17.

Szentimrey, T., 1997, Statistical procedure for joint homogenization of climatic time series, *Proc. 1st Seminar for Homogenization of Surface Climatological Data*, Budapest, Hungary 6-12 October 1996:47-62.

Szentimrey, T., 1999, Multiple analysis of series for homogenization (MASH). to be published in: *Proc. of 2nd Seminar for Homogenization of Surface Climatological Data*, held in Budapest, Hungary 9-13 November 1998.

Weber, R.O., Talkner, P., Auer, I., Gajic-Capka, M., Zaninovic, K., Brazdil, R., and Fasko, P., 1997, 20th-Century changes of temperature in the mountain regions of Central Europe, *Climatic Change* 36:327-344, Kluwer Academic Publishers, NL.

THE ONSET OF THE LITTLE ICE AGE

Jean M. Grove

Girton College
Cambridge, England

1. INTRODUCTION

The Little Ice Age was a period several centuries long, within the last millennium, during which glaciers enlarged and their fronts oscillated about forward positions without retracting as far as the positions they occupied before the initial advance. The term refers to the behaviour of glaciers, not directly to the climatic circumstances that caused them to expand. It is required as a collective term for the advances of the last few centuries which left many distinctive traces in the landscape, especially in the form of moraine sets.

It is argued that sufficient radiocarbon data are now available to allow critical examination of the suggestion by Porter (1986) that the Little Ice Age began in the 13th century. All the radiocarbon dates quoted here have been calibrated or, where necessary, recalibrated according to Stuiver and Reimer (1993). Data from mature soils or relating to surging glaciers are excluded. Ideally, data from all the dating methods available should be considered, but this would involve making the discussion extremely long. Little use is made here of dates obtained from lichenometry because it is doubted whether the accuracy of the method is sufficient to provide reliable results over a time interval of more than five centuries[1]

The most accurate reconstructions of glacier fluctuations come from the Swiss Alps, especially from the Grosser Aletsch, Gorner and Lower Grindelwald, where investigations have been unusually intensive and a variety of evidence has been employed (Holzhauser 1995, 1997; Holzhauser and Zumbühl, 1996). The exact ages of the outer rings of trees

---

[1] Several difficulties arise with the use of lichenometric dating, including differences in the methods used from one study to another, the sometimes unstated character of the substrate rock and, more particularly, the variations in growth rate which must be assumed to have taken place as climate varied over time. This last drawback is especially important here as this paper is much concerned with events which took place more than 500 years ago, that is before the period within which lichenometry is most useful. All lichenometric age estimates over about 300 years depend upon the extrapolation of growth curves (Luckman, 2000). In some regions serious difficulties have been shown to arise if lichenometry is used to date Little Ice Age events (Kirkbride and Dugmore, in press). A more detailed account of the complications involved in lichenometric dating, is given in Bradley (1999). He considers it is likely to continue to play a role in dating rocky deposits in arctic and alpine areas but indicates that caution is needed in using lichenometry as a dating method, even for establishing relative age.

History and Climate: Memories of the Future?
Edited by Jones et al., Kluwer Academic/Plenum Publishers, 2001

153

overrun by ice have been obtained by dendrochronological crossdating, a method also found valuable in the Canadian Rockies (Luckman 1996).[2].

A preliminary survey of evidence from regions round the North Atlantic (Grove, in press) showed that the Little Ice Age expansion was under way in the 13th century, and built up in the 14th century to the first of several culminations. Here, evidence from regions around the North Atlantic will be compared with that from other regions including the Himalaya, western North America, South America, and New Zealand. Data from all these regions, collected by Röthlisberger from glaciers of similar size and calibrated by Geyh in the same laboratory (Röthlisberger and Geyh 1986), are supplemented from other sources. Some relevant radiocarbon dates from lacustrine and marine sediments are taken into account in addition to limiting ages for moraine formation. Attention is paid to differences in the accuracy of data types and sets. Results from the Sub-Antarctic are mentioned, but only briefly because difficulties involved in estimating the magnitude of reservoir corrections for dates on marine samples allow only preliminary estimates of age to be made in the Antarctic region. Questions raised by differences in the dating of the Little Ice Age in ice-cores from high altitude and high latitude sites are discussed (Thompson *et al.*, 1995).

The Little Ice Age is a collective term for the period of glacial advances in the course of the last millennium which left behind many series of terminal and lateral moraines reflecting the influence of short term glacier fluctuations superimposed on longer term oscillations. The more recent of these have been recorded in considerable detail since the late 19th century[3] when deliberate monitoring of glaciers began in Switzerland. When Bradley and Jones (1993) advanced the view that understanding is not well served by the continued use of the term "Little Ice Age", they seem to have been influenced partly by the hypothesis that the event had to involve "a world-wide synchronous and prolonged cold period" and partly by disagreements in the literature about the time of its initiation and termination.

Differences in the dates attributed to the Little Ice Age could reflect genuine regional disparities but could also be accounted for in terms of the nature of the evidence available to authors at the time of writing, variations in the extent and type of research effort from place to place, the dating methods used, or uncertainties in the relations between radiocarbon and calendar dates particularly over the last five hundred years. Detailed evidence of glacier fluctuations has proliferated in recent years while the ability to present radiocarbon dates in the form of calendar years has much improved, particularly for the Holocene (Stuiver and Pearson, 1986; Stuiver and Reimer, 1986, 1993). This makes it easier to compare dates between one region and another, and to compare calibrated radiocarbon ages with absolutely dated evidence drawn from documents or other proxy sources.

It would be logical to expect that if the Little Ice Age were a global event then the dates of its initiation in different regions would be generally comparable, given that sufficient data could be found. It has been shown that Little Ice Age glacier expansions and an earlier phase of ice advances leaving multiple moraines bracketed an intervening period lasting from 900 to 1250 A.D. which is commonly known, and is subsequently referred to here, as the Medieval Warm Period (Grove and Switsur, 1994). The question as to whether the Little Ice Age was a global event is therefore linked to the question of the reality of the Medieval Warm Period and whether it too was global in extent. Use of the term has been questioned

---

[2] Estimates of moraine age based on the ages of the oldest trees growing on them cannot be very accurate on older moraines deposited during the early part of the Little Ice Age. The trees present may well not be the first generation and the lapse of time which occurred before the first trees grew is not necessarily known over a range of several centuries. There appear to be very few, if any, instances of dates of early Little Ice Age moraines being established on the basis of the age of the oldest tree found on them. Bradley (1999, p.123) cites an example where a tree about 750 years old according to dendrochronological dating, was found on a moraine dated by tephrachronology to older than 2500 years BP.

[3] Mass balance studies were initiated in Switzerland in 1882, with the installation of a stake network on the Rhône Glacier (Aellen 1996). International monitoring of glacier variations started in 1894, with the founding of the International Glacier Commission at the 6th International Geological Congress in Zürich (Haeberli *et al* 1989).

on the basis of the inadequacy of the data on which Lamb (1977) based his introduction of the term (e.g. Ogilvie and Farmer, 1997). The medieval period was not continuously warm, even in Europe. Winters varied markedly; those between 1091 and 1179 were as cold as in the Little Ice Age, while those between 1180 and 1209 were warmer than in the 19th century. Pfister *et al.* (1999), using numerous, verified European contemporary sources, have identified warm dry anomalies in the climate of western and central Europe in the medieval period which lack analogues even in the 20th century. Furthermore they found that a lower frequency of extreme negative anomalies was a feature common to the entire period, on the century time scale, and concluded that a climatic label is needed for the interval between the 9th and 13th centuries.

Hughes and Diaz (1994), surveying a wide range of data sets, based on a variety of evidence from many continents, found substantial decadal to multidecadal scale variability in regional temperatures over the last millennium. However, they were unable to demonstrate the occurrence of a global Medieval Warm Period, although more support for such a phenomenon could be drawn from high-elevation than low-elevation records. Other papers in the same number of *Climatic Change* showed that there had also been marked fluctuations in precipitation. The combined effect of fluctuations in temperature and precipitation, causing runs of negative mass balances, can provide as satisfactory an explanation of glacier diminution as a prolonged warm period. This may explain the gap which exists between the dates of moraines deposited during the Little Ice Age and during its Holocene predecessor in some regions of the world, including China (Zhang De'er, 1994) and western Alaska (Wiles and Calkin, 1994).

The fluctuations of the Grosser Aletsch and Gorner Glaciers in Switzerland have been traced in greater detail than any others (Holzhauser, 1995, 1997; Holzhauser and Zumbuhl, 1996). It is significant that the period from the 8th to the end of the 13th century during which these two glaciers were shrunken corresponds with the results of glacial geological evidence from elsewhere for the interval between the penultimate period of Holocene glacier expansion and the Little Ice Age (Grove and Switsur, 1994).

2.   LITTLE ICE AGE INITIATION IN REGIONS AROUND THE NORTH ATLANTIC

Evidence of the time at which the Little Ice Age began in the North Atlantic region has been considered at more length elsewhere and will not be repeated in full here (Grove, in press). Comparison of the glacial history of the last millennium in the Swiss Alps, where it has been traced in extraordinary detail (Holzhauser, 1995, 1997; Holzhauser and Zumbühl 1996), with other circum-Atlantic regions has shown that the Little Ice Age was under way in the 13th and 14th centuries (Grove, in press). In the Alps it is now seen as having commenced in the second half of the 13th century (Holzhauser, 1997). Its three culminations, including the first in the 14th century, were remarkably similar in scale. The fluctuation history of the Grosser Aletsch is known more completely than that of any other glacier in the world. Like those of the Gorner Glacier, its Little Ice Age advances were separated from the penultimate Holocene glacier expansion by a period lasting several centuries when retraction predominated (Holzhauser, 1997, Figures 1 and 2). The Little Ice Age histories of glaciers in other regions around the North Atlantic are compatible with those of the Grosser Aletsch.

In marine cores taken off the southeast coast of Greenland, variations with depth of sediment type and composition indicate warm, stable conditions between the 10th and 13th centuries, followed by cold conditions beginning about 1270 A.D. and culminating probably about 1370 A.D.,[4] leading on to even harsher conditions after 1630. A brief interruption between 1370 and 1470 did not involve reversion to the relative stability and warmth of the Medieval Warm Period (Jennings and Weiner, 1996). Isotopic values of deuterium from the GISP2 core suggest that the 14th century was the coldest in central Greenland in the last seven hundred years (Barlow *et al.*, 1997). Other physical evidence from ice cores demonstrates the occurrence of clusters of cold years in the 14th century (Meese *et al.*, 1994; Stuiver *et al.*, 1995; Barlow *et al.*, 1997).

In west Greenland, the rise of the snow line during the 20th century cut off some of the outlets from ice covered plateaux, leaving dead ice in the valleys below (Weidick *et al.*, 1992). Cairns in Godthåbsfjord, erected by deer hunters, have melted out of semiperennial snow cover on the plateaux only within the last decade. Radiocarbon dates on moss indicate that the cairns were built between 1290 A.D. and 1400 A.D. (Weidick *et al.*, 1992). This suggests that the snow cover in West Greenland was extending at the same time as Atlantic waters were cooling and Alpine glaciers extending.

In Iceland, 13th century moraines have been identified on the forefield of Svínafellsjökull, in Öræfi (Gudmundersson, private communication August 1997), while recent investigations around Myrdalsjökull show that one of the units forming the compound moraine of Gigjökull was formed around 700 BP, in the early Little Ice Age (Kirkbride and Dugmore, in press). Detailed examination of original documents led Ogilvie (1991) to conclude that climatic conditions in Iceland were generally mild up to the late 12th century, but that from then onwards until the 16th century short periods of harsh weather occurred intermittently during which mean annual temperatures may have fallen 1°C to 2°C below typical 20th century levels.

In western Spitsbergen, cores from Linnévatnet covering the last 700-800 years record a high influx of glacially derived sediments (Svendsen and Mangerud 1997). Radiocarbon dates and sedimentation rates indicate that the glacier expansion which culminated in the 19th century started in the 13th or 14th centuries. These results are in line with oxygen isotope analyses of an ice core from the Lomonosov plateau in central eastern Svalbard (Gordiyenko *et al.*, 1980) indicating lower temperature, interrupted by brief warmer intervals between the 12th and 20th centuries, with the coldest periods between about 1180 A.D. and 1500 A.D. and between about 1700 and 1900 A.D..

In both northern and southern Scandinavia the Little Ice Age culminated in the 17th and 18th centuries. In the south, glaciers reached their greatest extent since the Holocene Climatic Optimum, obscuring most of the evidence of earlier Little Ice Age events. In the north, radiocarbon and lichenometric dating of moraines[5] indicates that several earlier Holocene expansion phases were comparable in scale with those of the later Little Ice Age, or even larger. This is explicable in terms of the southerly position of the oceanic and atmospheric polar fronts over southern Norway during the Little Ice Age (Matthews, 1991). In northern Scandinavia, dates from peat beneath the outermost moraine of Fingerbreen, an eastern tongue of the Svartisen icecap, and a foreset bed of a delta in an ice-dammed lake suggest 13th or 14th century advances and accord with a date on soil from Ritajekna in northern Sweden (Karlén, 1991), but this evidence is not conclusive because of difficulties involved in the radiocarbon dating of peats (Grove and Switsur, 1994).

---

[4] Clusters of cold years recorded in ice cores from Greenland occurred in 1308-18, 1324-29, 1343-62 and 1380-84 A.D., abnormally cold summers in 1351 and 1352 A.D. and cold winters in 1351 and 1355 A.D. (Meese *et al.*, 1994; Stuiver *et al.*, 1995; Barlow *et al.*, 1997)

[5] The lichenometric dating of moraine surfaces in northern Sweden and Norway demonstrated effectively that older advances occurred before those of the Little Ice Age, but the reliability of the growth curves on which the dating was based was challenged by Innes (1982, 1983) on the grounds that they were based on radiocarbon dates from too short a period, and that many came from the surfaces of mining tips, the product of intermittent mining activity, difficult to date exactly. He also pointed to differences in the growth rates of *Rhizocarpon alpicola* and species in the *geographicum* group, which result in dating errors if they are not treated separately. But it must be noted that Karlen (1979) concluded that glaciers in northern Norway reached advanced positions in the early 1300s and subsequently retreated.

In southern Norway, the Little Ice Age has been identified as the most extensive Holocene glacier expansion "occurring between the 13th and the early 20th century" (Nesje and Dahl, 1991a). Its initiation has been dated to between 1220 A.D. and 1290 A.D. from sediment changes in Gjuvvatnet, a high altitude lake fronting a cirque glacier in mid-Jotenheimen (Matthews and Karlén, 1992). Other dates relating to early Little Ice Age initiation come from Sunndalen, northwest of Jostedalsbreen, where glaciation began after 1040-1220 A.D. (Nesje and Rye, 1993), and from Blåisen, a northeastern outlet of Hardangerjøkulen, where it occurred after 1300-1430 A.D. (Dahl and Nesje, 1994). In northwestern Jostedalsbreen, an initial Little Ice Age expansion has been recognized as occurring after 1040-1220 A.D. (Nesje *et al.*, 1991).

Small, high altitude glaciers, can react much more immediately to climatic fluctuations than larger, less sensitive valley glaciers and ice caps. Matthews and Karlén (1992) showed that in Norway, during the last few thousand years, glaciers have formed at different times at different sites, depending on critical altitudinal thresholds in relation to the scale of atmospheric variation. High altitude cirque glaciers may be expected to react rapidly to even a minor climatic fluctuation, sensitive valley glaciers somewhat later, and outlets from an icesheet are likely to react with substantial lag times. An attempt to discern whether or not initial Little Ice Age advances in other parts of the globe were contemporaneous with those in regions around the North Atlantic should be based as far as possible on comparisons of the advance histories of glaciers of similar size and type. In practice this favours the use of data from valley glaciers, as extremely few cirque glacier forefields have been studied. Data from several glaciers in the same massif should be used to obtain as complete a picture as possible because minor variations relating to differences in microclimate, aspect and height of accumulation basins are probable, and availability of evidence, such as sections in lateral moraines, is likely to vary from one forefield to another.

The relationship between climatic conditions and glacier fluctuations is complicated, depending upon the balance between accumulation and ablation. An advance phase occurs only after a sequence of years during which glacier volume has increased, melting and ablation having been insufficient to counterbalance accumulation. The time lag before frontal advance starts depends on the size and sensitivity of the glacier. While series of advances such as occurred during the Little Ice Age are clearly associated in general with colder periods, glaciers do not react to all marked temperature fluctuations; adequate accumulation is essential. This is exemplified in an extreme way by the inability of glaciers to form at high altitudes in hyper-arid regions, where temperature is not the limiting factor for glacier growth (Messerli, 1973), as in the Andes of the Atacama Desert where there are no glaciers despite elevations considerably above the continuous permafrost limit. In such conditions glacier advances are fully controlled by moisture availability (Grosjean *et al.*, 1998). The importance of accumulation is demonstrated in a less dramatic manner by the accordance of late Quaternary glacial advances with paleolake highstands in the Bolivian Altiplano (Clapperton *et al.*, 1997; Clayton and Clapperton, 1997). The coldest winters and springs in the Swiss Alps in the last 450 years, those between 1685 and 1715, failed to cause glaciers to advance because precipitation in winter was so low that it was not counterbalanced by reduced ablation in summer (Pfister, 1994). Nevertheless, it can generally be taken that recurrent decadal scale negative annual temperature anomalies, clustered on longer term climatic oscillations, will lead to glacier fluctuations, and can be seen to have done so on all continents during the Little Ice Age.

3.   ASIA

3.1   The Urals, Caucasus, Tien Shan and Altai

Very few of the many thousands of glaciers in Asia have been investigated in sufficient detail for relevant data to be available. Only a scatter of radiocarbon dates relating to glacial fluctuations has so far been obtained, no doubt in some regions because of lack of datable material. None are currently forthcoming from the Urals, the Pamirs or Tien Shan. In the Caucasus, moraines attributed to the last millennium have horseshoe shapes and are well spaced out, with the most recent close to the present ice fronts and those from the early Little Ice Age further down valley (Kotlyakov *et al.*, 1991). Glacier advances between the 13th and 15th centuries are said to be indicated by radiocarbon dates, but no details are available. Though systematic investigations of the moraines of the Tien Shan were made by the Institute of Geography of the Russian Academy of Sciences during the 1980s, dating was almost entirely by lichenometry (Serebryanny and Solomina, 1996, p. 160). This indicated that groups of moraines were formed about "A.D.: 1210-1215; 1340-1390; 1440; 1540-1550; 1590; 1650-1660; 1680-1710, and 1730-1910 (except for the 1800s)". However, Serebryanny and Solomina (1996, p. 60) admit that the lichenometric dating of Middle Asian moraines was of relatively low accuracy, and suggest that this is "partly compensated for by the abundance of age determinations"[6]. Several species of long-living trees in high mountain regions of Asia have already provided long dendrochronological series (Kotlyakov *et al.*, 1991). There are almost fifty tree ring chronologies from the upper tree lines of the FSU but only a few can be used as sources of proxy data for climate and glacier variations, mainly on account of lack of cross-dating procedures, the use of single samples for reconstruction, and low correlation of tree ring widths with meteorological parameters (Solomina, 1996) It is only possible to use two of the chronologies of *Larix sibirica* from the Polar Urals and Altai to estimate longer term variations and even here "as the data of the early parts of the chronologies are not very reliable because of the limited number of trees", it is still unclear whether the depressions before 1400 A.D. are real or not (Solomina, 1996, p. 190). The most detailed information about glacier history comes from the Aktru Valley (Figure 1) in the Severo-Chuisky Range of the Altai (87°45' E, 50°05' N). Measurement of lichen sizes on moraine surfaces indicated that the earliest Little Ice Age advances occurred in the 15th century (Serebryanny and Solomina, 1996). This is in line with the results of radiocarbon dating of buried wood from one of the moraines of the Bolshoy Aktru Glacier which has been dated to 1440-1480, although it is not clear whether the wood was found within or underneath the moraine. Several dates were obtained from wood found in proximity to the moraines of the Maly Aktru, but the species from which the wood came do not seem to have been identified (Serebryanny and Solomina, 1996, p. 161, quoting Ivanovsky and Panychev, 1982). Unfortunately the relationships of the samples with the moraines were not very clearly explained in the original publication. One of these, found "on the periphery of the Little Ice Age moraine" from a tree more than 60 years old gave an age of 1430-1490 cal A.D. Another, "at the contact of the left and right frontal ridges", gave the same age, but this tree was more than 210 years old, while a third sample, from a tree more than 180 years old, from "the steep slope of the right part of the terminal moraine", gave a date of 1440-1480. Two further wood samples near the upper edge of the left lateral moraine turned out to be modern, providing no further insight into the sequence of deposition. According to Kotlyakov *et al.* (1991), though no traces of glacier advance during the first phase of the Little Ice Age have been found, such evidence may lie buried

---

[6] The reasons for this lack of accuracy, include uncertainty in lichen species determination, lack of independently dated old surfaces, lack of lichens on some moraines, and construction of growth curves only on the basis of tree ring chronologies in the Altai. (Solomina, 1996).

The numbers in the margin give the laboratory identification number, the ¹⁴C age and the calibrated age (AD) for the 68% probability. Horizontal lines represent the calibrated ages.

> = advance after date
W = Wood

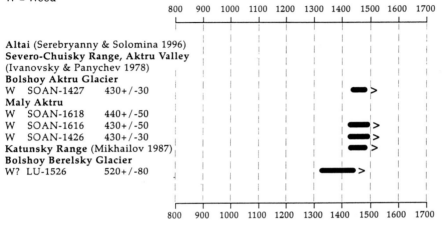

Figure 1. Key radiocarbon dates relating to Little Ice Age advances in the Altai.

beneath more recent sediments. They concluded that the earlier part of the Little Ice Age was characterized by smaller-scale advances than those occurring subsequently in both the Altai and the Tien Shan.

On the southern slopes of the Katunsky Range (86°45' E, 49°45' N) a date of 1330-1440 A.D., also on buried wood, comes from one of the moraines of the terminal complex of the Bolshoy Berelsky (Serebryanny and Solomina, 1996). In the latter case, the wood may have been *in situ* [7]. It is not clear whether the date relates to the oldest part of the complex.

Existing information demonstrates that glaciers in the Altai advanced in the 15th century, and perhaps also in the 14th century. The moraines are complex, containing material from many different advances and it is possible that that the earliest from the Little Ice Age have yet to be identified.

In Kamchatka, where material suitable for radiocarbon dating is scarce, the use of a recently compiled tephrachronological database (Braitseva *et al.*, 1997) is proving an important aid in dating moraines and unravelling of glacial history (Savoskul and Zech, 1997; Solomina *et al.*, 1995).

3.2 The Himalaya

The Himalaya stretch for 2000 km from the Hindu Kush and Karakoram and the sources of the Indus in the west to the headwaters of the Dikang-Brahmaputra in the east, dividing the monsoon lands to the south from the deserts and steppes of the Asian interior.

---

[7] According to a personal communication from the late Professor L. Serebryanny (25.8.1998) the original publication in Russian (Ivanovsky and Panychev, 1978) did not make this clear.

The main east-west arc is composed of several parallel ranges, the Lesser Himalaya rising to 4500 m and the Greater Himalaya to the north to over 5500 m. The ice-covered area is the largest outside the polar regions. Investigations of glacial history have been concentrated in the Everest region and especially on the moraines of the Khumbu glacier in the Sapte Kose drainage basin.

Röthlisberger examined the moraines of sixteen glaciers comparable in size with valley glaciers in the European Alps, along the length of the Karakoram and Himalaya, avoiding those in the arid zone, and those known to surge or to be affected by earthquakes (Röthlisberger and Geyh, 1986). He collected samples of palaeosols and *in situ* tree trunks overridden by ice and covered with till. Two sets of radiocarbon dates were obtained from the soil material; an older set given by material consisting mainly of lichen fragments, and a younger set consisting of humic acids, roots and *in situ* wood. It was argued that the lichen dates indicated the time when soil began to form after ice retreat, and the others referred to the time when soil formation ceased as readvance of the ice began (Geyh *et al.*, 1985). Röthlisberger took into account the information given by the heights at which samples were obtained and their positions up or down valley in assessing his findings, the approach he had earlier adopted in the Alps.[8]

Four of Röthlisberger's samples from the Khumbu region gave dates relating to the early Little Ice Age (see Figure 2). These can be compared with two others obtained by Muller (1961) and Fushimi (1978)[9]. Most of them indicate 14th century advances, but none can be rated as very close. His own dates, all from soil, suggest advances in the 14th or early 15th century, except for HV-12006 from Lhotse Shar which points to an advance between about 1160 and 1300 A.D.. Röthlisberger's single date from the Gilgit area, like (GaK-14029) providing a minimum date for the valley train of the Langshisa Glacier in the Langtang Valley, Nepal, taken to mark an early part of the Little Ice Age (Shiraiwa and Watanabe, 1991) also suggests expansion in the 14th or first half of the 15th century. Available evidence from the Himalaya is insufficient to prove that the Little Ice Age advances were delayed until the 14th or 15th centuries compared with those around the North Atlantic. Moreover data from the Dunde Icecap on the Tibetan Plateau do not support such a delay; neither does a scatter of dates on wood relating to moraines formed by glaciers in western China.

3.3 China

Most of the glaciers in China are in the northwest; in the Tien Shan, Kunlun Shan and Himalaya. The snowline rises towards the south from 2900-3400 m in the Altai, to 4600-6200 m in the Kunlun Shan, and 4600-6200 m in the northern Himalaya. In the southeast, the glaciers of the eastern Hengtuan Shan and the eastern Qilian Shan, fed by the southeasterly Pacific monsoon, are of maritime temperate type; those in the south, the Himalaya Nyenchintangla and western Hengduan Shan, fed by the southwest monsoon, are cold continental like those of the Karakoram, Kunlun Shan, Tien Shan, Altai and western Qilian Shan which receive snow from westerly airstreams. By 1992 it was estimated that about 45,375 glaciers covered a total area of 52,735 km^2 (Wang Songtai and Yang Huian, 1992). Glaciers are scattered across the Tibetan Plateau, where up to 80% of precipitation falls in the summer monsoon.

During recent years, understanding of the characteristics and history of the Chinese glaciers has increased with great rapidity, emphasis on their hydrological value extending to

---

[8] Full results of Röthlisberger's studies in the Himalayas, made with a view to comparing the timing of glacier advances in the northern and southern hemispheres, appeared in *1000 Jahre Gletschergeschichte der Erde*, a valuable book which has been sadly neglected (Röthlisberger and Geyh 1986).
[9] Röthlisberger and Geyh (1986,p.339) queried Fushmi's date for the Kyuwo, which he considered improbably young, though it is in line with Müller's date (B-174) from the Khumbu.

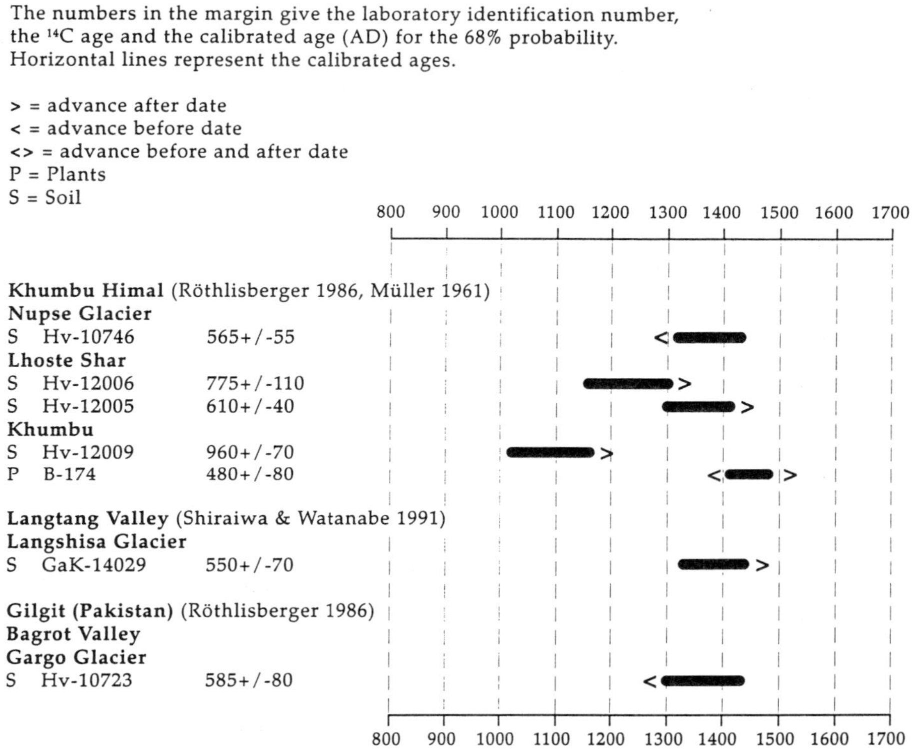

The numbers in the margin give the laboratory identification number,
the ^{14}C age and the calibrated age (AD) for the 68% probability.
Horizontal lines represent the calibrated ages.

> = advance after date
< = advance before date
<> = advance before and after date
P = Plants
S = Soil

Figure 2. Key radiocarbon dates relating to Little Ice Age advances in the Himalaya.

investigation of their variations and the relationship of these to climatic and environmental change (Xie Zichu, 1992). However, radiocarbon data relating to Holocene fluctuations have so far been obtained for only a handful of the thousands of glaciers. Dates relevant to the timing of Little Ice Age advances are even fewer (Figure 3).

The lower part of the Zepu, a 19 km long valley glacier in the southeast of the Nyenchintangla, is till covered and reaches into forest. A sample of wood (HR-566), found exposed in a landslide scar in the terminal area (Figure 2) dates a Little Ice Age advance to between 1290 and 1440 A.D. (Iwata and Keqin, 1993). This brought the Zepu front into close proximity with an older moraine, with a minimum age on charcoal (NUTA-1672) from a buried soil layer of 990-1220.[10] At least two of the older moraines were larger than those of the present millennium, causing the ice to reach as far as 5 km or more downvalley, compared with the 2 km of the Little Ice Age advances.

The Gongga Mountains, extending north to south between the southeastern periphery of the Tibetan Plateau and the Sichuan basin, carry glaciers small enough to be sensitive to

---

[10] The radiocarbon dates specified in Iwata and Keqin (1993,p.134) based on a paper by Li *et al* (1986), do not coincide with those published in Ageta *et al* 1991 p.32). The later set is used in Figure 2 as it is much more detailed and Iwata and Keqin were personally responsible for collecting the samples from which they came in 1989.

The numbers in the margin give the laboratory identification number,
the ¹⁴C age and the calibrated age (AD) for the 68% probability.
Horizontal lines represent the calibrated ages.

> = advance after date
< = advance before date
P = Plants
S = Soil
C = Charcoal

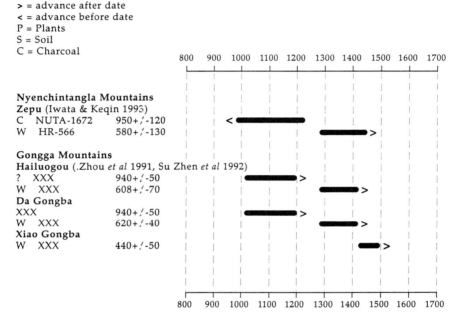

Figure 3. Key radiocarbon dates relating to Little Ice Age advances in China.

short period, low amplitude changes in climate (Su Zhen *et al.*, 1992). Precipitation comes
from both the southeast and southwest monsoons and is heaviest on the eastern side of the
mountains where the lower slopes are subtropical. Moraines, judged from their fresh
appearance to be Little Ice Age, are located 2 to 3 km from the termini on the eastern side of
the mountains, but only 0.2 to 1.5 km downvalley on the drier western side. The 13 km long
Hailuogou Glacier, on the eastern versant, is probably temperate. It is fronted, like many of
the Gongga glaciers, by three fresh-looking moraines. Decayed wood extracted from the
outermost gave a date of 1290-1410 cal A.D. Zhou *et al.* (1991) cite an advance of the
Hailuogou dated at 1020-1190 from Li and Feng (1984), which may have preceded the
Little Ice Age. Wood from the outermost moraines of the Da Gongba and Xiao Gongba,
which were once confluent, gave dates of 1290-1405 and 1430-1490 (Figure 2). These few
dates and accompanying descriptions suggest that the Little Ice Age had begun at least by
the 13th or 14th century, and was preceded by earlier Holocene advances on a similar or
larger scale.

No information about the timing of Little Ice Age fluctuations is available from
radiocarbon dating of moraines on the Tibetan Plateau. Information about contemporary
climatic conditions comes from ice cores extracted from both the Dunde Ice Cap, on the
northeastern margin of the Plateau, and the Guliya Ice Cap, 1400 km away on the
northwestern margin (Thompson, 1992; Thompson *et al.*, 1995; Lin *et al.*, 1995). The

accumulation histories of these two cores, both heavily dependent on summer precipitation, are similar, yet their isotopic histories, recording temperature, differ. The Dunde record shows a cool period, beginning in the 13th century, recognized by Thompson (1992) as representing the Little Ice Age. But there is no sign of a prolonged cool period in the Guliya core, except for a cool interval around 1650. The most negative isotopic values of $\delta^{18}O$, indicating lower temperatures, occurred at the beginning of the millennium at Guliya. The least negative values, indicating warmer conditions, are in the part of the core representing the period from 1300 to 1800 cal A.D. Between 1250 and 1500 cal A.D., warmth at Guliya appears to have been contemporaneous with cool conditions at Dunde.

The reason why the Dunde Ice core records cooling in the Little Ice Age period when the Guliya Ice Cap apparently experienced pronounced warming, despite the similarity in their accumulation histories, is not obvious.[11] The sequence recorded at Dunde is in general accord with documents from eastern China containing phenological information located in both space and time, showing that in the early 13th century crop boundaries lay further north than during the last few centuries (Zhang De'er, 1994). It appears that Xichuan and Henan experienced the Medieval Warm Period, with the Little Ice Age beginning sometime after 1264. If the isotopic records from the Guliya Ice Cap are correct in indicating a warm period from about 1300 to 1800 A.D. (Thompson, 1992; Mosley-Thompson et al., 1993), then the timing of medieval warmth was not coincident in central China, and the northeastern margin of Tibet. At Dunde, release of latent heat associated with cumulus-convective precipitation contributes to atmospheric heating, while around Guliya on the western side of the Plateau, heating is largely derived from vertical convection driven by surface heating. The difference in the $\delta^{18}O$ history could be related in part to the extent of localised snow drifting, to loss of mass by ablation, and especially to the addition of locally derived moisture from convective activity associated with intense radiative heating of the area round Guliya. Unfortunately no documentary evidence is available which might assist in understanding Guliya's climatic history.

4. NORTH AMERICA

4.1 Alaska

The extensive literature concerning glacier fluctuations in Alaska has been reviewed by Calkin (1988), Calkin and Wiles (1992) and also by Calkin et al. (in press).

In the interior of Alaska and adjacent Yukon Territory, where trees are sparse or non-existent, dating has depended mainly on lichenometry (Calkin and Ellis, 1980; Ellis and Calkin, 1984). In the Brooks Range (67-69°N), where Holocene moraines are 1000 to 2000 m above the upper level of spruce forest and to the north of its poleward limit, lichenometric dating of moraines has suggested a period of formation around 1200 A.D., in line with evidence of glacial expansion in the coastal regions (Calkin and Wiles, 1990). However Ellis and Calkin (1984) rated the reliability of their growth curve at only ± 20 per cent of the age.

Wiles and Calkin (1994) concluded that reliable records of Little Ice Age extension in maritime Alaska were separated from earlier Holocene episodes of expansion by an interval correlative with the Medieval Warm Period. During this interval, ice margins were at or behind their current positions as was clearly revealed in the coastal region by dendrochronological dating of forest beds overridden by expanding glaciers (Wiles et al., 1995; Wiles and Calkin, 1994; Wiles et al., 1999). Where many subfossil trees, affected by glacial processes, occur on glacial forefields, glacial history can be traced with great

---

[11] The impact of sublimation on the oxygen isotope signal is not known, but the firn to ice transition is so rapid that the snow is not open to sublimation for more than a year.

precision by cross-dating their ring characteristics with long dendrochronological series (Luckman, 1995, 1996). [12]

In the Kenai Peninsula (59-60°N), dendrochronological dating using cross-dating has been shown to be a useful and precise method of dating glacier fluctuations. An early medieval advance period centered around 600 A.D. has been identified at four glaciers. The Grewingk and Dinglestadt Glaciers both expanded to within a few kilometres of their Little Ice Age and Holocene limits by about 620 A.D. That a soil developed on till from this advance of the Dinglestadt is overlain by Little Ice Age deposits demonstrates that the ice receded during the Medieval Warm Period (Wiles and Calkin, 1994) The earliest Little Ice Age advances on the Peninsula registered by radiocarbon dates come from the Grewingk Glacier, the largest tongue from the Grewingk-Yalik Icefield. Here, dates on transported wood (BGS-1273 and BETA-2944) give maximum dates in the 15th or 16th centuries for the deposition of one of the many Little Ice Age moraines. A maximum date for the outermost Holocene moraine of the Tustemena, the largest glacier outlet from the Harding Icefield, (L-117) shows activity sometime between the early 15th and mid 17th century. Radiocarbon dated wood in sediments, in conjunction with tree-ring cross-dating, has given more precise information, for instance that the Grewingk Glacier had advanced by 1421 A.D. Dendrochronological investigations have shown that on the eastern, maritime side of the Kenai Mountains, general Little Ice Age advances took place from about 1420 to 1460 A.D., followed by further expansion phases between 1640 and 1670, around 1750, and between 1880 and 1910 A.D. Paleoclimate studies suggest that these were times of high winter precipitation. On the western side of the Peninsula, glaciers expanded between 1440 and 1460 A.D., 1640 and 1670, around 1750, and in 1880-1910. It appears that on this cooler and drier flank the glacier advances probably followed intervals with cooler summers (Wiles and Calkin, 1994).

Widespread advances of the glaciers of Prince William Sound in south-central Alaska (60-61°N) occurred in the 12th to 13th centuries (Calkin *et al.*, 1999), when low elevation coastal forests were overridden by the expanding ice (Barclay *et al.*, 1999). Cross-dating of living with glacially overrun trees on eight forefields in western Prince William Sound demonstrates that the Little Ice Age advances occurred in the 12th and 13th centuries, as well as the 17th and the early 18th centuries and more recently. Radiocarbon dating was used to provide independent verification of the dates assigned to glacial advances by tree-ring cross-dating (Wiles *et al.*, 1999). The earliest reliable kill dates from unrotted wood on the forefield of the Princeton Glacier, an outlet of the Sargent Icefield, suggest that the front was overriding forest by 1190, and that expansion continued at least until 1248. At Tebenkof Glacier, an outlet of the Spencer-Blackstone Ice Complex, evidence was found of an advance between 1289 and 1300 A.D. The record from Prince William Sound is consistent at the decadal level with that of the Sheridan Glacier further eastward on the Gulf of Alaska.

The Sheridan is a landbased glacier near Cordova, west of Malaspina, where numerous subfossil logs were exposed by a river cutting through the extensive terminal moraine complex (Yager *et al.*, 1998). Seventy three of them, mainly *in situ* stumps in four forest beds, were cross dated, providing a chronology from A.D. 911 to 1992. This chronology, which cannot be taken as exact as it was tied in by radiocarbon, suggests that aggradation occurred between 1182 and 1253 A.D., before three trees were crushed into bedrock in 1284. During this interval the front was 0.5 km from the outermost terminal moraine, but then retreated by over a kilometer. More outwash was deposited between 1491 and 1644. It is clear that the advance attributed to 1284 was in the early Little Ice Age and on a lesser scale than one or more of those which followed.

Röthlisberger (1986) used the same approach to date Juneau Icefield moraines to that he had employed in the Alps and Himalaya, avoiding those glaciers with floating tongues and

---

[12] An account of the procedures was given by Wiles and Calkin (1994).

those known to surge, which had attracted most attention from previous investigators (Röthlisberger and Geyh, 1986). He concentrated his investigations on the Juneau Icefield, finding indisputable evidence of early Little Ice Age initiation on the forefields of three of the Juneau outlet glaciers (Figure 4). The Gilkey Glacier overran forest during the first millennium, and either during the Medieval Warm Period or early Little Ice Age left an *in situ* stump dated 1000-1210 (Hv-11299). The Gilkey advanced into forest again around 1330-1430 (Hv-11301), and 1414-1440 (Hv-12094). The Llewellyn advanced about 1260-1390 (HV-11465) and again about 1445-1650 (Hv-11369), on each of these separate occasions killing trees and leaving *in situ* stumps. The second tree killed by the Llewellyn had 120 rings; evidently the climate was sufficiently improved between the two advances for trees to grow at the site. The Frontier Glacier advanced and killed an *Abies lasiocarpa* tree, for which the stump *in situ* has given a date of 1280-1390 (Hv-11363), and advanced again two centuries later, 1495-1565 (Hv-11362), as well as during the later part of the Little Ice Age (Hv-11361). Unfortunately tree ring studies which would have allowed identification of the calendar dates at which the Juneau outlets advanced into forest were not available to Röthlisberger.

It should be noted that all of the Alaskan glaciers discussed here emerge from icefields, and so their lag times are unlikely to be short. This has to be remembered when comparisons are made with the timing of the fluctuations of valley glaciers.

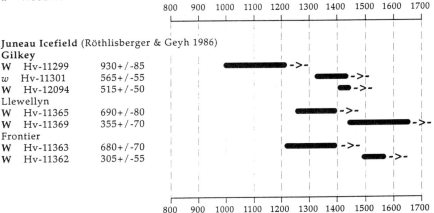

Figure 4. Key radiocarbon dates relating to glacial advances in Alaska.

4.2 The Canadian Rockies

The evidence of the Little Ice Age in the Canadian Rockies has been reviewed by Luckman (2000).[13.] As glacier extension was greater after 1700 A.D., unequivocal evidence of advances between 1200 and 1370 A.D. is fragmentary and comes from only four glacier forefields. Ring width and maximum density chronologies, developed from *in situ* stumps, crossdated with long, living tree chronologies, show that the Robson Glacier was advancing into forest between 1142 and 1150 and continued to do so until at least 1350 A.D., while the Peyto Glacier advanced between 1248 A.D. and about 1375 A.D. (Luckman, 1995), The French Glacier in the Peter Lougheed National Park, in the eastern Front Ranges of the Rockies, advanced about the same time (Smith *et al.*, 1995). The Stutfield Glacier, in Banff National Park, advanced into forest after 1270 (Osborn, 1993; Luckman, 2000)).

Annually resolved dendrochronological series (from *Picea engelmanni, Larix lyallii* and *Pinus albicaulis,* all of which can live for over 600 years) have been used to identify the incidence of temperature and precipitation conditions suitable for glacier advance in this region. The glacial evidence of advances early in the Little Ice Age is supported by summer temperature reconstructions from the tree ring series (Luckman, 2000). An April-August temperature reconstruction based on tree-ring densitometry from *Picea engelmannii* and *Abies lasiocarpa* growing near the Columbia Icefield demonstrates a strong agreement between periods of glacier advance and moraine building and periods of low summer temperature in the 1200s-1300s, late 1600s-early 1700s and throughout the 19th century, suggesting that summer temperatures are a primary control of glacier fluctuations in the region (Luckman *et al.*, 1997). The advances of the 12th and 13th centuries, bringing ice fronts to positions 0.5-1.0 km upvalley of more recent moraines, and recognized as representing the start of the Little Ice Age (e.g. Luckman, 1993; 1996a and b), were on a smaller scale than those which followed. In Canada multiple glacial advances became progressively more extensive over time, culminating in the 18th and 19th centuries (Luckman, 2000).

4.3 Coast Mountains of British Columbia

In the northernmost part of the Pacific Ranges of the Coast Mountains near Bella Coola, early Little Ice Age advances have been identified on the forefields of two glaciers, both outlets of the 350 km[2] Monarch Icefield (Desloges and Ryder, 1990). The Jacobson Glacier is judged to have advanced in 1440-1630 (S-2979), but this dating is tentative. It came from a succession exposed in the margin of a lobe of the left lateral moraine, and was derived from a composite sample of discontinuous Ah horizons found within a cumulic regosol developed within gritty sand, burying a mature podzolic soil, itself superimposed on till (Desloges and Ryder, 1990). A section through the main outer moraine of the Purgatory Glacier revealed three superimposed tills, separated by paleosols. Leaf litter, "clearly *in situ*", from the lower paleosol, (S-2977) records ice advanced over older till after 1210-1290, while litter from the upper paleosol (S-2978) indicates that the ice withdrew sufficiently for 2cm of leaf litter to accumulate again before the glacier readvanced after 1290-1410.

An advance of the Klinakini Glacier, an outlet of the Silverthrone Icefield, further south in the Coast Ranges (51°N-51°30'N), was recorded by a date (WS-1567) from an *in situ* tree trunk, but this gave a calendar age range of 1040-1210, too wide to reveal whether it relates to an interim advance during the Medieval Warm Period or at the beginning of the Little Ice Age (Figure 5). The trunk was found above the current ice and 150 m below the crest of the right lateral moraine. An advance of the Franklin/Confederation Glacier

---

[13] See also Grove (in press).

The numbers in the margin give the laboratory identification number, the ^{14}C age and the calibrated age (AD) for the 68% probability. Horizontal lines represent the calibrated ages.

->- = advance around or after indicated age
> = advance after date
W = Wood *in situ*
w = Wood not *in situ*
S = Soil
R = Roots *in situ*

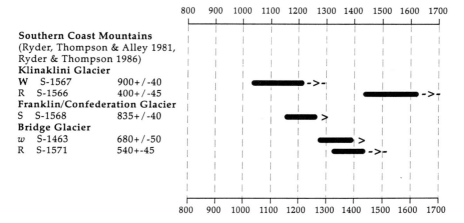

Figure 5. Key radiocarbon dates relating to Little Ice Age advances in the coastal range of British Columbia.

occurred sometime after 1160-1260, a date given by wood from a paleosol (S-1568) but there is no direct evidence of the length of the time lag involved. Recession of the Bridge Glacier, an outlet of the Lilloet Icefield has exposed a nunatak carrying overridden soil and trees about 100 m above the present ice level (Ryder and Thompson, 1986). Tree trunks and roots protrude from the highest moraine. Dates from a 1 m diameter tree trunk (S-1463) and a root (S-1571) indicate that the forest was overwhelmed sometime after 1280-1390 and 1330-1430 respectively. Though the age range is wide, it is clear that the Bridge Glacier advanced in the early Little Ice Age. Ryder and Thompson concluded from the history of a lake impounded by the glacier that the nunatak had not been overridden for a long period previously, perhaps as long as a thousand years.

Desloges and Ryder (1990) suggested that the Little Ice Age may have begun slightly later and ended generally later in the Coast Mountains of British Columbia than in the Rockies, noting that the mass balance of coastal glaciers is strongly controlled by precipitation, while temperature and solar radiation are important controls in inland regions.

5.   SOUTH AMERICA

Glaciers of every size, shape and type are scattered over the length of South America from tropical Venezuela to high subpolar Tierra del Fuego. The Cordillera Blanca[14] of Peru is the most heavily ice covered mountain range in the whole of the tropics; further south are the largest icefields in the mid-latitudes, the North and South Patagonian Icefields (the Hielo Patagonico Norte and the Hielo Patagonico Sur).[15] Many of the thousands of glaciers in South America are in very remote areas,[16] very few forefields have been investigated in detail. Knowledge of fluctuation history is founded to a great extent on the single-handed work of John Mercer (e.g. 1968, 1970, 1976), was much increased by that of Röthlisberger (Röthlisberger and Geyh, 1985), while more recently Japanese teams have been working on the Patagonian Icefields (e.g. Aniya, 1995, Aniya and Sato, 1995a and b). Data relevant to the start of the Little Ice Age is still sparse as well as scattered, but provides clear indications that in many parts of South America it began at much the same time as in regions around the North Atlantic. Particularly significant is the environmental information gained from ice cores extracted from the Qualccaya and Huascarán Icecaps in tropical Peru.

5.1 Tropical South America

Röthlisberger found evidence of the timing of Little Ice Age initiation on two of the five glacier forefields he investigated in the Cordillera Blanca (Röthlisberger and Geyh, 1985 and 1986). A river cut in the end moraine of Glaciar Huallcacocha, near Huascarán, revealed superimposed fossil soil 3 m below the crest. The soil depth indicated long term stability and no disturbance by the ice. A sample from the surface of the b-horizon (Hv-8703), indicated that the Little Ice Age advance occurred around 1280-1390 cal A.D. (Table 4, Figure 4). Another sample (HV-8704) from superimposed fossil soil on the right lateral moraine of Glaciar Ocshapalca indicated an advance after 1325-1650.

These results based on soil dating, though fragmentary and lacking in precision, give some support to Mercer's view (1977, p. 604) that "the most recent glacial expansion in the Cordillera Vilcanota-Quelccaya ice cap area culminated between about 600 and 300 BP". He noted a series of well-vegetated moraines occurring within 1200 m of the debris-covered terminus of a glacier at the head of the Upismayo Valley, Cordillera Vilcanota. A further 600m downvalley he found a moraine consisting of 'bulldozed' peat with its stratification intact, on top of coarse sand. A sample (DIC-678) of the uppermost peat gave an age range of 1290-1400 A.D. Mercer concluded that all the moraines closer to the glacier must have been deposited after 1290-1400, and evidently assumed that the initiation of the Little Ice Age was marked by the formation of the peat moraine. A moraine, about a kilometer from the margin of the Quelccaya Icecap also contained bulldozed peat, from which a sample (I-8441) gave an age range of 1020-1250 A.D. Mercer took it as coming from below the surface of the bulldozed bog, and therefore predating the Little Ice Age moraines. Dates from the Nevado Cuchpanga, obtained from peat samples taken from a fan inside the end moraine of the Glaciar Huatacocha, cannot be taken as reliable, in view of the uncertainties involved in dating peat by radiocarbon (Grove and Switsur, 1994).

---

[14] The Cordillera Blanca (77° 53'-77° 09' W, 8°08'-10° 02' S) is the largest of the twenty mountain groups carrying glaciers in Perú. Its 722 glaciers covered an area of 723 km^2 in 1920 and 58 km^2 in 1970. (Kaser and Ames 1996).

[15] Present knowledge of these icefields was reviewed in detail by Warren and Sugden (1993)

[16] The work of glacier inventory, nowadays assisted by use of satellite and airborne imagery (Aniya *et al* 1996), is well underway (e.g. Rabassa *et al.*, 1978, Corte and Espizua, 1981, Casassa 1995), but still incomplete.

The numbers in the margin give the laboratory identification number,
the ¹⁴C age and the calibrated age (AD) for the 68% probability.
Horizontal lines represent the calibrated ages.

> = advance after date
< = advance before date
P = Peat
S = Soil

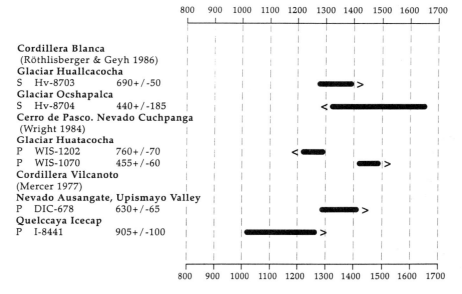

Figure 6. Key radiocarbon dates relating to Little Ice Age advances in tropical South America.

Cores from the Quelccaya icecap,[17] just south of the eastern end of the Cordillera
Vilcanoto, containing annual records covering more than a thousand years have provided
precise evidence of climatic conditions (Thompson, 1992; Thompson *et al.*, 1979 and 1986).
Dating, dependent upon a combination of annual dust layers, micro-particulate
concentrations, conductivity values, and identification of ash attributable to the eruption of
Huaynaputina (16°35' S, 70°52' W) in the spring of 1600, is rated as accurate to ± 2 years
back to A.D. 1500. Thompson (1992) placed the Little Ice Age in the period 1490 A.D. to
1880 A.D.,[18] during which microparticles and conductivities were 20% to 30% above the
averages for the entire core. This period stands out in the Quelccaya record as an important
climatic event, which evidently affected this part of the southern hemisphere as clearly as
the northern hemisphere (Thompson *et al.,* 1986). Perhaps this is not surprising as
Quelccaya's main source of precipitation is the Atlantic, 1800 km away, thunderstorms
recycling the water en route. Isotopically more negative values in the late 13th and early and
middle 14th centuries, indicating colder periods (shown on Figure 2 of Thompson *et al.*,

---

[17] The icecap covers about 55 km², with a central summit at 5650 m rising 400 m above the snowline.
Fortunately cores were obtained in the 1970s and 1980s; in 1991 it was found not only that the margins of
the ice cap had receded drastically, but also that the oxygen isotope profile had been smoothed as a result of
recent warming and percolation, and that annual variations could no longer be traced (Thomson *et al.*, 1993).

[18] Thompson *et al.* (1985, 1986) had initially dated the Little Ice Age at 1500-1900.

1993), may perhaps explain the difference between the period allocated to the Little Ice Age by Thompson and earlier dates of moraine formation, especially those based on radiocarbon ages of *in situ* trees.

In 1993 Thompson obtained two further cores from the col of Huascarán (9°06'41" S, 77°36'53" W), 970 km to the north of Quelccaya (Thompson, 1993; Thompson *et al.*, 1995; Davis *et al.*, 1995). At this site, at a height of 6048 m, the oxygen isotope profile had not been affected by recent warming; neither surface melting nor meltwater percolation have been observed, and the paleoclimate record is still intact.[19] No detailed results from the last few hundred years have been published, but it is already known that long term Holocene cooling here culminated in the 17th and 18th centuries, during the Little Ice Age period, and that the δ^{18}O decrease was similar to that recorded at Quelccaya.

5.2 Temperate South America

Little is known about the initial Little Ice Age advances of glaciers between 32°30 S and 35° S, in the most heavily glaciated part of the Andes outside Patagonia, apart from Röthlisberger's dating of some of the moraines of the Glaciar Cipreses in the Cordillera Central (Röthlisberger and Geyh, 1986). He dated a fossil soil with organic fragments (Hv-10915), superposed on bedrock 15m below the crest of the highest of its recent lateral moraines, to 1260 - 1440 cal A.D., also noting a high stand of the ice about 1858 A.D.

Between 37° S and 41°30' S, some 82 glaciers cover 266 km², the largest on Monte Tronador (3003 m). Recent moraines of the Frias Glacier on the northeast side were dated by Villalba (1990) using old drawings and photographs, ring counts of the oldest trees found on moraines, and the precise dates when the ice damaged living trees. He judged that the oldest moraine was deposited about 1638. On the basis of the similarity found between the timing of the Frias fluctuations and tree ring indices from *Fitzroya cupressoides*, Villalba suggested that Little Ice Age expansion of the Frias probably began between A.D. 1270 and 1380. He attributed the absence of dated moraines from this period to subsequent overriding during a longer interval of cool, wet summers and glacier advances from A.D. 1522 to 1664. Röthlisberger's dating (Hv-11800) of *in situ* tree trunks in the lateral moraine of the Rio Manso (or Ventisquero Negro) to 1000-1220 cal A.D. could relate either to an expansion during the Medieval Warm Period, or early Little Ice Age expansion (Röthlisberger and Geyh, 1986). A section in the left lateral moraine reveals that the ice advanced three times, overriding *in situ* trees of different generations between the 13th and early 15th centuries (HV-12865 and Hv-12864). A later advance occurred between late 15th and late 17th centuries, as well as in the 19th century (Röthlisberger and Geyh, 1986).

Evidence from the Patagonian icefields region is still insufficient to allow the earlier Little Ice Age culminations to be adequately traced, especially as topographical influences and glacier dynamics, which have caused recent fluctuations to vary in both timing and extent from one glacier to another, must be presumed to have operated in the past (Warren and Sugden, 1993). However, evidence that the Little Ice Age began in the 12th or 13th centuries is emerging. Aniya and Sato (1995b, p.321) have dated standing wood from a trimline of the Moreno in the Southern Icefield (NU-355) to the period 1160-1270, commenting that this indicates that "the glacier reached its recent maximum around the 12th century". They noted that their date is close to the one obtained by Mercer (1970), for the initial stage of the Little Ice Age advance of the Ofhido Norte (I-3827) of 1170-1290, a tongue known to have been land based for several thousand years.

In the Lago O'Higgins area, around 49° S, at the northern end of the Hielo Patagonico Sur, dates from *in situ* wood record advances in the 13th century and later in the Little Ice

---

[19] Though glaciers on Huascarán-Chopicalqui massif have been retreating in the present century, and equilibrium lines fell 95+/-5 m between 1920 and 1970 (Kaser and Ames, 1996).

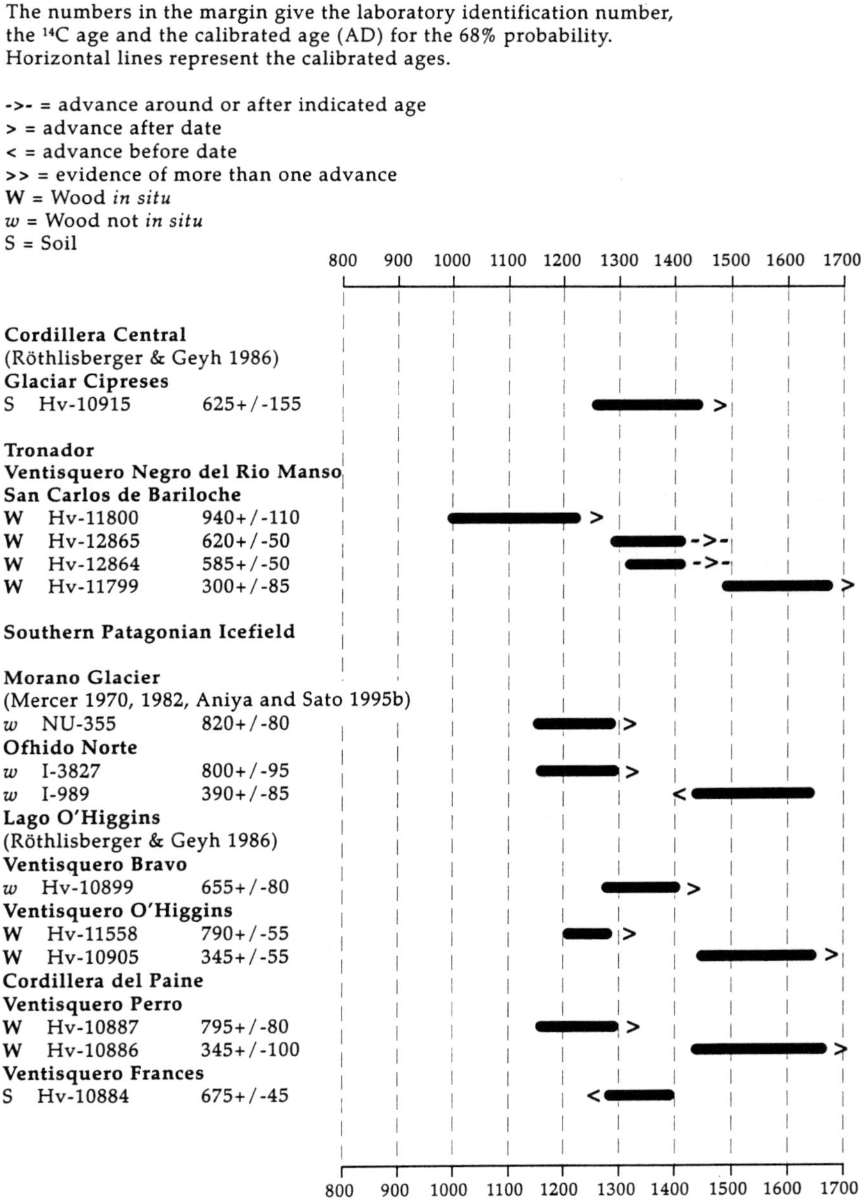

The numbers in the margin give the laboratory identification number,
the ¹⁴C age and the calibrated age (AD) for the 68% probability.
Horizontal lines represent the calibrated ages.

->- = advance around or after indicated age
\> = advance after date
< = advance before date
>> = evidence of more than one advance
W = Wood *in situ*
w = Wood not *in situ*
S = Soil

Figure 7. Key radiocarbon dates relating to Little Ice Age advances in temperate South America (30⁰ to 51⁰ south).

Age (Röthlisberger and Geyh, 1986). Röthlisberger dated wood from a fossil forest, growing from bedrock above the Ventisquero Bravo, a land based outlet of the Icecap (Hv-10899) to 1280-1400 cal A.D. (Figure 7).

At the southern end of the icefield in the Cordillera del Paine, around 51° S, clear evidence of advances in the 12th and 13th century comes from *in situ* wood (Röthlisberger and Geyh, 1986). Röthlisberger obtained radiocarbon dates relating to the initiation of the Little Ice Age from the forefields of three glaciers in the area. A date from an *in situ* tree (Hv-10887), growing near bedrock and overwhelmed by ice from the Ventisquero Perro, a small glacier separate from the main icefield, provided particularly clear evidence of an advance around 1160-1290 cal A.D., as well as in the period 1440-1660 A.D. The implication of dates from another small glacier, Ventisquero Francés (Hv-10884), indicating advance before 1285-1390 is less definitive as it came from a palaeosol in a lateral moraine.

Röthlisberger placed the start of the Little Ice Age at about 1350 BP, although in his summary he recognises two extension phases between 900 and 500 BP. He had himself taken part in the major research project which established that glaciers in the European Alps had advanced well before the 16th century (Röthlisberger *et al.*, 1980), but was then unaware of the more detailed investigations which have since shown that the 13th century advances in the Alps were also part of the Little Ice Age (Holzhauser, 1998).

6. NEW ZEALAND

New Zealand stretches from 33° to 47° S and 167° to 178°30' E. In North Island, only Ruapehu is high enough to support ice, but in South Island between latitudes 40°54' S and 44°53' S over 3,000 glaciers are scattered along the Southern Alps, the largest in the Mt Cook region around 43°35' S. [20] Whereas in the European Alps Little Ice Age advances were similar in scale to those earlier in the Holocene, in New Zealand the advances of the last few centuries have been smaller than their predecessors.

Evidence of early Little Ice Age initiation has been found on both sides of the Southern Alps (Gellatley, 1982 and 1985; Gellatley *et al.*, 1985 and 1988). A date on a fossil soil from the left lateral moraine of the Tasman Glacier (Hv-10500) giving a calendar age of 1220-1290 A.D. was one of a set of radiocarbon dates which led Gellatly *et al.* (1985) to recognize a close resemblance between the fluctuation histories of the Tasman and the Grosser Aletsch. The comparison was persuasive but the interpretation not entirely convincing because of the complexity of the moraine sections involved.

Dates from *in situ* stumps found emerging from the right lateral moraine of the Balfour Glacier in Westland, at several different heights, provide unmistakable evidence of 13th century and later fluctuations. Unfortunately, Hv-10521, from a stump 15 m below the crest, gave such a wide range of calendar dates (1030-1220 A.D.) that it is unclear whether it relates to an advance during the Medieval Period or the beginning of the Little Ice Age in the 12th or 13th century. A stump, 40 m below the crest was sheared off by the ice in the late 13th century (Hv-10524),[21] while another stump found 50 m below the crest (Hv-1127), was sheared off in the 14th or 15th century (1330-1440). These results demonstrate that at least two separate advances occurred in the early Little Ice Age (Röthlisberger, 1986). Evidently relatively benign conditions allowed the trees to grow between the advance episodes. [22]

---

[20] An inventory carried out by aerial survey gave a total of 3,153 New Zealand glaciers (Chinn, 1991).

[21] The lags between the dates given by the samples and the shearing episodes depend upon the unknown ages of the trees when they were killed.

[22] Dendroglacialogical approaches are more difficult than in North America, even on the densely wooded western versants because few species of trees with clear annual rings are found (Wardle, 1973). In *Weinmannia racemosa*, common at low altitudes, the central rings are invariably decayed in trees more than 300 years old. However *Nothofagus menziesii,* frequently found in the Hooker Range, has a life-span of about 600 years and "tolerably distinct rings". Some species, such as *Podocarpus furrugineas* and *Dacrydium cupressinum* do not enter the succession on Westland moraines during the first century of exposure. Further investigations would be worthwhile.

The numbers in the margin give the laboratory identification number, the ¹⁴C age and the calibrated age (AD) for the 68% probability. Horizontal lines represent the calibrated ages.

->- = advance around or after indicated age
> = advance after indicated date
< = advance before indicated date
S = Soil
W = Wood
P = Peat

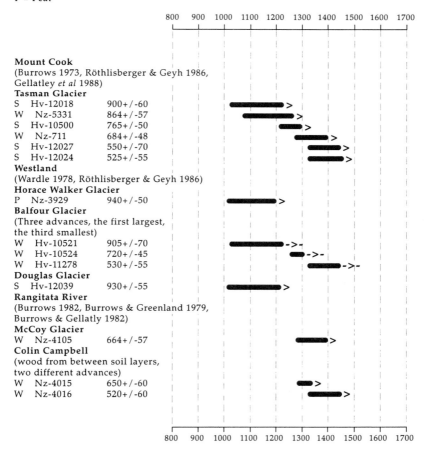

Figure 8. Key radiocarbon dates relating to glacial advances in New Zealand.

The evidence from the Tasman and Balfour Glaciers is reinforced by dates from other glaciers to both the east and west of the Southern Alps (Figure 8).

7. ANTARCTICA AND THE SUB-ANTARCTIC ISLANDS

Glaciers in the Sub-Antarctic Islands and small coastal glaciers in the Antarctic itself were affected by Little Ice Age fluctuations. Tracing the course of events in detail is hampered by the sparsity or absence of plant life and more particularly by difficulties involved in estimating the magnitude of the reservoir corrections required for calibrating radiocarbon dates of marine samples which form the major part of the available evidence.[23] However it may be hoped that tephra horizons of known date will mitigate problems of dating in the future.[24]

South of the Antarctic Circle (66°30' S), organic material for dating is lacking. To the north, moss and lichens occur in the South Orkney and South Shetland Islands, grass and moss communities in South Georgia. Reviewing the previous literature, Clapperton (1990) concluded that glaciers in South Georgia, the South Orkneys, South Shetlands and the Antarctic Peninsula have experienced at least two or three advance phases since the 12th century.

More than half of South Georgia (54-55°S) is covered by glaciers (Clapperton *et al.*, 1989). Advances during recent centuries have left multiple lateral and terminal moraines around the margins of land-based glaciers, the outermost moraines being more weathered and more densely covered with vegetation than those nearer the ice. The environmental succession recorded in peat bogs and lakes suggests that Little Ice Age advances began after the late 13th century, and peaked in the 18th, 19th and 20th centuries (Clapperton *et al.*, 1989).[25] The attribution of Little Ice Age initiation to the 13th century depends crucially on radiocarbon dating of whalebone.

Uncertainties involved in estimating the magnitude of the appropriate reservoir correction for dates on marine samples prevent attempts to correlate results with the ages of deposits elsewhere from being anything more than rough estimates. Nonetheless, placing the initiation of the Little Ice Age in South Georgia in the 13th or 14th century seems to be in line with the timing of events in South Shetlands. Here, glaciers were in contact with their terminal moraines when the 6m beach was formed. A beach at 2.5-3m lies outside but not inside moraines deposited by a later advance of up to 1 km outside present ice margins

---

[23] Variations from one area to another in the appropriate reservoir correction may be caused by several factors, including antiquity of reservoir ages near coasts due to glacial meltwater effects or reworking of old particulate organic carbon (Domack, 1992). Reservoir corrections used have varied from one study to another. On the basis of dates on marine creatures known to have died this century, it has been argued that a correction factor of 750 years, as made by Sugden and John (1973), may apply widely in the Antarctic region (Gordon 1987), but Stuiver *et al.* (1981) concluded that separate reservoir correction values were needed for different marine animals, including shells, penguins and seals. Clapperton and Sugden (1988) considered that a correction of 850 years might be applied in the case of the South Shetlands, stressing that this could only be a provisional guide, and that corrected dates might be a hundred years in error either way. Björck *et al.* (1991) put the true reservoir effect in the Peninsula area at 1200-1300 years. Domack (1992) also placed the most likely local reservoir age for the Antarctic Peninsula at around 1,200 years. Fitzsimmons and Domack (1993) used a provisional correction factor of 1300 years for the Vestfold Hills area of Antarctica. Orombelli and Baroni (1994) used 779±60 for Victoria Land, Antarctica. A useful review was provided by Gordon and Harkness (1992). Both the reality of regional variations and the importance of the influence of organic carbon from multiple sources has recently been demonstrated outside the Antarctic sector, as has the great utility of lipid biomarkers specific to individual organisms or a restricted range of them (Eglington *et al.*, 1997). It seems that there is room for much more intense investigation of the whole matter of reservoir corrections in Antarctic waters. Variations from region to region must be expected.

[24] The opportunity to obtain a tephrachronology which might be constrained by radiocarbon dating of mosses was recognised by Bjork *et al.* (1991). It involves establishing geochemical consistency between tephras deposited across wide geographical areas. The work was initiated by analyses of shards in sediment cores from Midge Lake, South Shetlands and Signy Lake, in the South Orkneys. Electron probe microanalysis of discrete tephra glass shards was used to identify the likely source volcanoes and to redefine tephrachronological data for the region (Hodgson *et al.*, 1998).

[25] Some of the impediments to radiocarbon dating of peats, such as the transport by water movement of humic and fulvic acids formed during decomposition, are discussed in Grove and Switsur (1994).

(Sugden and John, 1973). The two advances have been dated to the 13th and 18th centuries (Clapperton, 1990). A Southern beech log found on the 6m beach gave a date, not requiring correction, of 802 ± 43 (Birm-14), showing that the tree was growing between 1220 and 1280 cal A.D. But the sample was not from the outer layer of the trunk and the time taken for the log to drift across from South America can only be estimated. The assumption that that the log was stranded on the 6m beach round about 750 BP fits in with four radiocarbon ages from whalebone and seaweed, together suggesting ages of between 500 and 750 BP, i.e. some time between the late 13th and early 16th centuries A.D. for the 6m beach transgression and, by implication, the glacier advance.

Small glaciers near the coast of Antarctica itself are also believed to have advanced and retreated during the last millennium. The clearest example of probable Little Ice Age fluctuations comes from the Edmonson Point Glacier (165°08' E, 74°20' S) in the Terra Nova Bay area of Victoria Land, which apparently withdrew between the 10th and 14th centuries, and then advanced sometime later than the 14th, "in a time interval possibly corresponding to the Little Ice Age" (Baroni and Orombelli, 1994, p. 503). Radiocarbon dates from the Sub-Atlantic islands and the Antarctic coastal regions are not tabulated here because they are not sufficiently precise to allow meaningful comparison with those from other continents.

Reliable climatic information can be obtained from ice cores extracted from the icesheet itself, an approach that has revealed important regional differences in climatic history from one part of Antarctica to another. The South Pole core was the first in the Antarctic to be subjected to detailed analysis for the whole of the last millennium (Mosley-Thompson, 1992; Mosley-Thompson et al., 1982 and 1993). Isotopic values from the mid 16th century to the late 18th century were found to be generally below the long term mean, indicating colder conditions, especially in the mid to late 1500s. Before the mid 16th century, temperatures seem to have been higher, though interrupted by cold spells notably in the 13th century (Figure 5 in Mosley-Thompson et al., 1990). Accumulation was predominantly below the mean before 1700 A.D. and more frequently above the mean after that. Dust deposition between 1450 and 1850 was roughly double that before 1450, and from 1650 to 1850 was consistently above the long term average. Mosley-Thompson et al. (1993) date the Little Ice Age period at the South Pole to about 1450-1850 on the basis of this particulate evidence.

The characteristics of other Antarctic ice core records covering the last five centuries with sufficient time resolution and precision to allow comparison of environmental conditions over the continent, were surveyed by Mosley-Thompson (1992). At the South Pole, relatively warm conditions were found during the Medieval Warm Period and cool conditions from 1500 until the late 18th century. At the base of the Antarctic Peninsula, on the other hand, at Siple Station (84°15' W, 75°55' S), warm conditions predominated from the early 15th century to the early 19th century. Dust concentrations at Siple were below average from 1630 to 1880, except during a short cool spell around 1750, diminishing after 1880 as temperature fell. As with temperature, the dust sequence has been in antiphase with that at the South Pole.

Decadal averages of oxygen isotopes ($\delta^{18}O$) in cores from the Dyer Plateau in the middle of the Antarctic Peninsula (Thompson et al., 1994a) also fail to reveal any prolonged cold spell between 1450 and 1850, although marked cooling occurred between 1840 and 1850, initiating colder than normal conditions which lasted until the mid-20th century. Lack of any obvious period of significantly lower temperatures is common to the Siple and Dyer cores, as well as at several lower elevation and lower latitude Peninsula sites (Peel 1992). Dust deposition was consistently low from 1500 to the mid-19th century at both Siple and Dyer, and no long term trend in accumulation is recognizable at either.

The records from Mizuho and Law Dome in the East Antarctic, both register a short cool period from about 1650 to 1850, with the coldest conditions between 1750 and 1850, but as these records are less continuous than those from Siple and the South Pole, a comparison can only be general. It is significant that a smoothed deuterium record from Dome C indicates colder conditions from 1200 to 1800 A.D. (Benoist *et al.*, 1982). Unfortunately more detailed time resolution is not possible, but these data confirm that conditions over much of East Antarctica were cooler when those at Siple and on the Peninsula were warmer. It is also clear that a warming trend in East Antarctica since 1850 has coincided with cooling at Siple, on the Antarctic Peninsula, and on a part of the Ronne Ice Shelf.

The difference in dust concentration history at the South Pole and Siple may reflect different transport routes taken by dust originating from the various source regions: the South American Altiplano, the South African deserts and Antarctica (Mosley-Thompson, 1992). Intensified westerlies at lower latitudes could result in entrainment of more dust in the upper troposphere and lower stratosphere, increasing the amount ultimately deposited at the South Pole. The Antarctic Peninsula acts as a major climatic divide, extending northwards across the circumpolar trough, limiting advection of mild maritime air from the South Pacific and blocking the prevailing eastward flow of cold air around the edge of the ice sheet. Meteorological conditions are greatly affected by variations in the extent of sea-ice over the adjacent oceans and the Weddell Sea. Mosley-Thompson (1992) suggests that the similarity between the South Pole and Quelccaya records of both $\delta^{18}O$ and dust may be due to large scale upper atmosphere teleconnections between the South American Andes and the high East Antarctic Plateau. It seems probable that dust deposited at Siple, which is subject to frequent, severe storms throughout the year, comes via a lower tropospheric route associated with the passage of cyclonic systems.

SUMMARY

Moraine dating

The individual results shown in the various figures differ greatly in precision. Those depending on *in situ* wood are the most reliable because the position of the wood in relation to the ice is exactly known (see Figure 9).[26] Unequivocal interpretation of a date is only possible if the wood is found *in situ*.

Unfortunately, the necessary dendrochronological sequences allowing tree rings to be exactly placed are not available in many of the regions where *in situ* trunks have been radiocarbon dated. In spite of the variation in precision of dates and their closeness to the time of moraine formation, there are many indications that glaciers were advancing in the 13th and 14th centuries, and in some cases as early as the 12th century. It has to be remembered that the start of glacial advances must have preceded the deposition of the first Little Ice Age moraines by some decades, as Luckman (1995) has clearly demonstrated in Canada. Many early Little Ice Age moraines are extremely close to, even touching, those deposited in recent centuries. The data are as yet insufficient to compare the timing of the more recent culminations from all parts of the globe, although Luckman and Villalba (in press) have done so for sites along the length of the Pole-Equator-Pole transect of North and South America. While radiocarbon dating of some moraine sequences fails to indicate clearly whether the first advances were in the 13th or 14th centuries, where it is most precise it shows that the Little Ice Age was under way in the 13th century or earlier (Figure 10).

---

[26] Unequivocal interpretation of a date is only possible if the wood is found *in situ*. Alternative possible interpretations of dates on transported wood fragments are discussed by Ryder and Thompson (1994). Wood found within a moraine may provide maximum dates for its construction, ranging from precise limits, through broad estimates if it was lodged in snow before incorporation, to gross overestimates if reworking has occurred. If a wood fragment in a moraine was originally derived from overridden trees, its age may relate more closely to the start of an advance than to moraine deposition at the end of an advance.

The numbers in the margin give the laboratory identification number and the ¹⁴C age.
Horizontal lines represent the calibrated ages, 1σ = ━━━━ 2σ = ██████

->- = advance around or after indicated age range
>> = indicates evidence of more than one advance

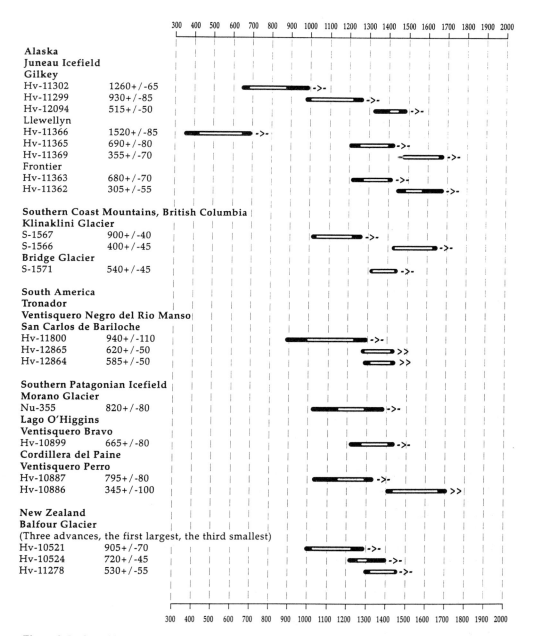

Figure 9. In situ evidence.

North Atlantic
European Alps   2nd half 13th century
Greenland   13th century
Iceland   13th century
Spitsbergen   13th century
Northern Scandinavia   13th or 14th century
Southern Scandinavia   13th century

Asia
Altai   14th-15th century
Himalaya   12th-14th century
Nyenchintangla   13th-15th century
Gongga   13th-14th century

North America
Juneau Icefield   13th-14th century
Canadian Rockies   12th century
British Columbia,   13th-14th century
Coast Mountains

South America
Tropical   13th-14th century
Temperate   12th-13th century

New Zealand   13th century

Figure 10. Time of Little Ice Age initiation indicated by radiocarbon dating of moraines.

The data provided by dendrochronological dating of trees overrun by ice are precise (Figure 11), and show definitively that in parts of North America, as well as in the Swiss Alps, the Little Ice Age was under way by the 13th century or earlier. The precision applies to the time when a tree or group of trees was killed or damaged; the advance of the ice is likely to have begun somewhat earlier (Luckman, 2000).

The combined evidence from radiocarbon dating of moraines and from dendrochronology confirms that the Little Ice Age was a global event. The exact time at which it began, as might be expected, differs somewhat from one location to another and from one region to another on account of differences in the morphology, aspect, size and sensitivity of the glaciers concerned, and their positions in relation to major climatic patterns and large scale wind fields. It would be possible to obtain a clearer view of such differences if more lengthy dendrochronological series suitable for cross-dating with sub-fossil wood were to become available, or if the intensity of the investigations carried out in the European Alps were to be matched elsewhere.

Kenai Peninsula　　　　　Wiles and Calkin 1994

Grewingk　　1422

Trustemena　1460

Princeton　　1190

Tebenkof　　1289-1300

Canadian Rockies

Robson　　1142-1350　Luckman 1995

Peyto　　1246-1375　Luckman 1996

Stutfield　1272　　Luckman 2000

Figure 11. Key dates relating to the initiation of the Little Ice Age given by dendrochronological dating of forest trees overridden by ice.

REFERENCES

Aellen, M., 1996, Glacier mass balance studies in the Swiss Alps, *Zietschrift für Gletscherkunde und Glazialgeologie* 20:159-68.

Adamson, D.A., and Colhoun, E.A., 1992, Late Quaternary glaciation and deglaciation of the Bunger Hills, Antarctica, *Antarctic Science* 4:435-46.

Ageta, Y., Yao, T., Jiao, K., Po, J., Shao,W., Iwata,S., Ohno,H., and Furukawa,T., 1991, Glaciological Studies on Qinzang Plateau, 1989, Part 2, Glaciology and Geomorphology. *Bulletin of Glacier Research* 9:27-32.

Aniya, M., 1995, Holocene chronology in Patagonia: Tyndall and Upsala Glaciers *Arctic and Alpine Research* 27:311-22.

Aniya, M., and Sato, H., 1995a, Holocene variations at Tyndall Glacier area, southern Patagonia, *Bulletin of Glacier Research* 13:97-109.

Aniya, M., and Sato,H., 1995b, Holocene chronology of Upsala Glacier at Peninsula Herminita, Southern Patagonia Icefield, *Bulletin.of Glacier Research* 13:83-96.

Barlow, L.K., Sadler, J.P., Ogilvie, A.E.J., Buckland, T.C., Amorosi, T., Ingimundarson, I., Skidmore, P., Dugmore, A.J., and McGovern, T.H., 1997, Interdisciplinary investigations of the end of the Norse Western Settlement in Greenland, *The Holocene* 7:489-99.

Barclay, D.J., Wiles, G.C., and Calkin, P.E., 1999, A 1119-year tree-ring-width chronology from western Prince William Sound, southern Alaska, *The Holocene* 9:79-84.

Baroni, C., and Orombelli, G., 1991, Holocene raised beaches at Terra Nova Bay, Victoria Land, Antarctica, *Quaternary Research* 36:157-77.

Baroni, C., and Orombelli, G.,.1994, Holocene glacier variations in the Terra Nova Bay area (Victoria Land, Antarctica), *Antarctic Science* 6:497-505.

Benoist, J.P., Jouzel, J., Lorius, C., Merlivat, L., and Pourchet, M., 1982, Isotopic climatic record over the last 2.5 ka from Dome C Antarctica, ice cores, *Annals of Glaciology* 3:17-22.

Björck, S., Hjort, C., Ingólfsson, Ó., and Skog, G., 1991, Radiocarbon dates from the Antarctic Peninsula region - problems and potential, *Quaternary Proceedings* 1:55-65.

Bradley, R.S., and Jones, P.D., 1992, *Climate since A.D. 1500,* Routledge. London and New York.

Braitseva, O.A., Ponomareva, V.V., Sulerzhitsky, L.D., Melekestsev, I.V., and Bailey, J., 1997, Holocene key-marker tephra layers in Kamchatka, Russia, *Quaternary Research* 47:125-39.

Bradley, R.S., 1999, *Paleoclimatology: Reconstructing Climates of the Quaternary, 2nd Edition.* International Geophysics Series, 64. Harcourt, San Diego, etc.

Bradley, R.S., and Jones, P.D., 1993, "Little Ice Age" summer temperature variations: their nature and relevance to recent global warming trends, *Holocene* 3: 367-76.

Broecker, W.S., 1963, Radiocarbon ages of Antarctic materials. *Polar Record* 11:472.

Buckland, P.C., Amorosi, T., Barlow, L.K., Mayewski, P., McGovern, T.H., Ogilvie, A.E.J. and Skidmore, P., 1996, Climate change, and the end of Norse Greenland, *Antiquity* 70:88-96.

Burrows, C.J., 1975, Late Pleistocene and Holocene moraines of the Cameron Valley, Arrowsmith Range, Canterbury, New Zealand, *Arctic and Alpine Research* 7:125-40.

Burrows, C.J., and Greenland, D.E., 1979, An analysis of the evidence for climatic change in New Zealand in the past thousand years: evidence from diverse natural phenomena and from instrumental records, *Journal of the Royal Society of New Zealand* 9:321-73.

Calkin,P.E., 1988, Holocene glaciation in Alaska (and adjoining Yukon Territory, Canada), *Quaternary Science Reviews* 7:158-84.

Calkin, P.E. and Wiles, G.C., 1992, Little Ice Age glaciation in Alaska: a record of recent global change, in: *The Role of the Polar Regions in Global Change,* G.Weller and C. Wilson, eds., Proceedings of the International Conference, Fairbanks, Alaska, UAF Geophysical Institute and Centre for Global Change, Fairbanks, Alaska, 617-25.

Calkin, P.E., Wiles, G.C., and Barclay, D.J., in press, Holocene coastal glaciation of Alaska, *Quaternary Science Reviews.*

Casassa, G., 1995, Glacier inventory in Chile: current status and recent glacier variations, *Annals of Glaciology* 21:317-22.

Chen, J., Nakawo, M., Ageta,Y., Watanabe, O., and Liu, L., 1989, Movement of Chongce Ice Cap and recent variations of some glaciers on the south side of West Kunlun Mountains, *Bulletin of Glacier Research* 7:45-8.

Chinn, T.J., 1991, *Glacier Inventory of New Zealand,* New Zealand Institute of Geological and Nuclear Sciences, unpublished.

Clague, J.J., and Matthews, R.W., 1992, The sedimentary record and Neoglacial history of Tide Lake, northwestern British Columbia, *Canadian Journal Of Earth Sciences* 19:94-117.

Clague, J.J., and Matthews, R.W., 1996, Neoglaciation, glacier-dammed lakes and vegetation change in northwestern British Columbia, *Arctic and Alpine Research* 28:10-24.

Clapperton, C.M., 1990, Quaternary glaciations in the Southern Ocean and Antarctic Peninsula area, *Quaternary Science Reviews* 9:229-52.

Clayton, J.D., and Clapperton, C.M., 1997, Broad synchrony of a late-glacial glacier advance and the highstand of palaeolake Tauca in the Bolivian Altiplano, *Journal of Quaternary Science* 12:169-82.

Clapperton, C.M., Clayton,J.D., Benn,D.I., Marden,J.C., and Argillo,J., 1997, Late Quaternary glacier advances and paleolake highstands in the Bolivian Altiplano, *Quaternary International* 38/39: 49-59.

Clapperton, C.M., and Sugden,D.E., 1988, Holocene glacier fluctuations in South America and Antarctica., *Quaternary Science Reviews* 7:185-98.

Clapperton, C.M., Sugden, D.E. Birnie,J., and Wilson, M.J., 1989, Late-Glacial and early Holocene fluctuations and environmental changes in South Georgia, Southern Ocean, *Quaternary Research* 31:210-26.

Clapperton, C.M., Sugden,D.E., and Wilson, M.J., 1989, Relationship of land-terminating and fjord glaciers to Holocene climatic change, in: *Glacier Fluctuations and Climatic Change,* J.Oerlemans, ed., Kluwer, Dordrecht, 57-75.

Corte, A.E., and Espinoza, L., 1981. *Inventario di Glaciares de la Cuenca del Rio.* IANIGLA-CONICT, Mendoza, Argentina.

Dahl, S.O., and Nesje, A., 1994, Holocene glacier fluctuations at Hardangerjøkulen, central-southern Norway: a high-resolution composite chronology from lacustrine and terrestrial deposits, *The Holocene* 4:269-77.

Dean, J.S., 1994, The Medieval Warm Period on the southern Colorado Plateau, *Climatic Change* 26:225-41.

Deslodges, J.S., and J.M., 1992, Neoglacial history of the Coast Mountains near Bella Coola, British Columbia, *Canadian Journal of Earth Sciences* 27:281-90.

Domack, E.W., 1992, Modern carbon-14 ages and reservoir corrections for the Antarctic Peninsula and Gerlache Strait area, *Marine Geology* 103: 63.

Eglington,T., Benitez-Nelson, B.C., Pearson, A., McNichol, A.P., Bauer, J., and Druffel, E.R.M., 1997, Variability in radiocarbon ages of individual organic compounds from marine sediments. *Science* 277: 796-9.

Fushimi, H., 1977, Glacial history in Khumbu region (1), *Seppyo* 39:60-67.

Fushimi, H., 1978, Glacial history in Khumbu region (2), *Seppyo* 40:71-7

Gellatley, A.F., 1982, *Holocene glacial activity in Mt Cook National Park, New Zealand*, PhD thesis, University of Canterbury, Christchurch, New Zealand, 218 pp.

Gellatley, A.F. 1985, Glacier fluctuations in the central Southern Alps, New Zealand: documentation and implications for environmental change during the last 1000 years, *Zeitschrift für Gletscherkunde und Glacialgeologie* 21:259-64.

Gellatley, A.F., Röthlisberger,F., and Geyh, M.A., 1985, Holocene glacier variations in New Zealand (South Island), *Zeitschrift für Gletscherkunde und Glacialgeologie* 21:275-81.

Gellatley, A.F., Chinn, T.J.H., and Röthlisberger,F., 1988, Holocene glacier variation in New Zealand: a review, *Quaternary Science Reviews* 7:227-42.

Gordon, J.E., 1987, Radiocarbon dates from Nordenskjöld Glacier, South Georgia and their implications for late Holocene glacier chronology, *British Antarctic Survey Bulletin* 76:1-5.

Gordon, J.E., and Harkness, D.D., 1992, Magnitude and geographical variation of the radiocarbon content in Antarctic marine life: implications for reservoir corrections in radiocarbon dating, *Quaternary Science Reviews* 11: 697-709.

Gordiyenko, F.G., Kotlyakov,V.M., Punning, Ya-K.M. and Vairmäe,R. 1980, Study of a 200 m core from the Lomonosov Ice Plateau in Spitsbergen and the paleoclimatic implications, *Polar Geography and Geology* 5:242-51.

Grosjean, M., Geyh, M.A., Messerli, B. Schreier, H. and Veit, H., 1998, A late-Holocene glacial advance in the south-central Andes (29° S), northern Chile. *The Holocene* 8:473-9.

Grove, J.M., 1985, The timing of the Little Ice Age in Scandinavia, in: *The Climatic Scene*, M.J.Tooley and G.M.Sheail, eds., Allen and Unwin, Hemel Hampstead, 133-53.

Grove, J.M., 1996, The century timescale, in: *Time-scales and Environmental Change*, T.S.Driver and G.P.Chapman, eds., Routledge, London and New York, 39-87.

Grove,J.M., in press, The initiation of the Little Ice Age in regions round the North Atlantic, *Climatic Change*

Grove, J.M., and Switsur, R., 1994, Glacial geological evidence for the Medieval Warm Period, *Climatic Change* 30:1-27.

Haeberli, W., Müller,P., Alean,A. and Bösch,H., 1989, Glacier changes following the Little Ice Age - a survey of the International Data Base and its perspectives, in: *Glacier Fluctuations and Climatic Change*, J.Oerlemans, ed., Kluwer, Dordrecht, 77-101.

Harkness, D.D., 1979, Radiocarbon dates from Antarctica, *British Antarctic Survey Bulletin* 47:43-59.

Heusser, C.J., 1956, Postglacial environments in the Canadian Rocky Mountains, *Ecological Monographs* 26:253-302.

Hodgson, D.A., Dyson, C.L., Jones, V.J., and Smellie, J.L., 1998, Tephra analysis of sediments from Midge Lake (South Shetland Islands) and Sombre Lake (South Orkney Islands), Antarctica, *Antarctic Science* 10:13-20.

Holzhauser, H., 1984a, Rekonstruktion von Gletscherschwankungen mit Hilfe Fossiler Holzer, *Geographica Helvetica* 39:3-15.

Holzhauser, H., 1984b, Zur Geschichte der Aletschgletscher und Fieschgletscher, *Physische Geographica* 13: (Geographisches Institut der Universität Zürich) 1-448.

Holzhauser, H., 1995, Gletscherschwankungen innerhalb der letzten 3200 Jahre am Beispiel des Grossen Aletsch-und des Gornergletschers. Neue Ergebnisse, in: *Gletscher im ständigen Wandel*. Publikattionen der Schweizerischen Akademie der Naturwissenschaften (SANW/ASSN 6, Zürich und Hochschulverlag AG an ETH.

Holzhauser, H., and Zumbühl, H.J., 1996, (Translated M.Joss, Checkmate Bern and T.Wachs) To the History of the Lower Grindelwald Glacier during the last 2800 years - paleosols, fossil wood and pictorial records - new results, *Zeitschrift für Geomorphologie* N.F. Suppl.Bd. 104:95-127.

Holzhauser, H., 1997, Fluctuations of the Grosser Aletsch Glacier and Gorner Glacier during the last 3200 years: new results, *Paläoklimaforschung/Palaeoclimate Research* 24:36-58.

Hughes, M.K., and Diaz, H.F., 1994, Was there a 'Medieval Warm Period', and if so, when and where, *Climatic Change* 26:109-42.

Innes, J.L., 1982, Lichenometric use of an aggregated *Rhizocarpon* "species" in lichenometry, *Boreas* 11:53-7.

Innes, J.L. 1983, Use of an aggregated *Rhizocarpon* "species" in lichenometry: an evaluation. *Boreas* 12:183-90.

Innes, J.L. 1985, Lichenometric dating of moraine ridges in northern Norway: some problems of application, *Geografiska Annaler* 66A, 341-52.

Ivanovskiy, L.N., and Panychev, V.A., 1978, Evolution and age of end moraines of the XVII-XIX centuries formed by the Aktru glaciers in the Altai, in: *Recent-relief forming.Processes in Siberia*. Irkutsk, (Protsessy sovremennogo reliefoobrazovaniya v Sibiri), Nauka, Irkutsk, 127-38 (in Russian).

Iwata, S. and Keqin, J., 1993, Fluctuations of the Zepu Glacier in late Holocene epoch, the eastern Nyainqêntanglha Mountains, Qing-Zang (Tibet) Plateau, in: *Glaciological Climate and Environment on Qing-Zang Plateau*, T. Yao and Y. Ageta, eds., Beijing Science Press. 130-9.

Jennings, A.E., and Weiner, N.J., 1996, Environmental change in eastern Greenland during the last 1300 years: evidence from foraminifera and lithofacies in Nansen Fjord, 68° N, *The Holocene* 6:179-91.

Jones, P.D., Marsh, R., Wigley, T.M.L., and Peel, D.A., 1993. Decadal timescale links between Antarctic Peninsula ice-core oxygen-18, deuterium and temperature, *The Holocene* 3:14-26.

Kaser, G., Ames, A., and Zamora, M., 1990, Glacier fluctuations and climate in the Cordillera Blanca, Peru, *Annals of Glaciology* 14:136-40.

Kaser, G., and Ames, A., 1996, Modern glacier fluctuations in the Huascarán-Chopicalqui massif of the Cordillera Blanca of Peru. *Zeitschrift für Gletscherkunde und Glazialgeologie* 32:91-9.

Karlén,W., 1991, Glacier fluctuations in Scandinavia during the last 9,000 years, in: *Temperate Paleohydrology. Fluvial processes in the temperate zone during the last 15000 years,* L. Starkel, K.J. Gregory, and J.B. Thornes, eds., Wiley, Chichester, New York, Brisbane, Toronto, Singapore, 395-412.

Karlén, W., and Matthews, J.A., 1992, Reconstructing Holocene glacier variations from glacial lake sediments: Studies from Nordvestlandet and Jostedalsbreen-Jotenheimen, southern Norway, *Geografiska Annaler* 74A:327-48.

Kirkbride, M., and Dugmore, A., in press, Can lichenometry be used to date the "Little Ice Age" glacial maximum ? *Climatic Change.*

Kotlyakov, V.M., Serebryanny, L.R., and Solomina, O.N., 1991, Climate change and glacier fluctuations during the last 1,000 years in the southern mountains of the USSR, *Mountain Research and development* 11:1-12.

Kotlyakov, V.M., Osipova,G.B., Popovnin,V.V., and Tsvetkov, D.G., 1996, Experience from observations of glacier fluctuations in the territory of the former USSR, *Zeitschrift für Gletscherkunde und Glazialgeologie* 32:5-14.

Li, J.J., and Feng, Z.D., 1984, On the remains of Quaternary glaciers in the Hengduan Mountains, *Journal of Lanzhou University* 6:61-72.

Li, J., *et al.*, 1986, Glaciers in Xizang, *Science press*, Beijing (in Chinese).

Lin, P.N., Thompson, L.G., Davis, M.E., and Mosley-Thompson, E., 1995, 1000 years of climatic change in China: ice-core $\delta^{18}O$ evidence, *Annals of Glaciology* 21:189-95.

Luckman, B.H., 1986, Reconstruction of Little Ice Age events in the Canadian Rockies, *Geographie physique et Quaternaire* 40:17-28.

Luckman, B.H., 1993a, Glacier fluctuations and tree-ring records for the last millennium in the Canadian Rockies, *Quaternary Science Reviews* 12:441-50.

Luckman, B.H., Holdsworth, G., and Osborn, G.D., 1993, Neoglacial glacier fluctuations in the Canadian Rockies, *Quaternary Research* 9:144-53.

Luckman, B.H., 1995,. Calendar-dated, early "Little Ice Age"' glacier advance at Robson Glacier, British Columbia, Canada, *The Holocene* 5:149-59.

Luckman, B.H., 1996a, Reconciling the glacial and dendrochronological records for the last millennium in the Canadian Rockies, in: *Climate variations and forcing mechanisms of the last 2000 years*, P.D. Jones, R.S. Bradley and J. Jouzel, eds.,. NATO ASI Series 14:85-108.

Luckman, B.H., 1996b, Dendrochroglaciology at Peyto Glacier, Alberta, in: Tree Rings, Environment and Humanity, J.S. Dean, D.M. Meko and T.W. Swetnam, eds. *Radiocarbon* 1996, 679-88.

Luckman, B.H., 1996c, Dendrochronology and global change I, in: Tree Rings, Environment and Humanity, J.S. Dean, D.M. Meko and T.W. Swetman, eds., *Radiocarbon* 1996 1-24,

Luckman, B.H., 2000, The Little Ice Age in the Canadian Rockies, *Geomorphology* 32:357-84.

Luckman, B.H., Holdsworth, G., and Osborn, D.G., 1993, Neoglacial glacier fluctuations in the Canadian Rockies, *Quaternary Research* 39:144-53.

Luckman, B.H., Briffa,K.R., Jones,P.D., and Schweingruber, F.H., 1997, Tree-ring based reconstructions of summer temperature at the Columbia Icefield, Alberta, Canada, A.D. 1073-1983, *The Holocene* 7:375-89.

Luckman, B.H., and Villalba, R., in press, Assessing the synchroneity of fluctuations in the Western Cordillera of the Americas during the last millennium, in: *Proceedings of the First Pole-Equator-Pole Paleoclimate of the Americas Conference*. Veneduala, 1998

Matthews, J.A., 1991, The late Neoglacial ("Little Ice Age") glacier maximum in southern Norway: new ^{14}C-dating evidence and climatic implications, *The Holocene* 1:219-33.

Matthews, J.A., and Karlén, W., 1992, Asynchronous Neoglaciation and Holocene climatic change from Norwegian glaciolacustrine sedimentary sequences, *Geology* 20:991-4.

Meese, D.A., Gow, A.J., Grootes, P., Mayewski, P.A., Ram, M., Stuiver, M., Taylor, K.C., Waddington, E.D., Zielinski, G.A, 1994, The accumulation record from the GISP2 core as an indicator of climatic change throughout the Holocene, *Science* 266:1680-2.

Mercer, J.H., 1968, Variations in some Patagonian glaciers since the Late Glacial, I. *American Journal of Science* 266:91-109.

Mercer, J.H., 1970, Variations of some Patagonian glaciers since the Late Glacial, II. *American Journal of Science* 269:1-25.

Mercer, J.H., 1976, Glacial history of southernmost South America, *Quaternary Research* 6:125-66.

Mercer, J.H., and Palacios, D., 1977, Radiocarbon dating of the last glaciation in Peru, *Geology* 5:600-604.

Messerli, B., 1973, Problems of vertical and horizontal arrangement in the high mountains of the extreme arid zone (central Sahara). *Arctic and Alpine Research* 5A:139-47.

Mikhailov, N.N., 1987, The dynamics of Belukha glaciers, Altai in historic time. *News from the Leningrad State University, Geology - Geography Series, (Vestnik Leningradskogo Gosudarstvennogo Universiteta, ser. geologiya-geografiya)* 3:100-103 (In Russian).

Mosley-Thompson, E., 1992, Paleoenvironmental conditions in Antarctica since A.D. 1500: ice core evidence, in: *Climate since A.D.1500*, R.S. Bradley and P.D. Jones, eds., Routledge, London, 572-91.

Mosley-Thompson, E., 1994, Holocene climatic changes in an East Antarctic core, in: *Climatic Variations and Forcing Mechanisms of the last 2000 years*, P.D. Jones, R.S. Bradley and J. Jouzel, eds., Springer, Berlin etc,. 263-79.

Mosley-Thompson, E., and Thompson, L.G., 1982, Nine centuries of microparticle deposition at the south Pole, *Quaternary Research* 17:1-13.

Mosley-Thompson, E., Thompson, L.G., Grootes, P., and Gunderstrup, N., 1990, Little Ice Age (Neoglacial) paleoenvironmental conditions at Siple Station, Antarctica, *Annals of Glaciology* 14:199-204.

Mosley-Thompson, E., Thompson, L.G., Dai, J., Davis, M., and Lin, P.N., 1993, Climate of the last 500 years: high resolution ice core records, *Quaternary Science Reviews* 12:419-30.

Müller, F., 1961, ^{14}C-Daten des Khumbu Glaciers, Mt Everest, *Radiocarbon* 3:16.

Nesje, A., and Dahl, S.O., 1991a, Holocene glacier variations of Blåisen, Hardangerjøkulen, central Southern Norway, *Quaternary Research* 35:25-40.

Nesje, A., and Dahl, S.O., 1991b, Late Holocene glacier fluctuations in Bevringsdalen, Jostedalsbreen region, western Norway ca. 3200-1400 BP, *The Holocene* 1:1-7.

Nesje, A., Kvamme, M., Rye, N., and Løvlie, R., 1991, Holocene glacial and climatic history of the Jostedalsbreen region, western Norway; evidence from lake sediments and terrestrial deposits, *Quaternary Science Reviews* 10:87-114.

Nesje, A., and Rye, R., 1993, Late Holocene glacier activity at Sandeskardfonna, Jostedalsbreen area, western Norway, *Norges Geografiske Tidsskrift* 47:21-30.

O'Brien, S.R., Mayewski, P.A., Meeker, L.D., Meese, D.A., Twickler, M.S., and Whitlow, S.I., 1995, Complexity of Holocene climate as reconstructed from a Greenland ice core, *Science* 270:1962-4.

Ogilvie, A.E.J., 1991, Climatic changes in Iceland A.D.c.865 to 1598, *Acta Archaeologica* 61:233-51.

Osborn, G.D., 1993, Lateral-moraine stratigraphy and Holocene chronology at Stutfield Glacier, Banff National Park, Alberta. *Geological Society of America Abstract Program* 25:157.

Osborn, G., and Luckman, B.H., 1986, Holocene glacial fluctuations in the Canadian Cordillera (Alberta and British Columbia), *Quaternary Science Reviews* 7:115-28.

Peel, D.A., 1992, Ice core evidence from the Antarctic Peninsula region, in: *Climate since A.D.1500*, R.S Bradley and P.D.Jones, eds. Routledge, London and New York, 549-71.

Pfister, C, 1981, An analysis of the Little Ice Age climate in Switzerland and its consequences for agricultural production, in: T.M. Wigley, M.J Ingram,. and G. Farmer, eds., *Climate and History: Studies on Past Climates and their Impact on Man*, Cambridge University Press, Cambridge, 214-48.

Pfister, C. 1994, Switzerland: The time of icy winters and chilly springs, *Paläoklimaforschung. Paleoclimate Research* 13:205-24.

Pfister, C., Luterbacher, L., Schwarz-Zanetti, G., and Wegman, M., 1999, Winter temperature in western Europe during the Early and High Middle Ages (A.D. 750-1300), *The Holocene* 8:535-52.

Porter, S.C., 1981, Glaciological evidence of Holocene climatic change, in: *Climate and History. Studies in past Climates and their impact on Man*, T.M.L. Wigley, M.J. Ingram and G. Farmer, eds., Cambridge University Press, Cambridge, 82-110.

Rabassa, J., Rubulis, S., and Suarez, J., 1978, Los glaciares del Monte Tronador Parque Nacional Nahuel Huapi (Rio Negro, Argentina), *Anales de Parques Nacionales* 14:259-318.

Ryder, J.M., and Thompson, B., 1986, Neoglaciation in the southern Coast Mountains of British Columbia: chronology prior to the late-Neoglacial maximum, *Canadian Journal of Earth Sciences* 23:237-38.

Röthlisberger, F., and Geyh, M. A., 1985, Gletscherschwungen der Nacheiszeit in der Cordillera Blanca (Peru) und den Südlichen Anden Chiles und Argentiniens. *Zentralblat für Geologie und Paläontologie* 11/12:1611-3.

Röthlisberger, F., and Geyh, M. A., 1986, *10,000 Jahre Gletschergeschicht der Erde*, Salxberg. Sauländer, Aarau, Frankfurt am Main.

Rull, V., and Schbert, C., 1989, The Little Ice Age in the tropical Venezuelan Andes, *Geologia* 70:71-3.

Savoskul, O.S., and Zech,W., 1997, Holocene glacier advances in the Topolovaya Valley, Range, Kamchayka, Russia, dated by tephrachronology and lichenometry, *Arctic and Alpine Research* 29:143-55.

Serebryanny L.R., and Solomina, O.N., 1996, Glaciers and climate of the mountains of the former USSR during the Neoglacial, *Mountain Research and Development* 16:157-66.

Shiraiwa,T., and Watanabe,T., 1991, Late Quaternary glacial fluctuations in the Langtang Valley, Nepal Himalaya, reconstructed by relative dating methods, *Arctic and Alpine Research* 23:404-16.

Smith,D.J., McCarthy, D.P., and Colenutt, M.E., 1995, Little Ice Age glacial activity in Peter Lougheed and Elk Lakes Provincial Parks, Canadian Rocky Mountains, *Canadian Journal of Earth Sciences* 32:579-89.

Solomina, O., 1996, Long-term variations of mountain glaciers in the former USSR (FSU), *Zeitschrift fur Gletscherkunder und Glazialgeologie* 32:197-205.

Solomina, O.N., Muraviev, V.D., and Bazanova, L.I., 1995, The Little Ice Age in Kamchatka, *Annals of Glaciology* 21:240-44.

Stuiver, M., and Reimer, P.J., 1986, A computer program for radiocarbon age calibration, *Radiocarbon* 28:1022-30.

Stuiver, M., and Pearson, G.W., 1993, High-precision bidecadal calibration of the radiocarbon timescale A.D.1950-500 B.C, *Radiocarbon* 35:323.

Stuiver, M., and Reimer, P.J., 1993, Extended ^{14}C data base and revised CALIB 3.0 ^{14}C age calibration program, *Radiocarbon* 35:215-30.

Stuiver, M., Grootes, P.M., and Braziunas, T.F., 1995, The GISP2 climate record of the past 16,500 years and the role of the sun, ocean and volcanoes, *Quaternary Research* 44:341-54.

Su Zhen, Liu Shiyin, Wang Ninglian, and Shi Aiping, 1992, Recent fluctuations of glaciers in the Gongga Mountains, *Annals of Glaciology* 16:163-7

Svendsen, J.I., and Mangud, J., 1997, Holocene glacial and climatic variations on Spitsbergen and Svalbard, *The Holocene* 7:45-57.

Tandong, Y., and Thompson, L.G., 1992, Trends and features of climatic changes in the past 5000 years recorded by the Dunde ice core, *Annals of Glaciology* 16:21-4.

Thompson, L.G., 1980, Glaciological investigations of the tropical Quelccaya Ice Cap, Peru, *Glaciological Journal* 25:69-84

Thompson, L.G., 1992, Ice core evidence from Peru and China, in: *Climate since A.D.1500*, R.S. Bradley and P.D. Jones, eds., Routledge, London and New York, 517-47.

Thompson, L.G., 1993, Reconstructing the Paleo Enso records from tropical and subtropical ice cores, *Bulletin de l'Institut francais d'Etudes Andines* 22:65-83.

Thompson, L.G., 1994, Climatic changes for the last 2000 years inferred from ice core evidence from tropical ice cores, in: *Climatic Variations and Forcing Mechanisms of the last 2000 years*, P.D.Jones, R.S.Bradley and J.Jouzel, eds., Springer. Berlin etc., 281-95.

Thompson, L.G., Mosley-Thompson, E., Bolzan, J.F., and Koci, B.R., 1985, A 1500-year record of tropical precipitation in ice cores from the Quelccaya Ice Cap, *Science* 229:971-3.

Thompson, L.G., Mosley-Thompson, E., Dansgaard, W., and Grootes, P.M., 1986, The Little Ice Age as recorded in the stratigraphy of the tropical Quelccaya Icecap. *Science* 234:361-5.

Thompson, L.G., Peel, D.A., Mosley-Thompson, E., Mulvaney, R., Dai, J., Lin, P.N., Davis, M.E., and Raymond, C.F., 1994a, Climate since A.D. 1510 on Dyer Plateau, Antarctica: evidence for recent climatic change, *Annals of Glaciology* 20:420-26.

Thompson, L.G., Mosley-Thompson, E., Davis, M.E., Lin, P.N., Dai, J., Bolzan, J.F. and Yao, T., 1995a, A 1000 year climate ice-core record from the Guliya ice cap, China: its relationship to global climatic variability. *Annals of Glaciology* 21:157-81.

Thompson, L.G., Mosley-Thompson, E., Davis, M., Lin, P.N., Henderson, K.A., Cole, J.-Dai, Bolzen, J.F. and Liu K,-b., 1995b, Late glacial stage and Holocene tropical ice core records from Huascarán, Peru, *Science* 69:46-50.

Villalba, R., 1990, Climatic fluctuations in Northern Patagonia during the last 1000 years as inferred from tree-ring records, *Quaternary Research* 34:346-60.

Villalba, R. 1994, Tree-ring and glacial evidence for the Medieval Warm Epoch and the Little Ice Age, *Climatic Change* 26:183-97.

Wang Zongtai and Yang Huian, 1992, Characteristics of the distribution of glaciers in China, *Annals of Glaciology* 16:17-19.

Wang Zongtai, 1992, Characteristics of the distribution of glaciers in China, *Annals of Glaciology* 16:17-20.

Warren, C.R., and Sugden, D.E., 1993, The Patagonian Icefields: a glaciological review, *Arctic and Alpine Research* 25:316-31.

Wardle, P., 1973, Variations of the glaciers of Westland National Park, and the Hooker Ranges, New Zealand, *New Zealand Journal of Botany* 11:349-88.

Wardle, P., 1978, Further radiocarbon dates from Westland National Park and the Omoeroa River mouth, New Zealand. *New Zealand Journal of Botany* 16:147-52.

Watson, H.M., 1986, *Little Ice Age glacier fluctuations in the Premier Range, B.C.* MSc Thesis, University of Western Ontario, London, Canada.

Wiedick, A., Boggild ,C.E., and Kudsen, N.T., 1992, Glacier inventory and atlas of West Greenland, *Gronølands Geologisk Unðersøelse* Rapport 158:1-194.

Wiles, G.C., Barclay, D.J., and Calkin, P.E., 1999, Tree-ring dated Little Ice Age histories of maritime glaciers from western Prince William Sound, Alaska, *The Holocene* 9:163-73.

Wiles, G.C., and Calkin, P.E., 1994, Late Holocene, high resolution glacial chronologies and climate, Kenai Mountains, Alaska, *Geological Society of America Bulletin* 106:281-303.

Wiles, G.C., Calkin, P.E., and Post, A., 1995, Glacier fluctuations in the Kenai Fjords, Alaska, U.S.A.: an evaluation of the controls on iceberg-calving glaciers, *Arctic and Alpine Research* 27:234-45.

Wright, H.E., 1984, Late Glacial and late Holocene moraines in the Cerros Cuchpanga, Central Peru, *Quaternary Research* 21:275-85.

Yager, E.M., Barclay, D.J., Calkin,P.E., and Wiles,G.C., 1998, Tree-ring based history of Sheridan Glacier, southern Alaska. *Geological Society of America* Abstracts with Programs: 30 (2).

Xie Zichu, 1992, Progress and prospect for research on mountain glaciers in China, *Annals of Glaciology* 16:107-11.

Yao Tangdon, and Agata, Y., 1993, *Glaciological climate and environment on Qing-Zang Plateau*, Science Press, Beijing.

Zhang De'er, 1994, Evidence for the existence of the MWP in China, *Climatic Change* 26: 289-97

Zhou, S.Z., Chen, F.H., Pan, B.T., Cao, J.X., Li, J.J., and Derbyshire, E., 1991. Environmental change during the Holocene in western China on a millennium timescale, *The Holocene* 1:151-6.

SEA ICE-CLIMATE-GLACIER RELATIONSHIPS IN NORTHERN ICELAND SINCE THE NINETEENTH CENTURY: POSSIBLE ANALOGUES FOR THE HOLOCENE

Maria Wastl[1,2], Johann Stötter[2] and Chris J. Caseldine[3]

[1]Abteilung Physiogeographie, Universität Bremen,
 Postfach 330440, D-28334 Bremen, Germany
[2]Institut für Geographie, Universität Innsbruck,
 Innrain 52, A-6020 Innsbruck,
[3]Department of Geography, School of Geography and
 Archaeology, University of Exeter, Amory Building,
 Rennes Drive, Exeter EX4 4RJ, U.K.

1. INTRODUCTION: APPROACHES TO PALAEOCLIMATIC RECONSTRUCTIONS IN NORTHERN ICELAND

Figure 1 shows the general approach which has been adopted here for palaeoclimatic reconstructions in northern Iceland, and which underlies the investigations presented in this paper. Iceland lies within the atmospheric and oceanic circulation of the North Atlantic. This determines the surrounding climatic and sea-ice conditions which, in turn, act upon indicators of the environmental system. Glacier behaviour (advance, standstill or retreat) is controlled by climate through the accumulation-ablation relation of the mass balance (Meier, 1965) and can be parameterized by the variations of the equilibrium line altitude (ELA). The approximately 250 small corrie and valley glaciers of the Tröllaskagi and Flateyjarskagi peninsulas in northern Iceland (Figure 2) react within 10^0-10^1 years to changes in temperature and precipitation conditions (see Häberle, 1991a; Stötter, 1991a; Caseldine and Stötter, 1993). This makes them very sensitive environmental indicators for climatic variations.

The approach used here encompasses a calibration period covering the time from the mid-nineteenth century, from which time continuous and homogeneous meteorological records in Iceland start (Sigfúsdóttir, 1969; see also Jónsson and Garðarsson, 2001), to the present. This period has shown climatic conditions close to, or even at, both Holocene pessimum and optimum levels. It is suggested here that the nineteenth century component of what has been termed the "Little Ice Age" represents a thermal minimum, whilst the warmest decades of the twentieth century are close to the estimated Holocene thermal maximum. By calibrating sea ice-climate-glacier relationships for this period it is possible to derive a

History and Climate: Memories of the Future?
Edited by Jones et al., Kluwer Academic/Plenum Publishers, 2001

187

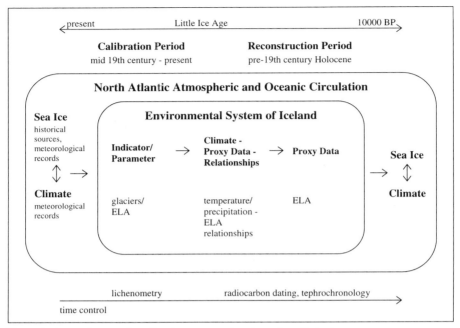

Figure 1. Approaches to palaeoclimatic reconstructions in northern Iceland.

basis for quantitative palaeoclimatic reconstructions in the pre-nineteenth century Holocene (= reconstruction period). The regional sea-ice and climate conditions thus determined for northern Iceland allow further inferences of the conditions of the larger-scale North Atlantic circulation system during the Holocene.

2. SEA ICE-CLIMATE-GLACIER RELATIONSHIPS IN NORTHERN ICELAND FROM THE MID-NINETEENTH CENTURY TO THE PRESENT

2.1 Glacier History

Based on detailed mapping of glacier forefields (*sensu* Kinzl, 1949) and lichenometric dating of moraines, the following phases of glacier advances on Tröllaskagi have been proposed for the calibration period (Kugelmann, 1989, 1990, 1991; Stötter, 1991a,b; see also Häberle, 1991a,b).

- 1840s/1850s
- 1870s/1880s
- 1890s
- 1920s
- 1940s
- 1970s/1980s

The extent of these advances decreases from the second half of the nineteenth century, when the glaciers in northern Iceland reached their "Little Ice Age" maximum, to the 1970s/1980s. Based on the investigation of 48 glaciers on Tröllaskagi (Kugelmann, 1989,

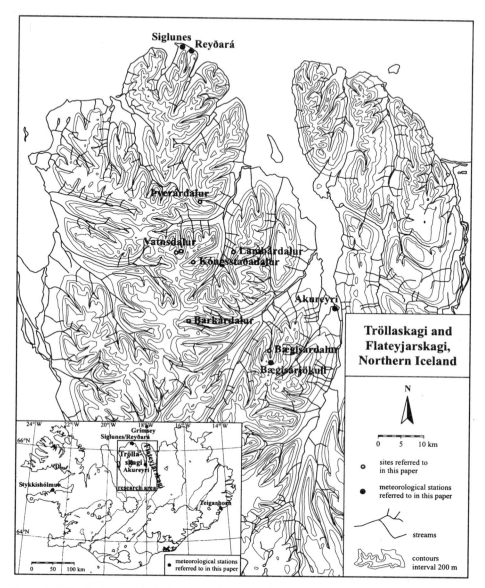

Figure 2. Tröllaskagi and Flateyjarskagi in northern Iceland.

1990, 1991; Stötter, 1991a; Caseldine and Stötter, 1993), the average ELA variation (*sensu* Gross *et al.*, 1976) since the nineteenth century maximum depression is 40 to 50 metres. For a detailed discussion of the method see Häberle, 1991a, 1994; Stötter, 1991a; Caseldine and Stötter, 1993.

2.2 Climate-Glacier Relationships

Simple models describing climate-glacier relationships have been established using temperature and precipitation as climate parameters, and the equilibrium line altitude as a measure for glacier response. The ELA is explained as a function of accumulation and ablation (Meier, 1965), with: (a) winter; or (b) glaciological-year precipitation representing the accumulation term, and summer temperature giving the ablation term (see, e.g., Greuell, 1989).

Temperature records for northern Iceland exist only for stations close to sea-level (Eyþórsson and Sigtryggsson, 1971). Continuous measurements go back to October 1881 (Jónsson and Garðarsson, 2001) at the Akureyri meteorological station (65°41'N, 18°05'W, 23 m a.s.l.; Figure 2), and a virtually continuous record extending back to 1873 exists for the island of Grímsey (66°32'N, 18°01'W, 15 m a.s.l.; Figure 2). Since 1943, the Siglunes meteorological station at the northern edge of the Tröllaskagi peninsula (Figure 2) has provided further temperature readings. Between 1943 and 1968 it was situated at 66°11'N, 18°51'W, 15 m a.s.l., from 1968 to 1981 at nearby Reyðará (66°11'N, 18°47'W, 8 m a.s.l.), and subsequently at Siglunes at 66°11'N, 18°50'W, 8 m a.s.l. Although Akureyri, Siglunes and Grímsey show a typical oceanic pattern in their annual temperature variations (Figure 3), an increasing impact of the Icelandic landmass can be seen at Akureyri, contrasting with the extremely exposed location of the other sites (see Figure 2). This supports the use of Akureyri as providing not only the longest continuous, but also the most representative record for Tröllaskagi.

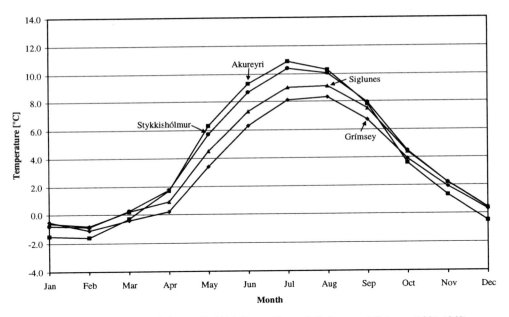

Figure 3. Annual temperature variations at Stykkishólmur, Akureyri, Siglunes and Grímsey (1931-1960).

As the Akureyri temperature record does not cover all of the calibration period it has to be extrapolated further back by use of the Stykkishólmur (65°05'N, 22°44'W, 16 m a.s.l.) temperature data, where continuous recording started in November 1845. Although Stykkishólmur is situated more than 100 km from Akureyri on the Snæfellsnes peninsula of western Iceland (Figure 2), the type of the annual temperature curve (Figure 3) and the monthly averages ± standard deviations of the two stations are similar (Table 1). With milder winters (only four months below zero) and cooler summers, the thermal situation at Stykkishólmur is slightly more controlled by oceanic influences. As changes of the recording technique and of the location of the meteorological station are well known at Stykkishólmur, it has been possible to produce a corrected homogeneous long-term temperature record (Sigfúsdóttir,

1969) which can be used to investigate temperature variations since the mid-nineteenth century. Based on a strongly significant correlation (Table 1; for all records: $p < 0.001$) with the Stykkishólmur temperature record, the Akureyri data set has been extrapolated back until 1846 (Stötter *et al.*, 1999).

Table 1. Comparison and correlation of monthly, annual, summer (May-September) and winter (October-April) temperatures at Akureyri and Stykkishólmur (1882-1992).

1882-1992	Jan	Feb	Mar	Apr	May	Jun	Jul	Aug	Sep	Oct	Nov	Dec	Year	Sum	Win
Akureyri Mean Temperature	-2.1	-1.9	-1.4	1.3	5.4	9.2	10.7	9.7	6.9	2.9	-0.2	-1.6	3.2	8.4	-0.4
Akureyri Standard Deviation	2.6	2.6	2.1	1.9	1.4	1.4	1.5	1.6	1.9	1.9	2.1	1.0	1.0	1.2	0.9
Stykkishólmur Mean Temperature	-1.2	-1.2	-0.8	1.3	4.9	8.2	10.1	9.6	7.2	4.0	1.2	-0.5	3.6	8.0	0.4
Stykkishólmur Standard Deviation	2.1	2.1	2.1	1.7	1.5	0.9	0.9	1.0	1.2	1.5	1.5	1.6	0.8	0.8	1.0
R^2	0.8760	0.8974	0.9259	0.9164	0.9028	0.6788	0.6135	0.7449	0.8670	0.8086	0.8391	0.8774	0.8865	0.8093	0.8837

5-year running means 1882-1992	Year	Sum	Win
Akureyri Mean Temperature	3.2	8.4	-0.4
Akureyri Standard Deviation	0.7	0.6	0.9
Stykkishólmur Mean Temperature	3.6	8.0	0.4
Stykkishólmur Standard Deviation	0.5	0.5	0.7
R^2	0.9035	0.8183	0.8877

The temperature record for northern Iceland since the middle of the nineteenth century (Figure 4) shows a generally low temperature level until around 1920, with minima in the 1860s and 1880s. In the 1920s there is a temperature shift of ca. 2 K in the annual mean, very rapid and pronounced in the winter temperatures and somewhat delayed and reduced in the summer temperature record. It is suggested here that this transition from a low to a following high temperature level marks the termination of the "Little Ice Age" in this area (however, see also Ogilvie and Jónsson, 2001). The late 1920s and early 1930s show the thermal maximum of the whole calibration period. Subsequently, there is a slightly negative trend with minima in the 1960s and late 1970s (see e.g. Ogilvie and Jónsson, 2001).

According to observations by Björnsson (1971), the temperature gradient between Akureyri and Bægisárjökull (65°36'N, 18°23'W, 1080 m a.s.l.; Figure 2) during the summers 1967 and 1968 was, on average, 0.65 K/100 m for prevailing southerly and westerly winds and lower for wind directions from the sea. Accepting a temperature gradient of 0.6 K/100 m, the average ELA variation of 40 to 50 metres since the "Little Ice Age" maximum of 48 glaciers on Tröllaskagi (Kugelmann, 1989, 1990, 1991; Stötter, 1991a; Caseldine and Stötter, 1993) does not agree with the observed change in summer temperature between the thermal minimum and maximum conditions. Thus the Akureyri temperature record (Figure 4) shows a variation of ca. 1.7 K from the minimum in the second half of the nineteenth century (1880s) to the 1940s, the warmest decade within the calibration period during which the formation of moraines indicates that glaciers were in equilibrium (Table 2).

This demonstrates that glacier behaviour in northern Iceland cannot be explained by a pure temperature-glacier relationship. On the basis of mass balance calculations of Norwegian glaciers (e.g. Liestøl, 1967), Sutherland (1984) demonstrates a close relationship between mean summer temperature and winter accumulation at the ELA for glaciers in southern Norway, assuming a state of equilibrium. Following these ideas, Ballantyne (1990) estimates the relationship between winter accumulation and summer temperature at the equilibrium line for glaciers in northern Norway from the following equation:

$$A = 0.915e^{0.339t}$$

where A is the winter accumulation [mm], and t the mean summer (May-September) temperature [°C]. Applying this model to 48 glaciers on Tröllaskagi in northern Iceland, Caseldine and Stötter (1993) show that the measured 1.7 K rise of summer temperature at Akureyri between the 1880s and 1940s corresponds to an average calculated increase in winter accumulation of 60%.

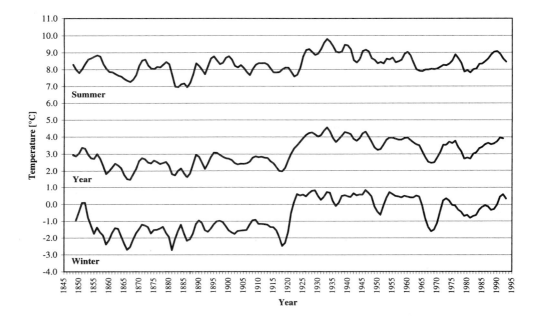

Figure 4. 5-year running means of reconstructed annual, summer (May-September) and winter (October-April) temperatures at Akureyri (1846-1995).

Table 2. Comparison of observed ELA variations of glaciers on Tröllaskagi from "Little Ice Age" maximum extents to the 1940s, and summer temperature change at Akureyri from the 1880s to the 1940s.

Geomorphological Observations	ΔELA 40-50 m	⟹	ΔT$_{Su}$ 0.3 K	
	≠		≠	
	ΔELA 280 m	⟸	ΔT$_{Su}$ 1.7 K	Meteorological Data (1880s - 1940s)

Due to the effects of avalanches and snow blowing from the plateau areas into the corries and valley heads, winter accumulation cannot immediately be equated with winter precipitation. Work by Dahl and Nesje (1992) in the Nordfjord area of southern Norway demonstrates that accumulation by wind must be taken into account in determining the ELA of cirque glaciers. There are no measurements quantifying the contribution of avalanches and wind drift to winter accumulation for glaciers in northern Iceland. By inference from investigations by Tangborn (1980) in the North Cascades, Caseldine and Stötter (1993) estimate that these factors may account for up to 35% of the total winter accumulation. On the other hand, the contribution from sources other than direct precipitation to the winter accumulation of the investigated glaciers on Tröllaskagi can be assumed to show no significant variation over time, as long as there is no change in the dominant wind direction (Caseldine and Stötter, 1993). This tends to be supported by the fairly constant relationship of September to May precipitation at Akureyri and the specific winter balance of Bægisárjökull in the two years of glacier mass balance observations (1966/67 and 1967/68) that are available for northern Iceland (Björnsson, 1971). Thus, while the calculated increase in winter accumulation does not provide absolute figures for the precipitation variation, it shows that there had to have been a major change in winter precipitation accompanying the recorded summer temperature change.

As the Akureyri precipitation record does not go back further than October 1927, there are no precipitation data for northern Iceland covering the "Little Ice Age" part of the calibration period. (The Grímsey meteorological station provides a long-term series since 1896, but according to Trausti Jónsson (1996 pers. comm.) before ca. 1935 these data are not reliable.) The Stykkishólmur meteorological data are therefore used as a basis for investigating precipitation variations from the middle of the nineteenth century. The Akureyri and Stykkishólmur annual variations of precipitation have the same pattern (Figure 5) with an early summer minimum (May) and a maximum in early winter (October). The Stykkishólmur record (Figure 6) shows at least a doubling in both annual and winter precipitation from the nineteenth century minimum (in the 1880s) to the twentieth century maximum (in the 1930s), while the variations of summer precipitation are reduced.

The major shift from the temperature level of the "Little Ice Age" (until about 1920) towards the twentieth century modern level (since the late 1920s) occurs in parallel with a step between two precipitation levels (see Figures 4 and 6). The correlation between temperature and precipitation, based on the 5-year running means from Stykkishólmur, is demonstrated in Figure 7, with both the annual and winter records showing significant positive correlations. This means that higher precipitation is linked to higher temperature, and lower precipitation to lower temperature. In summer this correlation is not so clearly developed. The positive correlation of temperature and precipitation at Stykkishólmur since the middle of the nineteenth century (Figure 7) is in good agreement with the pattern of temperature-precipitation relationships that has been established for northern Iceland on the basis of the observed glacier history in the nineteenth and twentieth centuries.

Based on these results, the climate-glacier relationships in northern Iceland can be summarized as follows (Table 3).

Table 3. Pattern of climate (temperature-precipitation)-glacier relationships in northern Iceland.

| Temperature | Precipitation | | Glacier | |
			Mass Balance	ELA
-	-	$\Rightarrow$	+	-
+	+	$\Rightarrow$	-	+

The direction of the behaviour (mass balance and resulting reaction) of glaciers in northern Iceland is predominantly governed by thermal conditions. Major advances occur when temperature is at a minimum level. The degree of glacier reaction is however strongly reduced by the precipitation variations which run counter to the temperature variations.

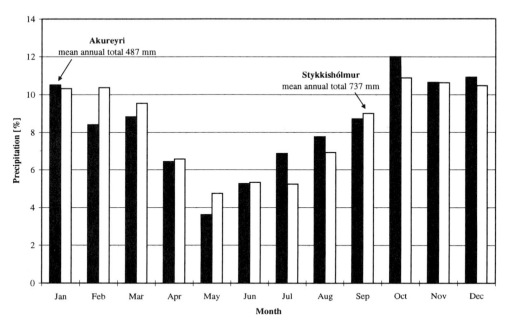

Figure 5. Annual precipitation variations at Akureyri and Stykkishólmur (1928-1992).

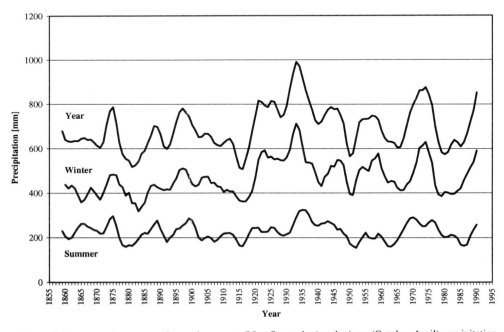

Figure 6. 5-year running means of annual, summer (May-September) and winter (October-April) precipitation at Stykkishólmur (1857-1992).

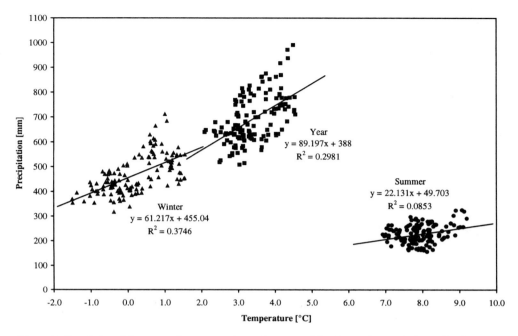

Figure 7. Correlation of 5-year running means of annual, summer (May-September) and winter (October-April) temperature and precipitation at Stykkishólmur (1857-1992).

2.3 Sea Ice-Climate-Glacier Relationships

The island of Iceland is affected by contrasting ocean currents. Warm water is brought primarily to the south and west of Iceland by the North Atlantic Drift (Irminger current) while cold water from the East Greenland and East Iceland currents mainly flow to the north and east of Iceland. Drift ice is brought to the coasts of Iceland by the East Greenland current, and most frequently affects the northern and eastern coasts. Because of the interplay of ocean currents it rarely, if ever, reaches the west coast, but in severe ice years it may, on occasion, reach the south coast (Ogilvie, 1992). Historical sources back to medieval times present abundant information on the occurrence of sea ice around Icelandic coasts (e.g. Thoroddsen, 1916-1917; Koch, 1945; Bergþórsson, 1969; Sigtryggsson, 1972; Ogilvie, 1984, 1991, 1992). For the period from A.D. 1600 these sources are prolific enough to permit the establishment of a reliable record of the sea-ice conditions around Iceland which may be quantified by use of indices (Ogilvie, 1996; Ogilvie and Jónsson, 2001).

Based on the observation that a fairly good correlation existed between sea ice and temperature, Bergþórsson (1969) was the first to use sea-ice information to reconstruct temperature variations. Furthermore, work by Kugelmann (1989, 1990, 1991) and Stötter (1991a) demonstrates that the occurrence of sea ice is clearly reflected in the temperature records from Akureyri, Teigarhorn and Stykkishólmur (see Figure 2). This shows that sea ice has a major control on thermal conditions in a large part of Iceland. Stötter (1991a) has demonstrated a further correlation between sea ice and precipitation for the calibration period.

The comparison of 5-year running means of both winter (October-April) temperature at Akureyri, and winter precipitation at Stykkishólmur with the sea-ice index after Sigtryggsson (Figure 8) shows a clear relationship. The highest incidence of sea ice occurs in parallel with minima in winter temperatures, whilst in times with no or only limited sea ice off the coasts of Iceland, temperature is relatively high, as shown by a significant negative correlation ($R^2 = 0.5500$, $r = -0.7416$, $p < 0.001$ for the period 1848-1970). The correlation of sea ice and winter precipitation ($R^2 = 0.4035$, $r = -0.6352$, $p < 0.001$ for the period 1858-1970) is only of slightly less statistical significance. During periods with the occurrence of sea ice, winter precipitation is low, in contrast to ice-free periods when precipitation levels are higher.

From these observations, the following pattern for the sea ice-climate-glacier relationships can be established for northern Iceland (Table 4). Periods with high occurrence of sea ice are characterized by low temperature and precipitation levels, with positive glacier mass balances leading to advances. Ice-free periods show warm and wet climate conditions, negative glacier mass balances and rising ELAs.

Table 4. Pattern of sea ice-climate (temperature-precipitation)-glacier relationships in northern Iceland.

Sea ice	Temperature	Precipitation		Glacier Mass Balance	ELA
+	-	-	$\Rightarrow$	+	-
-	+	+	$\Rightarrow$	-	+

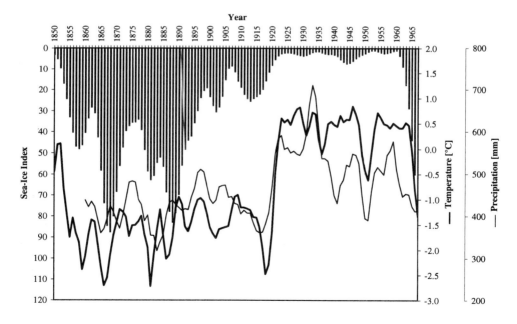

Figure 8. Comparison of 5-year running means of sea ice index after Sigtryggsson, reconstructed winter (October-April) temperature at Akureyri and winter precipitation at Stykkishólmur (1848-1970).

3. INFERENCES FOR HOLOCENE PALAEOCLIMATIC RECONSTRUCTIONS

The finding of the *Saksunarvatn* tephra dated to ca. 9200 radiocarbon years (Wastl, 2000) in sections immediately outside glacier forefields in northern Iceland indicates that, by this time, glaciers had reached positions at, or within, their "Little Ice Age" limits (Stötter *et al.*, 1999; Stötter and Wastl, 1999, in press; Wastl and Stötter, 1999; Wastl, 2000). Figure 9 summarizes the present dating control for post-Preboreal glacier advances

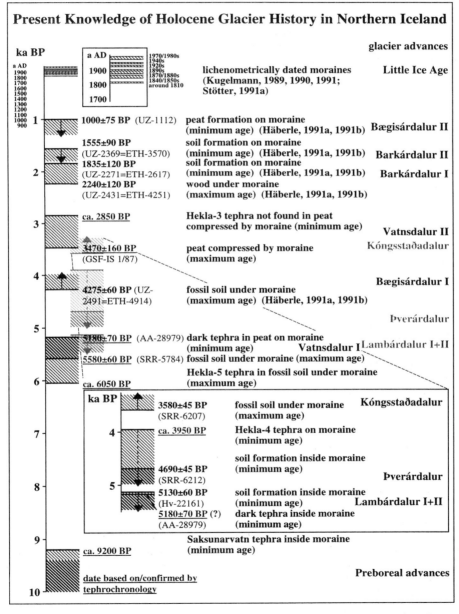

Figure 9. Present knowledge of Holocene glacier history in northern Iceland (Wastl, 2000). BP refers to uncalibrated radiocarbon years. The location of the sites is shown in Figure 2.

in northern Iceland (Wastl, 2000). These were, at most, only in slightly advanced positions relative to those of the "Little Ice Age". This is important for the quality of palaeoclimatic reconstructions based on glacial proxy data for the Holocene. Relationships between maximum glacier extents and climatic parameters in northern Iceland established for the second half of the nineteenth century, describe palaeoclimatic pessima since the early Holocene.

The sea ice-climate-glacier relationships established for northern Iceland (Table 4) allow a link between periods of glacier advances and thermal minima in the Holocene. The nature of the glacial evidence does, however, mean that only the extremes of negative climate variations will be recorded. On the other hand, present evidence does not give the absolute extent of positive temperature variation (see Stötter *et al.*, 1999). For the Holocene thermal maxima temperatures are, therefore, assumed to have been at least as high as during the 1930s (the warmest part of the calibration period). It should also be noted that the available dating control as yet only defines rather long intervals during which thermal conditions showed a minimum (see Figure 9). The temperature record since the middle of the nineteenth century (Figure 4) demonstrates, however, that negative and positive extremes can occur within few decades.

Holocene climatic pessima are comparable to extreme conditions in the second half of the nineteenth century, with an annual average temperature at sea-level of ca. 1.5°C, summer (May-September) temperature of ca. 7°C, and winter (October-April) temperature of ca. -2.5 to -3°C (see Figure 4). In these periods there was a high occurrence of drift ice at the Icelandic coasts. During periods of Holocene thermal optimum conditions annual average temperature at sea-level reached ca. 4.5°C, summer temperature almost 10°C, and winter temperature almost 1°C. Drift ice was absent from the sea around Iceland.

The range of Holocene temperature variations can thus be estimated to have been ca. 3 K for the annual average, slightly less for the summer and more for the winter. There was a doubling in both winter and annual precipitation from the thermal minima to the maxima, whereas summer precipitation showed no significant change.

4.   CONCLUSIONS

As glacier behaviour reflects a reaction to both temperature and precipitation the established sea ice-climate-glacier relationships in northern Iceland, and the palaeoclimatic reconstructions made on this basis, are only valid when the pattern of temperature-precipitation relationships is comparable to the calibration period. As changes in the temperature-precipitation relationships in northern Iceland are caused by shifts of major atmospheric boundaries in the North Atlantic, they are also reflected in the oceanic circulation system. Thus data produced by marine palaeoenvironmental and palaeoclimatic investigations in the North Atlantic is an important source of information to detect possible changes in the climatic preconditions for the reconstructions made on the basis of glacial proxy climate data in northern Iceland.

Acknowledgements

The authors wish to thank the Deutsche Forschungsgemeinschaft (German Research Council), Bundesministerium für Bildung und Forschung (German Federal Ministry of Education and Research) and the British Natural Environment Research Council (NERC) who have financed work in Iceland over several years. This paper has been supported further by the Deutscher Akademischer Austauschdienst (German Academic Exchange Service) through a PhD scholarship to Maria Wastl. We are most grateful to the Carl Kühne KG for its support of fieldwork in northern Iceland. Permission for these investigations was kindly granted by the Icelandic National Research Council. The help of Trausti Jónsson at the Icelandic Meteorological Office who has provided meteorological data and related information is gratefully acknowledged. Astrid E. J. Ogilvie made valuable comments on an earlier version of this paper. We thank Lene Zachariassen and Óskar Gunnarsson at Dæli in Skíðadal for their hospitality.

REFERENCES

Ballantyne, C. B., 1990, The Holocene Glacial History of Lyngshalvöya, Northern Norway, Chronology and Climatic Implications, *Boreas* 19:93-117.

Bergþórsson, P., 1969, An Estimate of Drift Ice and Temperature in Iceland in 1000 Years, *Jökull* 19:94-101.

Björnsson, H., 1971, Bægisárjökull, North Iceland, Results of Glaciological Investigations 1967-68, Part I: Mass Balance and General Meteorology, *Jökull* 21:1-23.

Caseldine, C., and Stötter, J., 1993, 'Little Ice Age' Glaciation of Tröllaskagi Peninsula, Northern Iceland, Climatic Implications for Reconstructed Equilibrium Line Altitudes (ELAs), *The Holocene* 3:357-366.

Dahl, S.O., and Nesje, A., 1992, Paleoclimatic Implications Based on Equilibrium-Line Altitude Depressions of Reconstructed Younger Dryas and Holocene Cirque Glaciers in Inner Nordfjord, Western Norway, *Palaeogeog., Palaeoclimatol., Palaeoecol.* 94:87-97.

Eyþórsson, J., and Sigtryggsson, H., 1971, The Climate and Weather of Iceland, *The Zoology of Iceland* 1:62.

Greuell, W., 1989, *Glaciers and Climates, Energy Balance Studies and Numerical Modelling of the Historical Front Variations of the Hintereisferner*, Broefschrift, Rijksuniversiteit te Utrecht (unpublished).

Gross, G., Kerschner, H., and Patzelt, G., 1976, Methodische Untersuchungen über die Schneegrenze in alpinen Gletschergebieten, *Zeitschrift für Gletscherkunde und Glazialgeologie* 12,2:223-251.

Häberle, T., 1991a, *Spät- und postglaziale Gletschergeschichte des Hörgárdalur-Gebietes, Tröllaskagi, Nordisland*, Dissertation, Universität Zürich, p. 191.

Häberle, T., 1991b, Holocene Glacial History of the Hörgárdalur Area, Tröllaskagi, Northern Iceland, in: *Environmental Change in Iceland: Past and Present*, J. K. Maizels and C. Caseldine, eds., Kluwer, Dordrecht:193-202.

Häberle, T., 1994, Glacial, Late Glacial and Holocene History of the Hörgárdalur Area, Tröllaskagi, Northern Iceland, in: *Environmental Change in Iceland, Münchener Geographische Abhandlungen* B12, J. Stötter and F. Wilhelm, eds.,:133-145.

Jónsson, T., and Garðarsson, H., 2001, Early Instrumental Meteorological Observations in Iceland, *Climatic Change* 48, in press.

Kinzl, H., 1949, Formenkundliche Beobachtungen in Vorfeld der Alpengletscher, *Veröffentlichungen des Museums Ferdinandeum* 26.

Koch, H., 1945, The East Greenland Ice, *Meddelelser om Grønland* 130,3:375.

Kugelmann, O., 1989, *Gletschergeschichtliche Untersuchungen im Svarfaðardalur und Skíðadalur, Tröllaskagi, Nordisland*, Diplomarbeit, Institut für Geographie, Universität München (unpublished).

Kugelmann, O., 1990, Datierung neuzeitlicher Gletschervorstöße im Svarfaðardalur/ Skíðadalur (Nordisland) mit einer neu erstellten Flechtenwachstumskurve, *Münchener Geographische Abhandlungen* B8:36-58.

Kugelmann, O., 1991, Dating Recent Glacier Advances in the Svarfaðardalur-Skíðadalur Area of Northern Iceland by Means of a New Lichen Curve, in: *Environmental Change in Iceland: Past and Present*, J.K. Maizels and C. Caseldine, eds., Kluwer, Dordrecht:203-217.

Liestøl, O., 1967, Storbreen Glacier in Jotunheimen, Norway, *Norsk Polarinstitutt Skrifter* 141:1-63.

Meier, M. F., 1965, Glaciers and Climate, in: *The Quaternary of the United States*, H. E. Jr. Wright and D.G. Frey, eds., Princeton University Press, Princeton:795-805.

Ogilvie, A.E.J., 1984, The Past Climate and Sea-Ice Record from Iceland, Part 1: Data to A.D. 1780, *Climatic Change* 6:131-152.

Ogilvie, A.E.J., 1991, Climatic Changes in Iceland A.D. c. 865 to 1598, in: *The Norse of the North Atlantic* (presented by G. F. Bigelow), *Acta Archaeologica* 61:233-251.

Ogilvie, A.E.J., 1992, Documentary Evidence for Changes in the Climate of Iceland, A.D. 1500 to 1800, in: *Climate since A.D. 1500*, R.S. Bradley and P.D. Jones, eds., Routledge, London, New York:92-117.

Ogilvie, A.E.J., 1996, Sea-Ice Conditions off the Coasts of Iceland A.D. 1601-1850 with Special Reference to Part of the "Maunder Minimum" Period (1675-1715), in: *AmS-Varia* 25, E.S. Pedersen, ed., Museum of Archaeology, Stavanger, Norway:9-12.

Ogilvie, A.E.J., and Jónsson, T., 2001, "Little Ice Age" Research: A Perspective from Iceland, *Climatic Change* 48, in press.

Sigfúsdóttir, A.B., 1969, Hitabreytingar á Íslandi 1846-1968, *Jökull* 19:7-10.

Sigtryggsson, H., 1972, An Outline of Sea Ice Conditions in the Vicinity of Iceland, *Jökull* 22:1-11.

Stötter, J., 1990, Neue Beobachtungen und Überlegungen zur postglazialen Landschaftsgeschichte Islands am Beispiel des Svarfaðar-Skíðadals, *Münchener Geographische Abhandlungen* B8:83-104.

Stötter, J., 1991a, Geomorphologische und landschaftsgeschichtliche Untersuchungen im Svarfaðardalur-Skíðadalur, Tröllaskagi, N-Island, *Münchener Geographische Abhandlungen* B9:166.

Stötter, J., 1991b, New Observations on the Postglacial Glacial History of Tröllaskagi, Northern Iceland, in: *Environmental Change in Iceland: Past and Present*, J.K. Maizels and C. Caseldine, eds., Kluwer, Dordrecht:181-192.

Stötter, J., and Wastl, M., 1999, Landschafts- und Klimageschichte Nordislands im Postglazial, *Geographischer Jahresbericht aus Österreich* 56:49-68.

Stötter, J., and Wastl, M., in press, Palaeoclimatic Investigations in Northern Iceland - Terrestrial Reference for the Variations in the System of Ocean, Atmosphere and Ice Distribution in the North Atlantic during the Holocene, *Paläoklimaforschung*.

Stötter, J., Wastl, M., Caseldine, C., and Häberle, T., 1999, Holocene Palaeoclimatic Reconstruction in Northern Iceland, Approaches and Results, *Quat. Sci. Revs.* 18:457-474.

Sutherland, D.G., 1984, Modern Glacier Characteristics as a Basis for Inferring Former Climates with Particular Reference to the Loch Lomond Stadial, *Quat. Sci. Revs.* 3:291-309.

Thoroddsen, Þ., 1916-1917, *Árferði á Íslandi í þúsund ár*, Hið Íslenzka Fræðafélag, Copenhagen.

Tangborn, W., 1980, Two Models for Estimating Climate-Glacier Relationships in the North Cascades, Washington, U.S.A., *J. Glaciol.* 25:3-21.

Wastl, M., and Stötter, J., 1999, Neue Ergebnisse zur holozänen Landschafts- und Klimageschichte Nordislands, *Norden* 13:181-195.

Wastl, M., 2000, *Reconstruction of Holocene Palaeoclimatic Conditions in Northern Iceland Based on Investigations of Glacier and Vegetation History*, PhD thesis, University of Innsbruck, p.176.

THE IMPACT OF SHORT-TERM CLIMATE CHANGE ON BRITISH AND FRENCH AGRICULTURE AND POPULATION IN THE FIRST HALF OF THE 18TH CENTURY

Axel Michaelowa[1]

[1]Hamburg Institute for International Economics
Neuer Jungfernstieg 21
20347 Hamburg
Germany
e-mail: a-michaelowa@hwwa.de

1. INTRODUCTION

The impacts of climate change on human societies have become a major political issue in the last decade. In this context, the study of the impacts of climate change in the past gains new importance as it can provide insights into societal adaptation processes. Economic history has so far played only a minor role in such analysis. Besides the work of Pfister (1984a, b, 1988, 1990) for Switzerland, Parry (1978) for the English-Scottish border and Lachiver (1991) and Lebrun (1971) for France, only a few scattered studies have appeared (see Galloway, 1994, for a thorough analysis of demographic impacts). This can only partly be explained by lack of data. For England, a wealth of demographic and climatic data exists back into the 17th century. The database for France is much scantier but has improved through recent research (Lachiver, 1991).

The methodological approach used in this chapter is semi-descriptive: it tries to isolate climatic impacts, describe them, and test their significance econometrically where data exist. Conceptually, it builds upon the work of Wigley *et al.* (1985). Climate has a direct effect on biological processes that are crucial for man's survival, such as growth and survival of foodplants and animals as well as pathogens. Moreover there are direct effects on the physical environment, such as flooding and storm impacts. These direct effects lead to second- and higher-order effects in the economic and social fabric. It is the latter which are the most difficult to analyze. Let us just take the different evaluation of impacts of population growth on agricultural yields by the followers of a Malthusian and a Boserupian (Boserup, 1981) approach. The former conclude that yields will rise underproportionally, leading to a vicious spiral while the latter contend that innovation will be spurred by population pressure and thus yields rise.

History and Climate: Memories of the Future?
Edited by Jones *et al.*, Kluwer Academic/Plenum Publishers, 2001

201

Data availability for England allows linkages to be made between climatic, agricultural and demographic data for the period 1700 to 1750. This period was characterised by a significant warming between 1700 and 1740 with subsequent cooling, a strong increase in agricultural production and a population growing quickly from the turn of the century. The results are compared to the French situation in the same period, via a literature review, since lack of demographic and meteorological data does not yet allow a quantitative analysis. For the first quarter of the century, though, French data might soon be sufficient to allow an analysis on the same lines as that for England.

2. DATA AVAILABILITY

Monthly temperature data for Central England exist from 1659 onwards (Manley, 1974). They make up the longest instrumental temperature series in the world. Seasonal rainfall data have been constructed using two precipitation series. The data of Lamb (1977) are used for the period from 1700 to 1726 and those of Craddock and Wales-Smith (1977) from 1726 onwards. The quality of these data is less reliable than the temperature series because they are based on single locations and precipitation data are less spatially coherent than temperature.

A homogeneous French temperature series only starts in the second half of the 18th century. Recently, however, a high-quality daily temperature series has been reconstructed for Paris for the period 1676-1713 (Legrand and LeGoff, 1992). A rainfall series also exists for the same time but it is not very reliable (Lachiver, 1991).

The demographic structure of England in the 18th century is well known (Wrigley and Schofield, 1981) even if there may be some biases because of the sample structure (Brown, 1991). French demography, on the other hand, suffers from a lack of data. Lachiver (1991) presents annual birth, marriage and death estimates until 1720. Dupaquier (1997) only adds an estimation for 1740, but states that data from an enquiry of the National Institute of Demographic Research might change the picture in the future.

Agricultural data are most hard to come by. There are no statistics on agricultural production either for 18th century Britain or France. Even the proportions of the different cereals cannot be estimated for the latter (Meuvret, 1977). Thus, food supply can only be estimated via indirect indicators such as agrarian prices where well-researched series exist (Thirsk, 1990, for England and Labrousse et al., 1971, for France). This is a major problem which has led to a number of studies trying to estimate agricultural production in Britain, and a major controversy on the timing and extent of production increases (for a discussion see Campbell and Overton, 1998).

A cautious summary of the debate suggests that production started to grow slowly from the 17th century. Growth accelerated around 1750 and fell back subsequently. In the second half of the 18th century yields rose decisively above historic levels. Grain prices show a marked decline until 1750 while livestock prices remained constant. Habbakuk (1987, p. 284) states that the low prices were due to productivity gains "either by human labour or a benign nature". Rising prices and a contemporary debate on productivity are indicators of the productivity improvement (Jackson, 1985 and Habbakuk, 1987). Grain exports from England rose almost tenfold until 1750 with the exception of the crises of 1709/10, 1727-30 and 1739-41 reaching around 13% of domestic food consumption (Bairoch, 1976). But already by 1766 imports had already become dominant and reached almost 5% of domestic consumption in the last quarter of the century (O'Brien, 1977).

An analysis using empirical values for income and price elasticities of demand leads to an annual production growth rate of 0.6% for 1700 to 1760 and a rate of 0.1% for 1760 to 1780 (Crafts, 1985). There are only a few yield values, which tend to show the greatest rises between 1700 and 1760 (Overton, 1979, Turner, 1982). It is unclear whether the share of cultivated land grew. Livestock numbers grew throughout the 18th century and the average weight of cattle rose (Deane and Cole, 1962).

In France, there is less current debate on the timing and extent of agricultural production increases while there was a heated discussion some decades ago (Kaplan, 1976). LeRoy Ladurie (1960, p. 4) stated bluntly that "there was no agricultural revolution" in the 18th century. He nevertheless concedes that production rose strongly from 1720 onwards and after 1750 passed a threshold that had not been reached in the preceding centuries. Morineau (1971) provocatively titled his study "The apparent signs of an economic take-off" stating that "there was neither a demographic nor an agricultural revolution" (p.334).

3. THE ROLE OF FOOD AVAILABILITY FOR DEMOGRAPHICS

The development of birth and death rates in preindustrial societies is often linked to domestic food availability which, in turn is influenced by climate. Among additional influences the virulence of epidemic diseases and social structure are important.

Malthus (1798) postulated a "preventive check", i.e. a reduction of the birth rate, to prevent a "positive check" through starvation and epidemic disease. Boserup (1981) comes to the opposite conclusion – population growth sustains itself by promoting innovation. Evidence suggests that a Malthusian situation (i.e. the preventive check) was prevalent in England in the 18th century (Galloway, 1994), where crisis mortality fell while in Scotland the positive check prevailed (Houston, 1986). Towards the end of the century a more Boserupian situation developed. In France, in contrast, the positive check prevailed until late in the 18th century (Appleby, 1979) which was punctuated by recurring mortality crises. But the intensity of these crises declined in France as well, and many observers state that the last real famine was in 1709/10 (Lebrun, 1971, 1980).

In Britain, a positive correlation between agrarian prices and population can be found over long periods showing that population growth led to reductions in food availability. The correlation weakens between 1700 and 1740 when agrarian prices fell while population grew slowly. Per capita food consumption rose. From 1740, the correlation once again holds: agrarian imports rose and per capita consumption fell (O'Brien, 1985).

No clear link exists between the nutritional status of a population and its direct susceptibility to epidemic diseases (Appleby, 1975). Only temporary migration to the cities and across the country enhanced the spreading of infectious diseases (Post, 1985). In England, there exists no correlation of the death rate with the real wage rate (Schofield, 1983), whereas nuptiality and fertility are strongly correlated, but lagging (Olney, 1983). Obviously, feedbacks have to be taken into account. For example, changes in the birth rate led to changes in the death rate as infant mortality was very high. Nuptiality could be positively influenced by a high death rate as more farms and jobs became vacant. Nevertheless, in England nuptiality and birth rates show a negative correlation with wheat prices with a lag of one or two years (Lee, 1981, p. 375). The proportion of married women rose 15% between the 1670's and the 1730's while the average age at marriage fell. The rapid growth of London led to a rise in average mortality in England as mortality in London remained high throughout the 18th century (Wrigley and Schofield, 1981). There is no evidence for a link between standard of living and mortality (Wrigley, 1982). Due to the relatively well-developed infrastructure, food distribution was not a problem in 18th century England. Growing consumption of imported foodstuffs is an indicator of a rising food availability. It seems that from 1700 to 1750 the average consumption of farmers and labourers rose and was higher than in the early 19th century (Grigg, 1980). Meat was eaten regularly even in middle and low income households by 1740. Beer and alcohol consumption in general rose and became a major social issue (Drummond and Wilbraham, 1957). In the second part of the century meat intake fell. Anthropometric studies such as Komlos (1993a) show that average heights of boys fell during this period. He concludes that

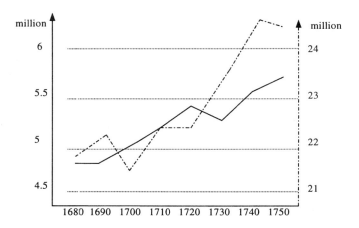

Figure 1: Population development in England and France.

dashed line: France (right-hand scale)

Source: Lachiver (1991), Grenier *et al.* (1988), Wrigley and Schofield (1981)

average food intake fell by 40% between 1760 and 1790 (Komlos, 1993b). His findings are disputed by Floud, Wachter and Gregory (1993) as being statistically biased. They see a continuous rise in adult height between 1740 and 1800 (Floud *et al.*, 1990).

In France, after the catastrophic decade of the 1690s, which had led to a major famine in 1692-94, food availability was high from 1700 until 1738 with the exception of the 1709-10 crisis. No war ravaged French soil; Louis XIV pursued his wars outside the French borders. Nevertheless, population did not grow throughout the period. The first two decades were punctuated by a series of epidemics at least partly due to hot, dry summers. 80% of the gain of 1.1 million people between 1700 and 1709 was lost through the 1709-10 famine, and 50% of the ensuing gain was wiped out by the dysentery epidemics in 1719 (Lachiver, 1991, p. 480). In early 1720 a plague outbreak in Provence led to 120,000 deaths (Lebrun, 1980).

Population (Figure 1) started growing quickly from 1720, gaining about 2.2 million people by 1740 (10%) (Dupaquier, 1997) while it had hovered around a level of 22 million between 1680 and 1720. The crisis in the first part of the 1740s disrupted this trend and by 1750 the 1740 level was barely held.

4.   CLIMATE CHANGE AND ITS EFFECTS ON AN AGRARIAN ECONOMY

Agrarian systems are directly dependent on climate. Short term climate shocks and long-term climate change lead to changes in agricultural production. If variability remains constant, a long-term change leads to a higher probability of extreme events, which will lead to crop losses. A rise in variability will always have a negative impact on agriculture. Temperature and precipitation are key variables for temperate agriculture. Accumulated warmth is a good indicator of production level (Parry and Carter, 1985). Not all yields, though, are positively correlated with temperature. The links can be complex and non-linear as threshold effects occur (Monteith and Scott, 1982, Wigley *et al.*, 1985). It is crucial to use monthly (Pfister, 1988) or at least seasonal climate data as the same average annual temperature or precipitation can mask completely different seasonal curves – one being favourable and another being less favourable. Higher temperatures in spring and autumn allow early sowing and reduce the risk of harvest losses through early frost (Parry, 1990).

There are huge differences between crops. While modern grain yields are negatively correlated with summer temperature, the slowly maturing varieties of the 18th century probably had higher yields at higher summer temperatures. Yields of potatoes, carrots and grass show a clear positive correlation with summer and winter temperature.

Drought effects are particularly severe in spring and early summer if germination is impeded. On the other hand, a moisture surplus in spring, summer and autumn disturbs sowing and harvesting and favours grain pests while clover, carrots and grass profit (Bourke, 1984). Generally, moisture surpluses had higher negative impacts in the English and French context than droughts (Le Roy Ladurie, 1983, vol. 2, Lebrun, 1971), especially if they occurred in spring and summer. "Excessive humidity was the great enemy and dry weather always welcome" (Desplat, 1987, p. 35). A cluster of dry or wet years led to cumulative effects (Pfister, 1984). Soils also play a role concerning climatic impacts. Heavy clay soils were very vulnerable to moisture surpluses while light limestone soils were not affected.

Pfister (1988) finds a clear link between both short-term and long-term climate change and cereal prices in Switzerland despite the lack of temperature and precipitation data before the 18th century. Calculations for England (see Appendix) confirm this result.

4.1 Climate and agricultural production in England and France

Climate historians (Wanner *et al.*, 1994 and 1995) have managed to shed light on the climatic conditions that prevailed in northwestern Europe in the late 17th and early 18th century. This period is known as the "Late Maunder Minimum" with low solar activity. The end of the 17th century had extremely cold and dry winter and springs as well as very wet and cool summers, especially the 1690s. In England, average decadal temperatures were a full 1.5° C lower than today, particularly during the winters (Manley, 1974). The snow on the Scottish peaks (altitude 1300 m) did not melt during the summers (Parry, 1978). French temperatures were also more than 1°C below today's (Lachiver, 1991).

From 1700 to 1740 average decadal temperature in England rose continuously by almost 2 degrees (see Figure 2) and reached the second highest value of the whole English series in the decade of 1730-1739. Only the decade of the 1990s has exceeded the warmth of this decade. Only two winters can be described as severely cold. Moreover, precipitation was below today's average by around 5 % (Brooks, 1928 and Craddock/Wales-Smith, 1977). High temperatures, especially in summer and autumn and low precipitation led to good grain harvests in all years except 1708-1709, 1711, 1713, 1715, 1725-1729 and 1735

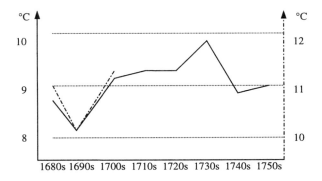

Figure 2: Decadal temperature averages in central England and Paris

dashed line: Paris (right-hand scale)

Source: Manley (1974), Lachiver (1991)

(Ashton, 1955a); the lack of freezing winters favoured fruit growing. Warm autumns lengthened the growing season.

This favourable situation ended abruptly when the year 1740 became the coldest of the entire Manley (1974) series (more than 2.5° C below today's average) and 1739/40 being an exceptionally severe winter. The decade 1740 to 1749 was 1.1° colder than its predecessor. Precipitation was also more than 15% below average (Lamb, 1965). Subsequently, livestock production fell sharply, while grain harvests remained above average. In the following two decades temperature rose by some tenths of a degree while precipitation, especially in summer, rose above average. This led to a sharp decline in grain harvests on clay soils (Ashton, 1955b).

In France, decadal temperature averages rose by 1.1°C from the famine decade of the 1690s to the 1700s. If one extrapolates the temperature data from 1713 (which is the last year of the Paris series) by using the mean of the Central England (Manley, 1974) and DeBilt series (Engelen and Nellesteijn, 1996) the temperature rise almost reaches 2°C by the 1730s and its fall from the peak in the 1740s amounts to 1°C. Precipitation declined by 7% in the 1700s and 13% in the 1710s compared to the 1690s (Lachiver, 1991). Unfortunately, no overall estimate of harvests exists despite the collection of tithe records by LeRoy Ladurie (1960)[1]. The fact that wheat prices fell from 1701 to 1730 with the exception of the two crises 1709-10 and 1712 (Meuvret, 1988) indicates that harvests must have been quite good. Lebrun (1971, p. 138) sees the whole period from 1726 to 1767 as a "meteorological success" for the whole South of France (with the exception of 1740 and 1741).

4.2   Increased Use of Marginal Land in the Early 18th Century in England

In the English lowlands, precipitation is more important than temperature for agricultural production while temperature is decisive for hilly regions as the climate is maritime. The tree line lies between 300 and 600 m above sea level while average annual temperature in the lowlands is 9.5°C (Manley, 1952). Grain yields at 350 m are 50% lower than at sea level. Thus, marginal hill land is very susceptible to climate change and changes can lead to an expansion or contraction of agricultural land. Parry (1978) studied the extension of agricultural land on a chain of hills near the southern Scottish border. He found an exponential correlation between temperature and crop failure. A temperature change of 1°C led to a change in the agricultural frontier of 140 m (Parry and Carter, 1985). Of course, the reaction of farmers involved a delayed response. Thus, comparing the decades 1690-1700 and 1730-1740, the agricultural frontier rose by 250 m. If one takes 30 year-intervals (1670-1700 compared to 1710-1740) the rise was 130 m.

The theoretical expansion of agricultural land can be underlined by historical examples. From the end of the 17th century, the number of new farms in the Lake District rose sharply (Manley, 1952). Cumbrian farmers doubled the number of livestock between 1660 and 1750, while there was no rise in livestock numbers in the lowlands. Concurrently, the wealth of the former rose more quickly than that of the latter. It peaked in the decade 1730-1739 (Marshall, 1980).

4.3   The Impact of Climatic Extremes in England and France

Besides decadal climate change, climatic extremes can have impacts on agriculture and population. An isolated climatic extreme does not lead to long-term economic damage if stocks exist. If extremes occur in clusters (Flohn, 1981) or high variability entails extremes of a different character, negative feedbacks occur such as the use of seeds for immediate food purposes or malnutrition that leads to a lower disease resistance.

---

[1] Direct wheat harvest statistics exist from Arles that indicate that the period between 1725 and 1740 was very favourable (Baehrel, 1961, p. 634).

The vulnerability to climatic extremes was markedly different in England and France. English agriculture was quite diversified, whereas northern France was mainly a wheat monoculture. Spring crops were neglected - barley and oats were not grown in large quantities. Cattle only played a role in marginal areas. This structure was not resilient to climatic shocks. In contrast to England (see below), prices of the different French cereals were strongly correlated leading to lower food availability for poor people in case of harvest failure (Appleby, 1979). In the following, two cases of 18th century climatic extremes will be discussed. The first is a single, but very marked extreme while the latter is a cluster of extremes. It is interesting to note the differences in impacts due to specific details:

The "Great Winter" of 1709. The first big climatic shock of the 18th century was the winter of 1708/09. Its impacts in France are described in great detail by Lachiver (1991, p. 268ff). Cold air arrived from the East on January 5, leading to a sudden drop in temperature of 15°C. In Paris, for the following two weeks minima reached about –20°C and daily maxima often stayed below –10°C, an extreme that has never since been repeated. The extreme cold touched the whole of France so that even shallow parts of the Mediterranean froze (Montpellier had a minimum of –18°C and 6 days below –10°C). People froze to death inside their homes, economic life was paralyzed and huge losses occurred due to death of domestic animals and bursting wine barrels. Most fruit trees were lost and the loss of olive and chestnut trees, in particular, deprived the poorer population of an important source of food and revenue for a long time; in the former case for decades. The same fate befell the vineyards. Wine harvests fell to less than 5% of normal values around Paris in 1709, reached 25% in 1710 and recovered only in 1711.

The real impact, though, did not come from this cold as deep snow cover protected the grain but from a second cold period in late February when temperature dropped to –14°C in Paris. Between the two cold periods, the snow cover had melted and in the plains the grain was lost completely. Paradoxically, regions above 500 m altitude did not suffer this loss as the snow cover there lasted for the full winter. When farmers in the plains realized the loss, they scrambled for alternatives. Everything that could grow was sown, even winter wheat that could not ripen. A royal edict issued in April stated that barley should be sown. Barley prices that usually were half those of wheat prices thus surpassed the latter level - multiplying by 8 compared to 1708. Due to favourable weather and the absence of weeds the barley developed into a bumper crop, three to four times the ordinary yield. Without this "barley miracle" the following famine would have been much worse.

The demographic impacts of the Great Winter were twofold. First, the severe cold killed directly or through respiratory diseases; Lachiver estimates these deaths at about 100,000. The longer-term impact was through the grain price rises between April and August 1709 that reduced food availability to the poor. Before the bumper barley crop came in, prices had reflected expectations of a complete harvest loss and thus the poor were completely priced out of the market. In the whole of France mortality rose and births fell by a third leading to a population loss of 1.2 million people, but this masks important regional variations. Regions with a diversified agriculture such as Brittany suffered much less. Much depended on the reaction of political authorities which, in some instances, was much better coordinated than in the famine years 1692-94 where mortality had been about double that of 1709 (Lebrun, 1971, Lachiver, 1991).

In England the winter did also bring severe cold but there was no catastrophic harvest loss as the snow cover lasted.

The crisis 1738 – 1742. At the end of the benign period a clustering of extremes occurred between 1738 and 1742. It had strong impacts both in Britain and France (see Table 1).

Table 1. Temperature (°C) and mortality (% of average 1735-1744) in England and Scotland and France 1735-1744.

	1735	1736	1737	1738	1739	1740	1741	1742	1743	1744
Temperature England	9.8	10.3	9.9	9.8	9.2	6.8	9.2	8.4	9.8	8.8
Mortality England	89	94	102	92	93	106	118	124	98	85
Mortality Scotland	88	96	109	88	93	122	116	107	90	91
Mortality France	80	93	93	96	109	123	116	111	94	83

Sources: Manley (1952, p. 394); Post (1985, p. 32)

The harvest of 1739 was damaged by a surplus of precipitation and autumn storms (Post, 1985). November 1739 as well as January and February of 1740 were extremely cold and strong easterly winds led to the freezing of many plants and trees. The coldest May of 300 years was a further blow. The grain harvest was only saved by the right amount of summer precipitation and a warm, dry September. Nevertheless, mortality doubled and tripled in the southwest of England (Post, 1985). Livestock prices rose sharply as fodder became scarce. In eastern England riots broke out and attempts were made to prevent grain exports (Charlesworth, 1983). As no freezing winter had occurred in England since 1708/9, the farmers were not adapted to the extreme situation the winter of 1739/40 posed. The anomalies prevailed until 1741 and mortality remained high (Wrigley and Schofield, 1981). This was not only due to food shortages but also to the confinement in unhealthy rooms during the cold winters. Droughts in spring and summer enhanced the prevalence of digestive illnesses (Post, 1985)

The year 1740 had a smaller temperature anomaly in France than in England. In Paris the departure from the mean was only -1.2°C compared to -2.5°C in England. While the winter anomaly was even sharper than in England (January – March -6°C compared to -4.6°C), the summer and rest of the year were also cooler than the 1961-1990 average (v. Rudloff, 1967). Nevertheless high precipitation in summer led to a crop failure (Post, 1985): the 1740's harvest brought only 30-50% of the average (Kaplan, 1982). Two preceding bad wheat harvests had already set the scene for a subsistence crisis. Due to a coordinated political response that entailed measures such as redistributionary emergency taxation (Appleby, 1979) a famine could be averted in contrast to the winter of 1709 described above. Still a mortality crisis similar to the English one followed. That was exacerbated by an epidemic of respiratory disease. Lachiver (1969) finds a mortality of 44% above average in 1738, 150% above in 1740 and 168% in 1743 for the village of Meulan. Lebrun (1971) decribes a twin-peaked subsistence crisis in 1739 and 1741/42 for Anjou, but states that it no longer led to a sudden mortality and birth crisis.

5.   STABILISING FACTORS IN ENGLAND

A number of developments that are commonly known as the "Agricultural Revolution" dampened the impact of climatic variability in the course of the 18th century, especially in its second half. French agriculture did not witness this progress and thus remained much more vulnerable to climatic shocks as the 1780s show (LeRoy Ladurie, 1983).

5.1 Agricultural Diversification

As different types of grain react in different ways to meteorological influence (Bourke, 1984, Pfister, 1984 and Lamb, 1982), growing diversification led to lower impacts of climatic extremes. As such a diversification took place in England with the growing cultivation of barley and oats from the 17th century, there were no famines compared to France, where wheat was dominant (Appleby, 1979). When autumn wheat was lost in the freezing winter of 1739/40, oats and barley were sown in spring and thus the harvest was saved (Post, 1985). In France, the only diversification was the use of maize in Southern regions. The potato was not yet accepted (Meuvret, 1977).

5.2 Agrarian Innovation

Farmers' calendars show that farming technology did not change much until the mid-18th century (Fussell, 1933), despite the availability of new knowledge since the late 17th century (Coleman, 1977). Nevertheless, innovation started to grow as investment in innovation could achieve high yields (Habbakuk, 1987). Only the period of rising agrarian prices from 1750 onwards led to a strong diffusion of innovation - it became fashionable to discuss agricultural technology and several "innovators" who just popularised older ideas became famous.

Fallowing had been reduced from the 17th century onwards through the introduction of nitrogen-fixing plants. This was accelerated by the period of low grain prices in the 1730s (Clark, 1992). Additionally, the combination of livestock and grain production allowed the use of manure as fertilizer more systematically than had been possible before. Thus, light limestone soils could be utilized for agriculture while heavy clay soils were used for livestock rearing. Irrigation of meadows led to higher availability of fodder (Jones, 1981). Carrots and potatoes were only used on a large scale in the 19th century.

Because many agricultural estates were not saleable, owners tried to achieve high income from renting their estates. This could only be achieved through renting to progressive tenants (Newby, 1987). Tenure was long-term and tenants had an incentive to introduce innovation. Sometimes, it was even stipulated in the tenure contract (Chambers and Mingay, 1966). Enclosures that are commonly seen as reasons for increases in yields were not relevant until 1750, while by 1800 84% of agrarian land had been enclosed (Wordie, 1983). Moreover, open-field agriculture was not generally less productive than enclosures (Turner, 1986, Havinden, 1961). Around 1730, the introduction of the triangular plough led to a doubling of productivity (Pierenkemper, 1989). Moreover, wooden parts were substituted by iron ones. The comprehensive use of these innovations was only achieved in the 19th century, as they were expensive at the start (Chambers and Mingay, 1966). Livestock productivity rose, particularly in the second half of the 18th century. Between 1732 and 1795 the average weight of livestock at the London market rose by 25% (Deane and Cole, 1962). In contrast, in France agricultural innovation only started after 1750. Before that date, almost no publications on agricultural matters existed (Meuvret, 1977).

5.3 Improvement of Infrastructure and Foreign Trade

In the 18th century, roads and canals were constructed on a large scale. Thus, local dearths vanished by 1720 and regional price variances fell (Schofield, 1983, Thirsk, 1990). Moreover, the incentive to change from subsistence to market-oriented production rose. Regional specialization developed (Kussmaul, 1990). Foreign trade in grains was able to lower price variability as supply, was lowered at times of bumper crops and enhanced times of crop failures. In 1740, the export of grain was suspended until late 1741 (Post, 1985).

The situation in France represents a complete contrast (Kaplan, 1982 and 1984). Transport infrastructure was non-existent and land transport extremely costly. There was no transparency of prices and available quantities. Grain trade was burdened by heavy political interference, especially in times of harvest crisis. Every region tried to look after itself in case of harvest failures. A chaos of differing weights and measures obstructed exchange. Imports were only possible for big cities using navigable rivers and the coastal areas.

6. RELATION BETWEEN CLIMATE CHANGE, AGRARIAN PRODUCTION AND POPULATION IN ENGLAND FROM 1700 TO 1750

Interactions between climatic, agrarian and population variables are analyzed using an econometric model (for data and numerical results see Appendix). The study uses seasonal data, based on the temperature series from Manley (1974), a constructed precipitation series (Craddock and Wales-Smith, 1977 from 1726, Lamb, 1977 from 1700 to 1726) and the population series from Wrigley and Schofield (1981). Agrarian prices are taken from Thirsk (1990). Export data come from Ashton (1955a). Real wages are taken from Wrigley and Schofield (1981). Regressions cover the period 1700 to 1749. An extension of the model using monthly data might be a promising further step. Hypotheses derived from the preceding discussion are tested using model results.

6.1 Impact of Climate on Agrarian Production

Using the literature, the following climatic influences on pre-modern agriculture can be assumed: a warm spring and summer led to high grain and livestock production while cold springs and high precipitation reduced production. Grain prices do not move in the same direction at the same time.

The quantitative analysis shows that wheat prices are indeed negatively correlated with summer and winter temperature and positively with summer precipitation. There is a one-year lagged negative correlation with autumn precipitation. As expected, hay prices are negatively correlated with precipitation. Cattle prices are negatively correlated with autumn and winter temperatures. Prices of rye, barley and oats surprisingly do not show significant correlations with climatic variables.

It is important to notice a strong positive correlation between wheat and rye prices (as suggested by Appleby, 1979) as well as between prices for barley and oats. Prices of hay, sheep and cattle are also highly correlated. Stock played an important role concerning wheat and barley prices as there is a positive correlation with prices of the previous year. The negative correlation of livestock and grain prices is somewhat unexpected. It could be explained by substitution of grain by meat in times of rising real incomes.

6.2 Links between climate and population

The literature suggests that climate can influence population through short-term shocks and through long-term changes.

Extreme heat and cold favours diseases, the former digestive ones, the latter respiratory ones (Galloway, 1985). London mortality peaks which occurred in summer in the 17th century were found in September and winter from 1725 to 1749 (Landers and Mouzas, 1988). Cold winters led to lagged deaths in the higher age categories in the long term while hot summers are positively correlated with mortality in medium age groups (Galloway, 1985).

Direct relations between temperature, precipitation and population variables have been tested by Lee (1981), although using an outdated precipitation series. He concludes that extreme temperatures in summer and winter raise mortality while this rise lagged by some months in summer, but was immediate in winter. This is due to the different type of diseases, respirative in winter and digestive in summer. No correlation with precipitation was found. A one degree rise in winter temperature and fall in summer temperature would have led to a rise of the population growth rate of 0.2%[2]. The regressions discussed below tend to underline Lee's values.

As expected, the calculations show that the death rate is negatively correlated with winter temperatures and positively with summer temperatures. Even the summer temperature of the previous year is significant! Precipitation in autumn and summer is negatively correlated with the death rate. This could be explained through a lower incidence of digestive diseases when propagation of bacteria was hampered through the absence of dust. This causal link can also be found in France where the dry summers 1705 to 1707 and 1719 (Lachiver, 1991) witnessed several epidemics, especially dysentery, that led to a high mortality (Lebrun, 1971, 1980, Lachiver, 1991) despite good harvests. In France, epidemics created by price hikes in certain regions also overwhelmed populations in areas where food availability was sufficient, such as port cities that could easily import wheat (Pousson, 1980). English mortality is strongly positively correlated with grain prices of the preceding year. The birth rate is less influenced by climatic variables. Autumn temperature and summer precipitation of the preceding year are negatively correlated with it. Correlation with winter temperature is positive. It is obviously positively correlated with the marriage rate of the preceding year and real wages of the last years. There is a negative correlation with marriage rates two years lagged and the death rate of the preceding year. Positive correlations exist between birth rates in winter and winter temperatures of the preceding year, while births in summer and autumn are negatively correlated with summer temperature of the preceding year. Marriage rates do not show any significant correlations. A significant positive correlation between real wage and annual temperature exists.

Galloway (1994) tries to derive quantitative results for the link between temperature and demographic variables in France in a similar way as Lee did for England. Galloway's temperature data, though, are derived from the Manley Central England series (he does not use the Legrand and LeGoff set) and must thus be treated with caution. His demographic data stem from unpublished data of the French enquiry. He concludes that births and marriages are negatively correlated with grain prices. In contrast to England, mortality remains positively correlated with grain prices for the whole 18th century. He finds a positive correlation between summer temperature and mortality and a negative one for winter temperature but no clear correlation between births and temperature.

7. CONCLUSION

Climate change is an additional explanatory variable for the course of English history in the 18th century. It was relevant in marginal agricultural areas. External shocks such as war, famine and epidemic diseases did not play a major role allowing the impact of climate change to be clearly distinguished (see Figure 3).

Farmers reacted to climate change, albeit with lags. Nevertheless, the effect of climate change cannot be quantified as there are no statistics of agrarian production. Demographic developments are induced by a large set of factors where climatic influences are indirect at best. The results of the descriptive and econometric analysis suggest that the favourable climate from 1700 to 1740 contributed substantially to the rise of agricultural production. The ensuing decline of prices led to efficiency improvements, additional investment and the acceleration of the spread of agricultural innovation. Because of stable meat prices the incentive to introduce a combination of livestock rearing and grain production grew. The

---

[2] Lee did not use a consistent methodology, though. He did not consider R^2 and used insignificant coefficients for his explanations. Moreover, his use of moving averages hides medium- and long-term effects.

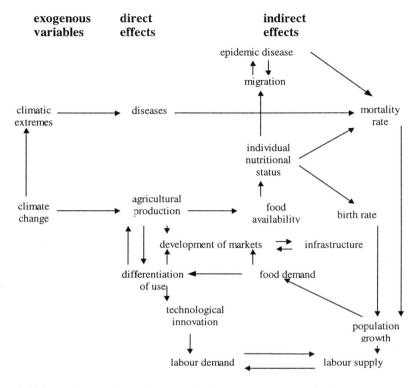

Figure 3. Linkages between climate change, agriculture, economy and population.

food intake of the lower and middle classes also grew. Because of more frequent epidemic diseases in warm and dry summers mortality remained high.

From 1740 onward the climatic trend was reversed. Yields fell when temperature fell and precipitation rose. Population started to grow quickly. This could be a lagged reaction to high real incomes of the preceding period which allowed earlier marriages for more people. As supply fell, demand for agrarian products rose. This led to a rise in prices and growing commercial interest in agriculture. Privatization of commons and elimination of fallowing were the consequences, while innovation became widespread. The availability of fertilizer was raised through the agricultural revolution, fallowing receded and new land could be taken into production. Specialization occurred. When the climate became colder and wetter, it was possible to shift grain production from heavy clay soils to limestone soils. As agrarian productivity rose, yields could be stabilized and rose even as the climate remained unfavourable. As the rate of yield rise remained below population growth food intake of lower classes was reduced.

England profited from the warm, dry climate of the early 18th century in the transition from pre-modern to modern agriculture. Climate change enhanced food availability in the early 18th century while it reduced it in the second half. Only the effects of the Industrial Revolution which allowed large-scale imports of foodstuffs finally reduced impacts of climate change.

In France, the impact of favourable climate change is harder to grasp. France suffered strongly from the Great Winter of 1709 due to the loss of the wheat crop. Moreover, dry summers such as 1706, 1707 and 1719 allowed epidemic diseases to spread. Nevertheless the period from 1700 to 1738 generally allowed good harvests. Wheat prices were low and population rose strongly from 1720 onwards. The climatic trend reversed in 1738 and a multi-peaked subsistence crisis followed. Generally, the impact of short-term extremes was higher than in Britain as there was no take-off of agriculture.

OLS-Regression on wheat prices

Variable	TSUM	TWIN	RSUM	RAUT (-1)	PWHE (-1)	PWHE (-2)	PRYE	PLIV	PLIV(-1)
Coefficient	-3.07	-2.15	0.047	0.028	0.34	-0.11	0.73	-0.42	-0.40
Std. error	1.69	1.09	0.025	0.015	0.1701	0.056	0.093	0.22	0.22
T-ratio	-1.81*	-1.97*	1.82*	1.81*	2.01**	-1.87*	7.81***	-1.89*	-1.83*

R^2= 0.94, D.W. (Durban-Watson statistic) = 2.13

OLS-Regression on rye prices

Only PWHE significant at 1% (T-ratio 6.87); R^2= 0.89

OLS-Regression on barley prices

Variable	PRYE	POAT	PBAR (-1)
Coefficient	0.21	0.59	0.31
Std. error	0.12	0.093	0.063
T-ratio	1.71*	6.38***	4.86***

R^2= 0.90

OLS-Regression on oat prices

Only PBAR significant at 1% (T-ratio 5.42) and REALWAGE at 5% (T-ratio 2.47); R^2=0.80

OLS-Regression on hay prices

Variable	RSPR	RAUT (-1)	PCAT (-1)	PSHE	PSHE(-1)
Coefficient	-0.20	-0.063	0.43	0.47	0.43
Std. error	0.069	0.037	0.23	0.199	0.17
T-ratio	-2.80***	-1.71*	-1.84*	2.36**	2.54**

R^2= 0.45, D.W. = 1.48

OLS-Regression on sheep prices

Only PHAY significant at 1% (T-ratio 2.75) and PCAT at 5% (T-ratio -2.48); R^2= 0.27, D.W. = 2.31

OLS-Regression on cattle prices

Variable	TWIN	RAUT (-1)	PCAT (-1)	PSHE	PSHE(-1)
Coefficient	-0.20	-0.063	0.43	0.47	0.43
Std. error	0.069	0.037	0.23	0.199	0.17
T-ratio	-2.80***	-1.71*	-1.84*	2.36**	2.54**

R^2= 0.57, D.W. = 1.56

OLS-Regression on births

Variable	TSPR	TAUT (-1)	TSPR (-1)	TWIN (-1)	RAUT (-1)	RWIN (-1)	MWIN	MAUT (-1)
Coefficient	2.42	-2.64	-3.67	2.77	0.040	-0.059	1.55	4.55
Std. error	1.34	1.49	1.48	0.92	0.015	0.025	0.81	1.01
T-ratio	-1.80*	-1.78*	-2.49**	3.00***	2.66**	-2.34**	1.93*	4.50***

OLS-Regression on births (continued)

Variable	MSPR (-1)	DAUT	DAUT (-1)	PGRA (-1)	PLIV (-1)	REAL (-1)	REAL (-2)
Coefficient	-0.32	0.49	-0.56	-0.19	0.33	0.064	0.055
Std. error	0,18	0.16	0.19	0.10	0.16	0.026	0.023
T-ratio	-1.80*	3.05***	-2.94***	-1.84*	2.06*	2.37**	-2.38**

R^2= 0.93, D.W. = 1.89

OLS-Regression on deaths 1725-1750

Variable	TSUM	TWIN	TSUM (-1)	RAUT	RSUM	PWHE (-1)	PWHE (-2)
Coefficient	10.77	-6.35	9.35	0.066	-0.12	0.84	0.52
Std. error	2.12	1.26	2.39	0.021	0.049	0.19	0.13
T-ratio	5.08***	-5.06***	3.9₁***	-3.12***	-2.56**	4.31***	3.99***

R^2= 0.94, D.W. = 1.77

OLS-Regression on real wage

Variable	POP (-1)	TYEA	REAL (-1)
Coefficient	-0.085	26.32	0.61
Std. error	0.029	12.69	0.15
T-ratio	-2.90***	2.07**	4.13***

R^2= 0.67, D.W. = 1.60

REFERENCES

Appleby, A.B., 1975, Nutrition and Disease - the Case of London 1550-1750, *J. of Interdisciplinary Economics* 6:1-22.

Appleby, A.B., 1979, Grain Prices and Subsistence Crises in England and France 1590-1740, *J. of Economic History* 39:865-887.

Ashton, T., 1955a, *Economic Fluctuations in England 1700-1800*, Clarendon, Oxford.

Ashton, T., 1955b, *An Economic History of England - the 18th Century*, Methuen, London.

Baehrel, R., 1961, *Une croissance. La basse-Provence rurale (fin XVI siècle – 1789)*, S.E.V.P.E.N, Paris.

Bairoch, P., 1976, Die Landwirtschaft und die Industrielle Revolution 1700-1914, in: Cipolla, C.M. and Borchardt, K., Europäische Wirtschaftsgeschichte, vol. 3: *Die Industrielle Revolution*, UTB, Stuttgart, 297-332.

Boserup, E., 1981, *Population and Technological Change: A Study of Long-Term Trends*, University of Chicago Press, Chicago.

Bourke, A., 1984, The Impact of Climatic Fluctuations on Agriculture, in: Flohn, H. and Fantechi, R., eds., *The Climate of Europe - Past, Present and Future*, Kluwer, Dordrecht, 269-314.

Brooks, C.E.P., and Glasspole, J., 1928, *British Floods and Droughts*, Brill, London.

Brown, R., 1991, *Society and Economy in Modern Britain 1700-1850*, London.

Campbell, B., and Overton, M., 1998, L'histoire agraire de l'Angleterre avant 1850. Bilan historiographique et état actuel de la recherche, *Historie et Sociétés Rurales* 9:77-105.

Chambers, J.D., and Mingay, G.E., 1966, *The Agricultural Revolution 1750-1880*, Batsford, London.

Charlesworth, A., 1983, *An Atlas of Rural Protest in Britain 1548-1900*, London.

Chartres, J., ed., 1990, *Agricultural Markets and Trade 1500-1750*, Cambridge University Press, Cambridge.

Clark, G., 1992, The Economics of Exhaustion, the Postan Thesis and the Agricultural Revolution, *J. of Economic History* 52:61-84.

Coleman, D.C., 1977, *The Economy of England 1450-1750*, London.

Craddock, J.M., and Wales-Smith, B.G., 1977, Monthly Rainfall Totals Representing the East Midlands 1726-1975, *Meteorological Magazine* 106:97-111.

Deane, P., and Cole, W.A., 1962, *British Economic Growth 1688-1959*, Cambridge University Press, Cambridge.

Desplat, C., 1987, The climate of 18th-century Béarn, *French History* 1(1):27-48.

Drummond, J.C., and Wilbraham, A., 1957, *The Englishman's Food*, 2nd ed., London.

Dupaquier, J., 1997, La France avant la transition demographique, in: Bardet, J-P. and Dupaquier, J., eds., *Histoire des populations de l 'Europe*, vol. 1, Flammarion, 435-452.

Grenier, J.-Y., Dupaquier, J., and Burguière, A., 1988, Croissance et déstabilisation, in: Dupaquier, J., ed., *Histoire de la population Française, De la Renaissance à 1789*, Presses Universitaires de France, Paris, 437-498.

Engelen, A., and Nellestijn, J., 1996, *Monthly, seasonal and annual means of the air temperature in tenths of centigrade in De Bilt, Netherlands, 1706-1790*, mimeo, De Bilt.

Flohn, H., 1981, Short-Term Climatic Fluctuations and their Economic Role, in: Wigley, T.M.L., Ingram, M.I. and Farmer, G., eds., *Climate and History*, Cambridge University Press, Cambridge, 310-318.

Floud, R., Wachter, K., and Gregory, A., 1990, *Height, Health and History,* Cambridge University Press, Cambridge.

Floud, R., Wachter, K., and Gregory, A., 1993, Measuring Historical Heights - Short Cuts or the Long Way Round: A Reply to Komlos, *The Economic History Review* 46:145-154.

Fussell, G.E., 1933, Farmers' Calendars from Tusser to Arthur Young, *Economic History* 1:521-535.

Galloway, P.R., 1985, Annual Variations in Deaths by Age, Deaths by Cause, Prices and Weather in London 1670-1830, *Population Studies* 39:487-505.

Galloway, P.R., 1994, Secular changes in the short-term preventive, positive and temperature checks to population growth in Europe, 1460 to 1909, *Climatic Change* 26:3-63.

Grigg, D., 1980, *Population Growth and Agrarian Change - a Historical Perspective*, Cambridge University Press, Cambridge.

Habbakuk, H.J., 1987, The Agrarian History of England and Wales - Regional Farming Systems and Agrarian Change 1640-1750, *The Economic History Review* 40:281-296.

Havinden, M.A., 1961, Agricultural Progress in Open-Field Oxfordshire, *Agricultural History Review* 9:73-83.

Houston, R.A., 1986, British Society in the Eighteenth Century, *Journal of British Studies* 25:436-466.

Jones, E.L., 1981, Agriculture 1700-1800, in: Floud, R. and McCloskey, D.N., *The Economic History of Britain since 1700*, vol. 1, 1700-1860, Cambridge University Press, Cambridge, 66-86.

Kaplan, S., 1976, *Bread, politics and political economy in the reign of Louis XV*, 2 vols, Martinus Nijhoff, The Hague.

Kaplan, S., 1982, *Le complot de famine – histoire d'une rumeur au XVIII siècle*, Armand Colin, Paris.

Kaplan, S., 1984, *Provisioning Paris*, Cornell University Press, Ithaca.

Komlos, J., 1993a, The Secular Trend in the Biological Standard of Living in the United Kingdom 1730-1860, *The Economic History Review* 46:115-144.

Komlos, J., 1993b, Further Thoughts on the Nutritional Status of the British Population, *The Economic History Review* 46:363-366.

Kussmaul, A., 1990, *A General View of the Rural Economy of England 1538-1840*, Cambridge University Press, Cambridge.

Labrousse, E., Romano, R., and Dreyfus, F., 1970, *Le prix du froment en France du temps de la monnaie stable (1726-1913)*, S.E.V.P.E.N., Paris.

Lachiver, M., 1969, *La population de Meulan du 17 au 19 siècle (vers 1600-1870)*, S.E.V.P.E.N, Paris.

Lachiver, M., 1991, *Les années de misère. La famine au temps du Grand Roi 1680-1720*, Fayard, Paris.

Lamb, H.H., 1965, Britain's Changing Climate, in: Johnson, C.G. and Smith, L.P., eds., *The Biological Significance of Climatic Change in Britain*, London, 4-31.

Lamb, H.H., 1977, *Climate - Present, Past and Future*, 2 vols., Methuen, London 1977.

Lamb, H.H., 1988, *Weather, Climate and Human Affairs*, Routledge, London.

Landers, J., and Mouzas, A., 1988, Burial Seasonality and Causes of Death in London 1670-1819, in: *Population Studies* 42:59-83.

Lebrun, F., 1971, *Les hommes et la mort en Anjou au 17e et 18e siècle*, Mouton, Paris.

Lebrun, F., 1980, Les crises démographiques en France aux XVII et XVIII siècles, *Annales E.S.C.*, 35, 2, 205-234.

Lee, R., 1981, Short-Term Variation - Vital Rates, Prices and Weather, in: Wrigley, E. and Schofield, R., *The Population History of England*, Edward Arnold, London, 356-401.

Legrand, J.-P., LeGoff, M., 1992, *Les observations météorologiques de Louis Morin*, Monographie 6, Direction de la Météorologie Nationale, Paris.

LeRoy Ladurie, E., 1960, *La production agricole en France (15-18 siècle) notamment d'après les dimes*, mimeo, Paris.

LeRoy Ladurie, E., 1983, Histoire du climat depuis l'an mil, 2 vols, Flammarion, Paris.

Malthus, T., 1798, *An Essay on the Principle of Population*, London.

Manley, G., 1952, *Climate and the British Scene*, Collins, London.

Manley, G., 1974, Central England Temperatures - Monthly Means 1659-1973, *Quarterly Journal of the Royal Meteorological Society* 100:389-405.

Marshall, J.D., 1980, Agrarian Wealth and Social Structure in Pre-Industrial Cumbria, *The Economic History Review* 33:503-521.

Meuvret, J., 1977, *Le problème des subsistances à l'époque Louis XIV, La production des céréales dans la France du XVII au XVIII siècle*, Mouton, Paris.

Meuvret, J., 1987, *Le problème des subsistances à l'époque Louis XIV, La production des céréales et la société rurale*, EHESS, Paris.

Meuvret, J., 1988, *Le problème des subsistances à l'époque Louis XIV, Le commerce des grains et la conjoncture*, EHESS, Paris.

Monteith, J.L., and Scott, R.K., 1982, Weather and Yield Variation of Crops, in: Blaxter, K. and Fowden, L., eds., *Food, Nutrition and Climate*, London, 127-153.

Morineau, M., 1971, *Les faux-semblants d'un demarrage économique: agriculture et démographie en France au 18e siècle*, Cahiers des Annales 30, Armand Colin, Paris.

Newby, H., 1987, *Country Life - A Social History of Rural England*, Totowa.

O'Brien, P.K., 1977, Agriculture and the Industrial Revolution, *The Economic History Review* 30:166-181.

O'Brien, P.K., 1985, Agriculture and the Home Market for English Industry 1660-1820, *English Historical Review* 10:771-798.

Olney, M.L., 1983, Fertility and the Standard of Living in Early Modern England - In Consideration of Wrigley and Schofield, *The Journal of Economic History* 43:71-77.

Parry, M.L., 1978, *Climatic Change, Agriculture and Settlement*, Dawson, Folkestone.

Parry, M.L., 1990, *Climate Change and World Agriculture*, Earthscan, London.

Parry, M.L., and Carter, T.R., 1985, The Effect of Climatic Variations on Agricultural Risk, *Climatic Change* 7:95-110.

Pfister, C., 1984a, *Bevölkerung, Klima und Agrarmodernisierung*, Haupt, Berne.

Pfister, C., 1984b, *Das Klima der Schweiz von 1525-1860 und seine Bedeutung in der Geschichte von Bevölkerung und Landwirtschaft*, Haupt, Berne.

Pfister, C., 1988, Fluctuations climatiques et prix céréaliers en Europe du XVIe au XXe siècle, *Annales E.S.C.* 43(1):25-53.

Pierenkemper, T., 1989, Englische Agrarrevolution und preußisch-deutsche Agrarreformen in vergleichender Perspektive, in: Pierenkemper, T., ed., *Landwirtschaft und industrielle Entwicklung*, Stuttgart, 7-25.

Post, J.D., 1985, *Food Shortage, Climatic Variability and Epidemic Disease in Preindustrial Europe*, Cornell University Press, Ithaca.

Pousson, J.-P., 1980, Les crises démographiques en milieu urbain: l'exemple de Bordeaux (fin XVII – fin XVII siècle), *Annales E.S.C.* 35(2):235-252.

v. Rudloff, H., 1967, *Die Schwankungen und Pendelungen des Klimas seit Beginn der regelmäßigen Instrumenten-Beobachtungen*, Braunschweig.

Schofield, R., 1983, The Impact of Scarcity and Plenty on Population Change in England 1541-1871, *The Journal of Interdisciplinary History* 14:265-291.

Thirsk, J., 1990, *Agricultural Change - Policy and Practice 1500-1750*, Cambridge University Press, Cambridge.

Turner, M., 1986, English Open Fields and Enclosures - Retardation or Productivity Improvements, *The Journal of Economic History* 36:669-692.

Wanner, T., Pfister, C., Brazdil, R., Frich, P., Frydendahl, K., Jónsson, T., Kington, J., Lamb, H.H., Rosenørn, S., and Wishman, E., 1995, Wintertime European Circulation Patterns During the Late Maunder Minimum Cooling Period (1675-1704), *Theoretical and Applied Climatology* 51:167-175.

Wanner, T., Brazdil, R., Frich, P., Frydendahl, K., Jónsson, T., Kington, J., Pfister, C., Rosenørn, S., and Wishman, E., 1994, Synoptic interpretation of monthly weather maps for the Late Maunder Minimum (1675-1704), in: Frenzel, B., Pfister, C. and Gläser, B., eds., *Climatic trends and anomalies in Europe 1675-1715*, Stuttgart, 401-424.

Wigley, T.M.L. *et al.*, 1985, Historical Climate Impact Assessments, in: Kates, R.W., Ausubel, J.H. and Berberian, M., eds., *Climate Impact Assessment - Studies of the Interaction of Climate and Society* (Scope 27), John Wiley & Sons, New York, 529-564.

Wordie, J.R., 1983, The Chronology of English Enclosure 1500-1914, *The Economic History Review* 36:483-505.

Wrigley, E., 1982, The Growth of Population in 18th Century England - A Conundrum Resolved, *Past and Present* 98:121-150.

Wrigley, E., 1989, Some reflexions on corn yields and prices in preindustrial economies, in: Walter, J. and Schofield, R., eds., Famine, disease and the social order in early modern society, Cambridge University Press, Cambridge, 235-278.

Wrigley, E., and Schofield, R.: 1981, *The Population History of England*, Edward Arnold, London.

APPENDIX: QUANTITATIVE ANALYSIS OF ENGLISH CLIMATE, AGRICULTURE AND DEMOGRAPHIC DATA

Variables used (covering the period 1700-1750):

Climate:

TYEA (average annual temperature), TWIN (average temperature [°C] of December of the previous year, January, February), TSPR (average temperature of March-May), TSUM (average temperature of June-August), TAUT (average temperature of September-November), RWIN-RAUT (average precipitation [mm] of the seasons defined as above). Sources: Manley (1974) for temperature, Craddock and Wales-Smith (1977) and Lamb (1977) for precipitation.

Population:

POP (population of England and Wales at 1st July [1000s], BWIN-BAUT (births [1000s] of the seasons as defined above), DWIN-DAUT (deaths [1000s] of the seasons as defined above), MWIN-MAUT (marriages of the seasons [1000s] as defined above), REAL (real wage index [1500=1000]) . Source: Wrigley and Schofield (1981).

Agriculture:

PWHE (price of wheat), PBAR (price of barley), PRYE (price of rye), POAT (price of oats), PGRA (average grain price), PHAY (price of hay), PSHE (price of average sheep), PCAT (price of average cattle), PLIV (average price of livestock (cattle, sheep, horses, pigs)) (all as % of average 1640-1749), EXWHE (export of wheat [1000 quarters]). Source: Thirsk (1990) for prices, Ashton (1955a) for exports after 1750.

*: significant at 10%, ** significant at 5%, *** significant at 1%. All insignificant variables are not shown.

BONS BAISERS D'ISLANDE: CLIMATIC, ENVIRONMENTAL, AND HUMAN DIMENSIONS IMPACTS OF THE *LAKAGÍGAR* ERUPTION (1783-1784) IN ICELAND

G.R. Demarée[1] and A.E.J. Ogilvie[2]

[1]Royal Meteorological Institute, Ringlaan 3,
 B-1180, Brussels, Belgium
[2]INSTAAR, University of Colorado, Campus Box 450,
 Boulder, Colorado, 80309-0450, USA

1.   INTRODUCTION

Fires from beneath, and meteors from above,
Portentous, unexampled, unexplain'd,
Have kindled beacons in the skies; and th' old
And crazy earth has had her shaking fits
More frequent, and foregone her usual rest.
Is it a time to wrangle, when the props
And pillars of our planet seem to fail,
And Nature with a dim and sickly eye
To wait the close of all?

(William Cowper, *The Task*, Book II: *The Time-Piece*, 1802).

During the year 1783, a haze was spun out like a large veil over much of the Northern Hemisphere, persisting for periods of up to three months. In particular, the summer in Europe and elsewhere was characterized by the appearance of the phenomenon described by many contemporaries as the "great dry fog". The origin of this was the *Lakagígar* volcanic eruption (1783-1784) in Iceland. In the non-Icelandic literature, this eruption has often been referred to as the "Laki" eruption. Strictly speaking, this is a misnomer as the eruption did not occur on Mount Laki, but on either side of it. In Iceland it is often called *Lakagígar*, the "Laki fissure". (The total length of the *Lakagígar* crater row, from one end to the other, is 27 km, and it is divided by the Laki mountain into two nearly equal parts.) It is also called *Skaftáreldar*, the "Skaftá fires" from the river Skaftá in southeast Iceland. One of the two main lava flows from the eruption travelled along the canyon through which the river runs, causing it to dry up. The eruption began in Iceland in early June 1783, and the effects were

History and Climate: Memories of the Future?
Edited by Jones *et al.*, Kluwer Academic/Plenum Publishers, 2001

219

soon noticed in Europe. However, news of the eruption itself did not reach the continent of Europe until 1 September when it was brought to Copenhagen from Iceland, carried by the vessels of the Danish trading monopoly.

The connection between volcanic eruptions and climate seems to have been noted first by Benjamin Franklin (1785, 1786) who, at the time of the *Lakagígar* eruption, was the United States Ambassador to France. Since that time, this relationship has been researched extensively (see e.g., Lamb, 1970; Schneider, 1983; Self and Rampino, 1988; Bradley and Jones, 1992; Jones *et al.*, 1995; Stothers, 1996, 1998; Briffa *et al.*, 1998; Crowley, 2000; Robock, 2000). It has also become generally accepted that large explosive volcanic eruptions are followed by a global-scale lowering of temperatures during the ensuing seasons. (*Lakagígar* was not an explosive eruption, however; an argument considered e.g., by Wigley, 1996.) The magnitude and spatial character of such coolings appears to be dependent on a number of factors such as: the latitude of the volcano and the precise time of the eruption within the year (see, e.g., Stommel and Stommel, 1979; Self *et al.*, 1981; Self and Rampino, 1988; Harington, 1992; Ogilvie, 1992). The climate response to volcanic eruptions may, in some cases, be modulated or even marked by factors internal to the climate system (such as ENSO; see e.g., Wigley, 2000; Wigley and Santer, 2000). Furthermore, there is strong evidence that over the land areas of the Northern Hemisphere the response in winter may actually be a warming (Robock and Mao, 1995; Robock, 2000). The interest in the influence of volcanic eruptions on climate has grown considerably in recent years as a result of the increasing scientific and public awareness of a possible "global warming" due to anthropogenic activities (see e.g., Santer *et al.*, 1996), not least because the natural volcanic input signal could obfuscate an increased "greenhouse" effect (Tett *et al.*, 1999). However, in spite of the considerable research effort expended on the issue of volcanoes and climate, many questions concerning the relationship still remain unanswered.

Pioneering geological research on the *Lakagígar* event may be traced to the time around the first centenary of the eruption (see, e.g., Helland, 1882/84, 1886; Thoroddsen, 1879). It is also possible that a renewed interest in the eruption at this time may be due partly to the eruption of another Icelandic volcano, Askja, in 1875, which, like *Lakagígar,* was followed by an ashfall in Europe (specifically, in this case, in Scandinavia; Daubrée, 1875; Nordenskiöld, 1876; Mohn, 1877, 1877/78; Thórarinsson, 1981). The second centenary of *Lakagígar* in 1983-84 saw the publication in Iceland of a comprehensive book dealing with hitherto unpublished Icelandic historical sources on the event, as well as details of impacts in all the districts of Iceland (Gunnlaugsson *et al.*, 1984).

Subsequently, a number of researchers have considered the effects of the *Lakagígar* eruption in various parts of Europe. Thus, for example, Grattan and co-authors analysed the spreading, as well as the environmental and social impacts, of the "great dry fog" (Grattan, 1994, 1998; Grattan and Brayshay, 1995; Grattan and Charman, 1994; Grattan and Gilbertson, 1994). Stothers (1996) thoroughly reviewed the contemporary literature regarding the appearance of the dry fog, studied its horizontal and vertical spreading, and estimated its optical thickness and total mass. Demarée (1997, 1999, 2000) placed the appearance of the "great dry fog" in the southern Netherlands in the context of the scientific knowledge of the eighteenth century (at that time incomplete understanding caused it to be linked to: earthquakes; the warm weather followed by heavy thunderstorms; epidemics, etc.) and also assessed the human dimensions of these phenomena. As regards the spreading of the "great dry fog" of 1783-84 to regions other than Europe, more information has recently come to light. Demarée *et al.* (1998) were able to demonstrate that the fog had been witnessed by a missionary of the Moravian Brethren located at Nain in Labrador. Further-more, Jacoby *et al.* (1999) used tree-ring data and Inuit oral history to link the *Lakagígar* eruption to a climatic event in Alaska. In addition to this specific work on the *Lakagígar* event, Camuffo and Enzi (1994, 1995) have produced valuable chronologies of dry fog episodes in Italy during the last seven centuries.

In this paper, a number of the climatological, phenological and environmental impacts in Europe of the *Lakagígar* eruption are presented and considered. It is clear that the coincidental occurrence of unusual phenomena such as the Calabria and Messina earthquakes, the emergence of a volcanic island off the southwest coast of Iceland, the hot summer weather, and severe thunderstorms all over western Europe, as well as the concurring outburst of epidemics, all had a profound impact on the population of the time. This "human dimensions" aspect of the eruption is discussed in the context of these other phenomena. Further information regarding the spreading of the "great dry fog" is also given. Specifically, the geographical distribution of this fog is updated by many new findings in hitherto unused documents from several regions of Europe and elsewhere. It is also speculated that the adverse weather conditions experienced worldwide may, in fact, be manifestations of a concurrent ENSO warm event (i.e., an *El Niño* phase) within the period 1782-1786.

2. CONTEMPORARY HISTORICAL SOURCES OF THE "GREAT DRY FOG"

The data presented here are derived from the reports of climatological, oceanographic, geophysical and other environmental phenomena taken from a wide variety of contemporary European documentary records. In particular, a large number of early newspapers, ranging in location from the Iberian peninsula to central and eastern Europe, were scrutinized for information. A list of these newspapers, and the places where they were consulted, follows the references. A number of extracts from these newspapers are given below. They give information on phenomena which may be attributed to the eruption, such as the dry fog, and also on a number of other interesting natural phenomena such as earthquakes and unusual weather events which were reported during the *annus mirabilis* of 1783. Taking into account the status of eighteenth-century science, and the rather slow rate of distribution of the news, these sources have been evaluated and are judged to be reliable (see e.g., Bell and Ogilvie, 1978; Ingram *et al.*, 1981). Another important source consists of the scientific journals of the time. In addition to this, more than a dozen pamphlets were published on the subject of the *Lakagígar* eruption within a year of the event. This illustrates the lively scientific discussion that the natural occurrences such as the dry fog and the earthquakes initiated at the time. In this regard, the pioneering role of early researchers such as Abbot Giuseppe Toaldo (1719-1797) in the observation of the dry fog at Padua may be noted (Fig. 1). Toaldo established, *inter alia*, a chronology of similar phenomena, and the distribution of his writings on the subject has also been investigated (Demarée, 2000). The specific scientific discourses of Toaldo (1784) and others on the "great dry fog" will be not discussed here, but will be considered further in a subsequent paper.

The striking events of the year 1783 were not only noticed by scientists, however. Other contemporary writers also commented on the phenomena they observed. Thus, for example, the stanza from the work by William Cowper (1731-1800) quoted at the start of this paper describing the meteor or "fireball" (see below) of 18 August 1783 ('...meteors from above') and the "dry fog" during the summer of 1783 ('And Nature with a dim and sickly eye') gives us an insight into the fact that the natural events of 1783 were observed by poets as well as scientists.

OBSERVATIONS

E T

MÉMOIRES

S U R

LA PHYSIQUE,

SUR L'HISTOIRE NATURELLE,

ET SUR LES ARTS ET MÉTIERS.

OBSERVATIONS MÉTÉOROLOGIQUES

FAITES A PADOUE AU MOIS DE JUIN 1783;

Avec une Diſſertation ſur le Brouillard extraordinaire qui
a régné durant ce temps-là (1);

Traduites de l'Italien de M. Toaldo *, & accompagnées de nouvelles Vues ſur
l'origine de ce Brouillard , de l'Académie de Turin ;*

Par M. le Chevalier DE Lamanon *, Correſpondant de l'Académie des
Sciences de Paris.*

Baromètre. Sa plus grande élévation a été de 28 pouces 4 lignes le
24 & le 25 du mois. Son plus grand abaiſſement a été de 27 pouces

---

(1) Ces Obſervations du célèbre Profeſſeur de Padoue ſont inſérées dans un Journal
Tome XXIV , Part. I, 1784. *JANVIER.*   A 2

Figure 1. Title page of the dissertation on the extraordinary fog and the meteorological observations for the
month of June 1783 made at Padua, Italy, by Giuseppe Toaldo, augmented with the views of de Lamanon,
Knight, published in the *Observations et Mémoires sur la Physique* of the Abbot Rozier at Paris (see Toaldo,
1784, pp.3-18.)

3. NEWS FROM ICELAND: SKAFTÁRELDAR

> Around midmorn on Whitsun, June 8th of 1783, in clear and calm weather, a black haze of sand appeared to the north of the mountains nearest the farms of the Síða area. The cloud was so extensive that in a short time it had spread over the entire Síða area and part of Fljótshverfi as well, and so thick that it caused darkness indoors and coated the earth so that tracks could be seen. The powder which fell to earth looked like the burnt ash from hard coal. A light drizzle, which fell from that black cloud that day over the Skaftártunga region, turned this powder into black, inky liquid. A southeasterly sea breeze drove the dark cloud back inland later that day so that, like other clergymen here, I was able to celebrate the day with religious services under clear skies. The joy of those celebrations was turned to sorrow soon enough. That night strong earthquakes and tremors occurred. (This contemporary account by Jón Steingrímsson, one of the main Icelandic chroniclers of the eruption, whose work was probably begun in 1784, has been translated into English by Kunz, 1998, p.25)

The *Skaftáreldar* eruption occurred at a time when the full effects of the Enlightenment movement were being felt in Iceland. In consequence, there was a great general interest at this time in the observation of nature and natural events. In the year following the beginning of the eruption, six major contemporary descriptions of the event were published in Iceland (for a list of these accounts see Gunnlaugsson *et al.*, 1984. See also Thordarson and Self, 1993; and Steingrímsson, 1993, 1998). Some of these publications were rapidly translated from Icelandic into other European languages (see e.g., Holm, 1784a,b, 1785, 1799; Stephensen, 1786a,b). Interestingly, Icelandic references tend not to mention the text by Sæmundur Magnússon Holm contained in the *Ephemerides Meteorologicae Palatinae* series. This is a collection of the data from the embryonic worldwide meteorological network, edited by Johann Jacob Hemmer (1733-1790) at Mannheim. The volume for the year 1783 includes detailed information on the appearance of the dry fog (Hemmer, 1785) as well as Holm's text translated into Latin by Hemmer (Holm, 1785).

The context of the impact *Skaftáreldar* had in Iceland is outlined briefly below; however, the social and environmental aspects of the eruption in Iceland are not discussed here in any detail as this topic has already been covered extensively (see Gunnlaugsson *et al.*, 1984, and references cited therein). *Skaftáreldar* (the "River Skaftá fires") which lasted from June 1783 until February 1784 was, without doubt, one of the most noteworthy and largest fissure eruptions in historical times. One of the largest basaltic lava flows (14.7 km^3), it covered an area of 580 km^2 while the total volume of tephra produced is estimated at 0.4 km^3 dense rock equivalent volume (DRE), or 2.65% of the total erupted volume. Maximum fire fountain heights reached 800 to 1400 m. Maximum effusion rates occurring in the first two episodes (8-12 June 1783) were 8700 m^{3}s^{-1}. This is about double the discharge of the river Rhine near its mouth (data from Thórarinsson, 1969; and Thordarson and Self, 1993). The *Lakagígar* eruption should not be disassociated from the neighbouring Grímsvötn eruption which lasted from May 1783 until May 1785, and the two eruptions should therefore be considered as one two-year-long vulcano-tectonic episode.

The *Skaftáreldar* eruption had a catastrophic effect in Iceland, not because the eruption caused direct loss of life, but because of the indirect effects as a result of the emission of volcanic gases and ashes, distributed by wind. The grass, the basic food supply for the grazing livestock, became polluted and fluorine poisoned (Thórarinsson, 1979). Within a year of the eruption, 53% of the cattle, 80% of the sheep and 77% of the horses died. The Lakagígar eruption must be seen as the primary cause of the ensuing famine which came to be known as *Móðuharðindin* or "famine of the mist" (Bjarnar, 1965). However, the severe weather of the years 1782-84 also undoubtedly played a part in negatively impacting the human as well as the livestock population (Ogilvie, 1986). Difficulties were further compounded by an epidemic of smallpox. It is estimated that the total death toll of *Móðuharðindin* ultimately represented 19 to 22% of the Icelandic population, or approximately 10,000 people (Stephensen, 1808; Hooker, 1813; Henderson, 1819; Thoroddsen, 1879, 1882; Bjarnar, 1965; Gunnlaugsson *et al.*, 1984; Jackson, 1982; Vasey, 1991, 1997, 2001; Steingrímsson, 1998). Vasey (2001) found, however, that the excess mortality and decline in the human population in Iceland did not begin until losses in livestock dropped to a level of approximately one quarter of the usual.

The consequences of the *Lakagígar* eruption were not confined to Iceland, however (Thoroddsen, 1925; Thórarinsson, 1981; Bradley and Jones, 1992). The fine ash and volcanic dust that rained down on most parts of that country were also reported in many regions of northern Europe. Examples are: the Faeroe Islands; Caithness[1] in Scotland, Copenhagen; Friesland (Brugmans, 1783); Bergen in Norway; and northern Germany (Holm, 1784a,b). The dry fog, observed in many parts of Europe also appears to have been witnessed in areas as far apart as Labrador, Newfoundland, the Tunisian coast, Asia Minor, and possibly China (Demarée *et al.*, 1998). This spreading of the fog is discussed below. First, however, some of the other natural phenomena that occurred during the year 1783 are described.

4.   NEWS FROM ELSEWHERE: *ANNUS MIRABILIS* 1783 - CHRONICLE OF AN EXTRAORDINARY YEAR

4.1. Introduction

In order to understand the human responses in western Europe to the *Lakagígar* eruption, it is necessary to mention the other geophysical, meteorological, and biologically-related events that made the year 1783 so extraordinary. They include earthquakes, other volcanic eruptions, and unusual weather events, as well as illnesses and epidemics. The spirit of eighteenth-century science was such that the borderlines between these phenomena were not yet well defined. It will thus be seen how earthquakes, in particular, are frequently described by contemporary observers as being connected with the "great dry fog". Typically, natural scientists of the time also provided detailed descriptions of the weather conditions during earthquake episodes (see, e.g., von Hoff, 1840/1841). Usually, this type of information also included a description of the appearance of dark clouds and fog.

Many of these natural phenomena were frequently regarded as harbingers of doom and disaster, and hence caused much consternation among the general populace. This viewpoint is in accord with the prevailing Renaissance-influenced ideas on astro-meteorology of eighteenth-century scientists, who frequently interpreted the occurrence of other natural events as having a direct influence on meteorological events. The anthology of newspaper extracts and other contemporary accounts given below illustrate the links that the general populace tried to establish between all the different natural occurrences which they observed. Where necessary, the original texts used here have been translated by the present authors from several different European languages into modern English. All the newspaper reports cited below are from the year 1783 or, in some cases, from 1784.

---

[1] Some doubts remain about this particular occurrence since its main source is Geikie's non-contemporary *Text-book of Geology*. The event described may refer to the Icelandic Katla eruption of 1755 (see also Thórarinsson, 1981) or other nineteenth-century Icelandic volcanic events.

4.2. Earthquakes in Europe

4.2.1. Messina and Calabria. The large earthquake which laid waste Messina (in Sicily) and Calabria (in what is now southern Italy) occurred on 5 February 1783. Further tremors were experienced over the rest of the year. These earthquakes were perceived by contemporaries as overshadowing the remarkable events of the year 1783, and as being closely connected with them. The news of the earthquakes was reported in all European newspapers, and, from week to week, news - some clearly exaggerated - was distributed through these channels. Several Italian scientific reports were published on the earthquakes, and Lord Hamilton reported on them to the Royal Society (Hamilton, 1784). In contemporary accounts the earthquakes were frequently placed in the context of related geophysical, oceanographic and meteorological events such as the account below from the *Gazette van Gent*:

> Recorded at Naples: 11 February 1783. One has received the sad news that on the 5th at 19 hours according to the Italian clock, a heavy earthquake occurred at Messina in Sicily...The earthquake is ascribed to the bursting out of the Etna mountain which is situated 25 miles away. However, others assure [us] that this accident is the consequence of a most terrible thunderstorm...It is said that more than 20,000 inhabitants lost their lives in the waves or by the collapsing of their houses...One has counted more than 30 different earthquakes which were preceded by a heavy storm at sea and on land and accompanied by excessive rain with thunder, lightning and a terrible darkness. For the rest, it is mentioned from Naples that the mountain Vesuvius throws out such a thick and uncommon smoke that one fears for many catastrophes [occurring]...Letters from Palermo and Messina contain [the information] that the burning mountains in the small [Liparian] islands Vulcano and Stromboli throw out a large mass of fire (*Gazette van Gent*, XIX, 6 March 1783).

The following account, which almost certainly refers to the dry fog from the *Lakagígar* eruption (the eruption *per se* was still unknown in Europe) was published in numerous newspapers all over the continent and stated that:

> Since the last earthquakes in Upper and Lower Calabria on 8, 11 and 12 June, our coastal pilots have noticed a phenomenon, which makes [us] fear for some sad news. They attest that the atmosphere was heavily obscured and the sea has been in a permanent state of unrest, and the waves have receded further from the shore than usual (see e.g., *Journal historique et politique de Genève*, No. 30, Du Samedi 26 Juillet 1783).

4.2.2. Franche-Comté, the Jura, Burgundy and Geneva. On 6 July 1783, a further earthquake occurred in the region of: Franche-Comté, the Jura, Burgundy, and Geneva. Fortunately, there was not too much material damage; however, the populace was alarmed due to the continuing disturbing news about the earthquakes in Messina and Calabria. For many, that earthquake was obviously connected with the ongoing series of earthquakes in other areas. Furthermore, the fact that the earthquake in eastern France occurred during the height of the dry fog was, for many, a strong argument to link both phenomena. The account below illustrates this contemporary view.

A letter from Dijon of the 7ᵗʰ of this month [July] contains [information] that, the day before, while the atmosphere was covered, as usual, with a thick fog, an earthquake was felt from the east to the west, lasting three seconds. This earthquake, associated in the imagination with the catastrophes in Calabria, and with the dreadful subterranean noise, has caused a general panic among the inhabitants of Dijon...From Geneva, it was reported that, at the time, it was thought that the city would be destroyed. The waters of the lake were very swollen, and the waves generated by the movement of the earth, were crashing against the shores of the town...so that a few houses did collapse (*Gazette van Gent*, LVII, 17 July 1783).

4.2.3. Aachen, Maastricht, Cambrai and Northern France. In the neighbourhood of Aachen and Maastricht, at 3 o'clock in the morning of the night from 7 to 8 August, a light earthquake was felt (*Wiener Zeitung*, No. 70, 30 August 1783; *Esprit des Gazettes*, No. 21, 23 August 1783). On 9 December there were tremors at Cambrai as well as in other towns of northern France (*Feuilles de Flandres*, No. 42, Lille, 23 December 1783).

4.2.4. Tripoli in the Middle East. A further earthquake occurred in Tripoli [Lebanon] on 30 July 1783. This event is noted together with what is presumably the dry fog reported elsewhere, and the sun is described as being obscured.

[On] the 30ᵗʰ of this month [July] one has experienced here an earthquake that was felt on two occasions. The tremors succeeded each other rapidly, lasting together...8 to 10 seconds. They had been preceded by a muffled rumble, similar to the one of the roaring of the waves which one hears from afar. [In] the evening there was a pouring rain which was extraordinary for the season. For more than one month a heavy fog covered the land and the sea, the winds blew with equal violence as in the winter, the sun appeared only seldom and always with a bloody colour, unknown phenomena until now in Syria. The earthquake was also felt in Lebanon. An entire village near Naplouse [Palestine] was buried under a rock that had collapsed. The Turks, being informed about the disaster of Messina, are in a state of great consternation (*Journal historique et littéraire*, Luxemburg, 1 November 1783)

4.3. General Weather Conditions of 1783

Unusual weather conditions were described during 1783 in many places throughout Europe and elsewhere; in areas as far apart as, for example: Sweden; Malta; Istanbul; Croatia; Turkey; the northeastern United States; Egypt; and Japan.

Stockholm: 30 April 1783. The frost continues to be very strong during the night. The ice has not yet melted nor consequently [has] any ice flow occurred which means that commerce still remains interrupted, mostly in the north of the country. It [also] results in such high prices of all foods necessary for the provision of this capital. One does not remember ever to have known such high prices (*Journal politique, ou Gazette des Gazettes*, Bouillon, Mai 1783).

On Malta, on 1 July there was a storm lasting 24 hours during which many lightnings struck, and an abundant rain fell which caused great damage. No such storm is remembered on this small island that has no mountains nor forests which are the principal causes of the storms that disturb other countries. On the days 28 and 29 August were stormy winds (Reported in a letter from Naples[2], 5 August, 1783, *Gazeta de Barcelona*, 1783, p. 801).

Pera [a suburb of Istanbul]: 6 August 1783. The wind has always been westerly since the beginning of this month. We cannot see Scutari from here nor Constantinople [Istanbul], so thick is the fog. The northern and northeasterly winds which are continuous during the summer in these provinces, have been replaced by southerly winds. It is not yet sufficiently warm to put on our summer clothes. The inhabitants from Angoury [Ankara] and of Smyrna [Izmir] make the same complaints. So far [this summer] we have not had one day where we did not experience stormy weather accompanied by abundant rains. The Turks conclude from all this that the end of the world or, at least, of their empire, approaches (*Journal historique et littéraire*, Luxemburg, 15 September 1783).

Fiume [Rijeka] Croatia: 15 August 1783. Upon the fog which has equally covered our horizon for such a long time, have followed very heavy thunderstorms which are still continuing. The frequent thunders were causing great damage to our vineyards, vegetable gardens and fruit trees. Since then such a cool weather set in [that it was] like in the middle of the Autumn (*Wiener Zeitung*, No. 65, 13 August 1783).

In contrast, good weather was reported from the northeastern United States:

Hartford, Connecticut: 5 August 1783. We have not experienced finer weather for agriculture for some time past. From almost every part of the country we have the most satisfactory intimation that all appearances indicate a year of astonishing plenty (*The Continental Journal and Weekly Advertiser*, Boston, No. 390).

As regards Egypt, the low Nile level in 1782 might, albeit speculatively, be in part the result of an ENSO warm event. Certainly, on the basis of all available information, the period 1782-1786 has been classified as a general *El Niño* episode (Allan and D'Arrigo, 1999).

Cairo, Egypt: 29 November 1782. We are presently treated, in all these fertile provinces, to high prices. [This is] because the River Nile has not given enough water this year, and [because] of a coming war which is believed to be inevitable (*Gazette de Cologne*, XXV, 27 March 1783).

---

[2]News was transferred from one place to another in Europe in the form of letters. These letters usually start with the location and the date as here: Naples, 5 August 1783. In this case the news concerns information from the island of Malta.

In Japan, the year 1783 has been regarded as "The Year Without a Summer[3]". It was extremely cool and moist in the summer of 1783, which caused heavy damage to rice crops, and led to the great *Tenmei* famine (Mikami, 1982). It has been suggested that the Japanese Asama volcanic eruption in August 1783 might have contributed to this climatic anomaly (Mikami and Tsukamura, 1992). However, there may also be an ENSO association since Allan and D'Arrigo (1999) classify the year 1783 as part of the protracted 1782-1786 *El Niño* event. Wet/flood conditions in southern Japan and cool summer conditions in western and central Japan are characteristics of *El Niño* phases (Allan *et al.*, 1996).

For northern China, the chart of dryness/wetness indices for 1783 presents an amalgam of: grade 3 (normal); grade 4 (dry); and grade 5 (very dry), while for southern China; grade 3 (normal); grade 2 (wet); and grade 1 (very wet) classifications are given (Chinese Academy of Meteorological Sciences, 1982, p.162). This is in line with "normal" *El Niño* conditions for that region (Halpert and Ropelewski, 1988). In the light of the above evidence, especially that for Egypt, Japan and China, it is possible that the worldwide adverse weather conditions around 1783 are more likely to be associated with an ENSO warm event than with the *Lakagígar* eruption.

4.4. The Summer Weather in Europe in 1783

The summer of 1783 is known to have been one of the warmest of the last three centuries in western Europe (Kington, 1980). According to a contemporary meteorological overview from Roesbrugge in Flanders, the Abbot van der Meulen stated: "The year with the nicest weather was 1783; it was permanently or nearly [always] sunny from 8 o'clock in the morning till evening, during 103 days" (quoted in Blondeau, 1981). During that hot summer, numerous heavy thunderstorms, some of them followed by inundations, occurred in nearly the whole of Europe. They were promptly ascribed by contemporaries to the extraordinary and long-lasting foggy state of the sky. Many of these thunderstorms were reported in detail in European newspapers. Thus, for example, accounts are given below from Prague, Silesia, Frankfurt, Paris, Regensburg, Milan, and Livorno:

> Prague: 5 July 1783. The town of Klattau [Klatovy] has experienced on the 29th of June a terrible thunderstorm (*Gazette van Antwerpen*, No. 58, 22 July 1783).

> Silesia: 6 July 1783. Lightning struck 8 times in Schazlar [Zaclar, now in the Czech Republic]. It still thunders in a dense and persisting fog (*Gazeta Warszawska*, No. 59, 23 July 1783; *Gazety Wilenskie* [Vilnius, Lithuania], No. XXX, Saturday 26 July 1783).

> Frankfurt: 11 July 1783. The letters from many places of the German Holy Roman Empire scarcely speak of anything other than the damage which has been...caused by the thunderstorms, the downpours and the inundations. Among them, it is written from Lautern in Swabia that there, on the 27th of last month, the most dreadful weather that ever has been seen has been experienced there. In that region, the thunder has fallen at least twelve times, and has caused fires in some places and has killed a few people. In Upper-Saxony, it has also been experienced that the atmosphere has been covered for two weeks by a thick fog, and [there was] a heavy thunderstorm on the 26th and 27th June. In upper Austria so much

---

[3] The epithet "The Year Without a Summer" has frequently been applied to 1816, after the eruption of Tambora. See, e.g., Harington, 1992.

rain has fallen that the roads are under the water and may not be used (*Gazette van Gent*, LVII, 17 July 1783).

Paris: 11 July 1783. One has experienced thunderstorms in many regions of the Kingdom [of *France*]. These seem to be the result of the...general state of the atmosphere which has continued for some time (*Gazette van Antwerpen,* No. 57, 18 July 1783; *Journal politique, ou Gazette des Gazettes*, Bouillon, Première Quinzaine Août 1783).

Regensburg: 14 July 1783. Thunderstorms and inundations have, on the 4th of this month, terribly shaken several areas of Bavaria (*Berlinsche Nachrichten*, No. 50, 29 July 1783).

Milan: 30 July 1783. In these last days, the thunderstorms were more frequent, lightnings having struck in various places...

Livorno: August 1783. After midday of the 30th of last month, arose another fierce thunderstorm with rain and lightning...(*Gazzetta universale*, Florence, No. 62).

In many other parts of Europe, from Scandinavia to the Mediterranean region, unusual summer weather conditions were experienced. In most cases, contemporaries either ascribed these adverse weather conditions to the persisting dry fog, or sought to relate them in some way to this phenomenon.

4. 5. Other Unusual Events

4.5.1. Fireballs and similar phenomena. In the evening of 18 August, a hissing fireball, or meteor, was said to have sped through the air above England, across the Channel, through northern France to Burgundy, and then exploded in seven to eight pieces.

According to the account, one has seen at Ostend on the 18th of this month at half past nine in the evening, rising northerly from the sea, a fireball equal to the disk of the full moon. Its brightness [gave the sense of] a long trail of light coupled with this phenomenon...At the time of the peculiar event, the evening was in brightness equal to the daylight; the moon appeared red-coloured, but took its usual appearance again immediately afterwards. Many persons [state that they] have heard four explosions. It was result of the extraordinary state of the sky, with which the atmosphere has been fogged since approximately the middle of June. One does not remember having seen anything similar (*Gazette van Antwerpen*, No. 69, Friday 29 August 1783).

A similar phenomenon was witnessed in Antwerp on 4 October 1783:

One has seen on the 4th of this month at half past seven in the evening, north of this town, the sky being clear under the light of the stars, an appearance in the atmosphere of a sphere with a diameter of 9 to 10 inches. It seemed to rise high from the stars, and it described at high speed an oblique line from above to below, and [then] it disappeared from the view westwards around the Lower Scheldt in the space of 2 minutes (*Gazette van Antwerpen*, No. 80, Tuesday 7 October 1783).

4.5.2. A New Volcanic Island near Iceland. Although the birth of a new island off Iceland may well have been related to the volcanic activity in Skaftafellsýssla in Iceland, it is included in this section because it was the subject of unusual news in European newspapers. In the neighbourhood of Reykjanes, at Fuglasker (the "Bird Skerries") off the southwest tip of Iceland, a volcanic island rose out of the sea. This event was observed by a Danish boat skipper, Pedersen, and his assistant, Svendborg, of the *Forsken,* and written about on 22 and 24 May 1783. Their account formed the "headline" news of many European newspapers around that time.

> On our return trip, we discovered at 7 or 8 miles from the furthest rocks a small island which we sailed around at half a mile distance with nice weather [at the same time]...It was burning so strongly such that we could see the thick smoke which obscured the sky, at more than 6 miles distance...The inhabitants report that around Easter [20 April] they have seen something burning at sea, south of Grindavík, without knowing what it could be. It is noticed that this event occurred at the same time that Calabria was devastated by an earthquake (*Gazette de Leyde*, Supplément LVII, 18 July 1783).

A manuscript preserved in the Icelandic National Library in Reykjavik reports the discovery by a Captain Mindelberg of this new volcanic island (see also Thórarinsson, 1967; Wood, 1992). The birth of the temporary volcanic island named at the time in Danish *Nye-Ø* (New Island) remained in the "headlines" in the entire European press during the year 1783. It was related in the popular discussions to similar events near Santorini in 1707, and the Azores in 1720. The event further recalls the story of the voyage of a Venetian ship in the fourteenth century visiting the mythical island of Friesland. This island appeared on maps as late as the sixteenth century (de Pingré, 1784). Curiously enough, the birth of a new volcanic island and a similar story was repeated in 1963-65 when the island of Surtsey rose out of the sea near the Westman Islands (Vestmannaeyjar) off the south coast of Iceland (Thórarinsson, 1967).

5. THE GEOGRAPHICAL SPREADING OF THE "GREAT DRY FOG"

The dry fog of the year 1783 lasted for approximately three months in Europe and elsewhere, starting around mid June, with most of the last descriptions dating from September 1783. However, a brief note from Mannheim mentions a late appearance in January or February of the year 1784. It is unclear if this can be ascribed to the continuing *Lakagígar* eruption, which was still in action, or to normal weather phenomena in the winter season.

> Mannheim: 1 February 1784. Notwithstanding the rigour of the season and the continuing snowfall, [it is] believed [that one has] observed that the atmosphere has been nearly always enveloped by the vapours that have obscured it during the course of the summer. It is therefore concluded that the cause that produced them that time, still exists (*Gazette de Cologne*, XII, 9 February 1784).

Stothers (1996) discussed in great detail the geographical spreading of the "great dry fog" of the year 1783. Additional information on the spreading for remote areas was given by Demarée *et al.* (1998). Using contemporary sources, these authors showed that the fog was certainly witnessed in Labrador, Newfoundland, the Azores, the Tunisian coast, Asia Minor, and possibly China. In this section, further new information on the spreading is brought to light. Examples are given of descriptions of the "great dry fog" from newspaper and other accounts from northern, western and southern Europe as well as from Croatia, Romania, Istanbul (formerly Constantinople) and Syria.

One of the first to notice the appearance of the dry fog in western Europe appears to have been the Belgian agro-meteorologist Baron de Poederlé who noted at his estate:

> Saintes, Hainault: 17 June 1783. The day has been cold and very rainy with large bursts [*sic*] since noon, and the southwesterly wind raging the whole day. Before 6 o'clock in the morning, a light fog, and the sun was red (de Poederlé, 1784).

As regards northern Europe, a weather journal from Åbo (Turku) in Finland reports unusual weather and fog in June 1783:

> The wind was strange the whole month, but did not blow away the dry fog (Anonymous, 1783).

Accounts from Vienna stress the dry fog:

> Vienna: 5 July 1783. The same fog which fills the atmosphere in many other places, and hides the sunshine, which is so bright at this season of the year, is noticed here. During some days it has been as thick as in the month of October (*Journal historique et politique de Genève*, No. 30, Du Samedi 26 Juillet 1783).

> Vienna: 6 August 1783. All other messages from Hungary, Siebenburgen and from the other to-the-Crown incorporated provinces, state clearly that also there (as in the other countries of Europe) a thick dry fog has covered the horizon for a long time, which has also had here no other consequences than numerous and unusually heavy thunderstorms, which mostly were accompanied by heavy showers and hail, and in many places have caused greater or lesser damage (*Wiener Zeitung*, No. 63, 6 August 1783).

Describing the circumstances in Lisbon, a dry summer was reported (in contrast to wet conditions reported elsewhere):

> Nonetheless, what did make this year more remarkable than the many past years, was the hazy weather in the summer. From 22 June until 6 July there was a permanent haze by day and by night, and thereupon from 12 until 20 July the same happened during these 8 days. What was even more remarkable is that the same weather was prevalent in the larger part of the northern hemisphere. Notwithstanding this continuous haze, and a little bit of fog, the whole summer was very dry. One can easily consider it exceptional that during the 70 days from 19 June until 27 August, it did not rain except for the little above-mentioned moisture, although the shortage

of rain [at this time] was truly compensated for afterwards (Pretorius, 1785).

For Barcelona it was noted:

> At the end of June, a white and thick fog started to be observed in this capital [Barcelona] which did not allow [anyone] to see the nearest mountains and through which the sun seemed to be of a bloody colour, particularly at sunset. At night it seemed to have been absent since the stars were seen brightly, but at sunrise it returned...and continued the whole day, with the winds unable to prevent it (Salvà, 1783).

In Italy and Hungary also, the fog is described:

> Livorno: 11 July 1783. The fog of which one complains in nearly all of Europe, also affects Italy. It is so thick on the Adriatic Sea that vessels are obliged to make signals in order not to damage each other (*Journal historique et politique de Genève*, No. 32).

> The fogs of which one complains everywhere, prevails in several districts of Hungary, and they are so thick that they nearly obscure the daylight...Another singular phenomenon, noticed in all the places where are warm baths, is the baths are presently one degree warmer than they are normally (*Journal historique et politique de Genève*, No. 33).

Further accounts of the spreading of the fog come from areas as far apart as Malta, Romania, Istanbul, Croatia, and Syria:

> Malta: 5 July 1783. ... until 20 of the last month [June] such a thick fog was experienced on that island that the sun was obscured (*Gazeta de Barcelona*, p. 801).

> Hermannstadt [Sibiu, Romania]: 14 July 1783. After a long rainy weather, an extraordinary heat followed which lasted approximately 8 days. During that period the sun set every evening in a bloody red colour behind dense foggy exhalations. These exhalations remained through the whole night in the upper atmosphere. The sun that appeared similarly bloody coloured had to work itself every morning through these heavy exhalations. This appearance also called the attention of learned people. In the countryside, as has been reported, in particular in the neighbourhood of Klausenburg [Cluj, Romania] this appearance gave rise to different explanations. The imagination confined to the fog a stinking smell of sulphur, and everybody here was in fear about those things which might come. The heavy thunderstorms and numerous rainfalls diminish these fears (*Pressburger Zeitung* [Bratislava, Slovak Republic], 30 July 1783).

> Constantinople [Istanbul]: 23/27 July 1783. Since the beginning of the current month, one suffers here, in Smyrna [Izmir] and Angora [Ankara], from a thick fog with westerly winds (*Gazeta de Madrid*,

No. 76, 23 September 1783; *Notizie del mondo ossia Gazzetta*, Venezia, No. 73, 10 September 1783).

Fiume [Rijeka], Croatia: 15 August 1783. Upon the fog which has equally covered our horizon for such a long time, have followed very heavy thunderstorms which are still continuing (*Wiener Zeitung*, No. 65, 13 August 1783).

An account from Tripoli in the Middle East has already been mentioned above with regard to an earthquake. The account also describes the spreading of the fog in this region.

Tripoli, 30 July 1783: For more than one month a heavy fog covered the land and the sea, the winds blew with equal violence like in the winter, the sun appeared only seldom and always with a bloody colour, unknown phenomena until now in Syria (*Journal historique et littéraire*, Luxemburg, 1 November 1783).

6. ENVIRONMENTAL AND PHENOLOGICAL IMPACTS OF THE *LAKAGÍGAR* ERUPTION

Many harmful effects of the dry fog on the environment were noted in certain European sources. Specifically, these negative effects seem to be mainly limited to the banks of the North Sea and Baltic Sea region. Other sources report its beneficial influence on the vegetation and on the harvest. These include accounts from regions in: Germany; Austria; Hungary; Livonia (comprising nearly all of modern-day Latvia and Estonia); and Upper-Hungary (modern Slovakia). In the following discussion, the negative impacts are noted first. Some observers put their accounts in the context of what was occurring with the fog in Europe as a whole. Thus, for example:

Embden: 12 July 1783. The dry and thick fog which has reigned for a long time seems to have spread over the whole surface of Europe. Several seamen have also observed it at sea. During the day, the fog veils the sun, near the evening it takes on a vile smell. At some places it dries out the leaves; nearly all trees on the banks of the Ems have been denuded of their leaves in one single night (*Gazette de France*, No. 60, 29 July 1783).

Another account comments further on the Ems river region:

Munster: 30 July 1783. Nearly all the trees on the banks of the Ems have been denuded of their greenery in one night. One complains about having all the appearances of the autumn without having enjoyed the summer. The farmers say that the harvest will be very bad if these fogs do not change into rain, and they are afraid of its influence on the horned animals (*Journal historique et littéraire*, Luxemburg, 1 August 1783).

Others comment specifically on local conditions. Thus, for example, the Baron de Poederlé (1784) noted at his estate in Saintes in Hainault that:

24 June: The day has been warmer. The penetrating, unhealthy, stinking fog, like a light smoke [in colour] has lasted the whole day

> with a smell of sulphur (of burnt powder said the countryfolk)...Shortly after the appearance of this fog, I have noticed that several plants, and the leaves of several trees and shrubs were burnt and had become completely red. I also have noticed that after the so-much-expected rain, which fell on the last days of May, the earth was largely spotted with greenish and red spots (de Poederlé, 1784).

In the French region of Pas-de-Calais, the Abbot of Bazingham noted in his journal:

> The fog made the corn rusty in some parts of the province which made the bishop call for three days of prayers (*Journal d'Abot de Bazingham*, coll. partic., Archives du Pas-de-Calais).

In southeast Hungary, dry and warm weather changed to wet and cold:

> Vienna: 10 August 1783. One conveys from Szegedin [Szeged, southeast Hungary] the news that the dry and warm fogs were replaced there on 15 July by wet vapours, so cold that the plants and fruit trees have suffered a lot (*Journal Politique ou Gazette des Gazettes*, Bouillon, Septembre 1783).

Later, in the autumn, regarding "Upper Hungary" (modern Slovakia) it is stated:

> Vienna: 17 September 1783. One writes from Upper-Hungary [Slovakia] that the grain harvest has not been too plentiful there because of the excessive heat. One hopes that the wine harvest will be better (*Gazette van Antwerpen,* No. 78, 30 September 1783; *Journal historique et littéraire*, Luxemburg, 15 October 1783).

In a discussion of the possible implication of the *Lakagígar* eruption in a narrow tree-ring for the year 1784 in southwest Norway, Kalela-Brundin (1996, pp. 115) mentions two contemporary accounts by Holm (1784a) and Brun (1786). Referring to Trondheim in central Norway, Holm (1784a) wrote:

> Because of the acid rain the leaves on the trees were partly burnt through and the grass on the ground turned almost black.

Brun, writing from Bergen in southwest Norway, stated:

> The fog fell onto the leaves of various [types of] vegetation which withered.

Many of the contemporary accounts refer to the strong smell of sulphur:

> From the Lower Rhine...the people, when leaving their houses, from the most heavy smell of sulphur, had to put on sheets (*Koblenzer Intelligentzblatt*, No. 57, 18 July 1783).

From other regions, more typically in eastern and southern Europe, there are reports of a beneficial effect on agriculture, both on the grape harvest and other crops.

Upper Rhine: 6 July 1783. From nearly all regions of Germany come messages of an unknown fertility with which fields and meadows, vineyards and gardens are blessed this year. In Hungary and in Austria the vines are so loaded with grapes that a great part of them must be cut away so that the sticks become aerated and light. It is also so in Franconia, one expects from the vine press such streams of blessings that it will not be possible to contain them. It is possible that this general unusual fertility originates from an exceptional state of the sky which through its composition and modification on all kinds of vegetation and life has a very great influence (*Koblenzer Intelligenzblatt*, No. 56, 14 July 1783).

Pressburg [Bratislava]: 23 July 1783. The letters from Rust [Neusiedler See, Austria] and from the whole of Hungary, proclaim that the vineyards there promise a most abundant grape harvest (Gazette *van Antwerpen*, No. 64, 12 August 1783; *Schlesische privilegirte Zeitung*, No. 90, 2 August 1783).

Hungary: 8 October 1783. The wine harvest which started in all areas of this Empire by the end of the last month, has been nearly everywhere very successful as one had hoped. At many places it has exceeded expectations (*Wiener Zeitung*, No. 81, 8 October 1783).

Giovanni Lapi, a Florentine *georgophile*[4], describes the vigorous vegetation of the year 1783 which he ascribes to the beneficial influence of the dry fog (Lapi, 1783). Other accounts of fruitfulness come, for example, from the Banat (an ethnically mixed historical region, now divided among the modern states of Romania, Hungary and Serbia): "there is an abundance of fruits of all kinds" (*Das Wienerblättchen*, 3 August 1783). Also, according to a letter from Warsaw, "it is stated that already in the beginning of July the corn was cut, an unheard of event there, and oats and barley started to ripen" (*Das Wienerblättchen*, 10 August 1783). In contrast, in Stockholm in Sweden, an account for 16 September 1783 notes that "the grain harvest has failed in several provinces, and the hay has become so scarce that the inhabitants fear they will be forced to get rid of a part of their livestock" (*Journal politique, ou Gazette des Gazettes*, October 1783).

Fischer (1730-1793), a German-Livonian natural scientist reported as follows:

In Livonia [nearly all of modern Latvia and Estonia] the fog was as advantageous to the wheatfields and the plants in proportion to its covering the sun, so that the heat could not dry out the earth too much, which otherwise with the shortage of rain would have been disastrous for plants. Only fruits in sandy soil did badly, on heavy soils [they were] better. Vegetables did rather well; although near Riga there was a shortage of crops of fruits and on old trees the leaves dried out and shrivelled together. We have noticed also a notorious smell of sulphur and also the barometer was very high (Fischer, 1791, pp. 106-107).

---

[4] A *georgophile* was a member of the *Accademia dei Georgofili*, founded in 1753 in Florence. It is the earliest scientific institution in Europe whose explicit goal was the improvement of agriculture. The name is derived from "The Georgics" by the poet Virgil.

7. HUMAN DIMENSIONS ASPECTS OF THE *LAKAGÍGAR* ERUPTION

7.1. Epidemics and Illnesses in Europe

The earliest known account that explores the influence upon human beings of the natural environment is that encompassed in the fourth century BC work *Airs, Waters and Places* by Hippocrates. Further to this, according to the neo-hippocratic hypothesis formulated, among others, by Thomas Sydenham (1749) in the mid-eighteenth century, an interaction exists between climate, health and environment. Following from Hippocrates, this hypothesis states that it is the air that a person breathes, the water that he drinks and the topography of the place where he lives, together with the dry/wet and cold/warm succession of the four seasons that affects his health. The establishment of meteorological networks in the eighteenth century by scholarly societies in a number of different European countries may well have been sparked by a curiosity among contemporaries regarding the current unusual meteorological and geophysical phenomena, as well as the occurrence of epidemics.

A number of epidemics and pandemics prevailed during the 1770s and early 1780s throughout western Europe. These phenomena were monitored by the French *Académie de Médecine* at Paris and the Dutch *Geneeskundige Correspondentie Societeit* at the Hague (see Anonymous, 1793, for the observations related to the year 1783). The French network also included a number of Belgian, Dutch, Catalan and German correspondents. Indeed, within the framework of the neo-hippocratic hypothesis prevalent at the time, mentioned above, a link was clearly made by contemporaries between the appearance of epidemics and the extraordinary state of the atmosphere, as witnessed through the long-lasting dry fog, the hot summer and the heavy thunderstorms (Demarée, 1996). Thus, for example:

> In the summer of 1783 putrid illnesses occurred due to the great heat of the summer, and one counted a larger number of deaths than births during the year, in particular due to the dysentery in the villages near to the swamps (*Journal d'Abot de Bazinghem*, coll. partic., Archives du Pas-de-Calais).

> At that time, the great heat and the extreme droughts turned into a constellation of illnesses which were malignant fevers, from which no few persons died. It was compounded by a strong shortage of provisions in many regions (Maldà, 1988).

Mathias van Geuns, a member of the *Correspondentie Societeit* and Professor of Medicine at the University of Harderwijk, the Netherlands, wrote:

> One has probably to seek the causes of this illness in the state of the sky, the temperature of the air and the season of the year. One observes that the dysentery generally occurs in the great heat of the summer, and that it continues, but always declining, until the autumn, and that finally it disappears in the winter (van Geuns, 1784, pp. 64ff.).

Other accounts focus on direct adverse affects on human health. Thus, for example, from Dillenburg (Rhineland):

> The fogs which are covering us, seem to be here more noxious than elsewhere. Several persons have eye pains. One notes also that the leaves of the trees, the grass, etc., start to dry out (*Journal politique, ou Gazette des Gazettes*, Bouillon, Première quinzaine août 1783).

De Lamanon, a natural scientist at Sallon-de-Crau, in Provence, wrote extensively on the dry fog (see de Lamamon, 1784, 1799) stating, for example:

> One has noticed that this fog sometimes has a stinking smell...It causes a light pain in the eyes and people who have a delicate chest, are unpleasantly affected (*Journal historique et politique de Genève*, No. 31, 2 août 1783).

7.2. Human Perceptions and Responses

Several cases are known where the extraordinary events of the summer of 1783 became the subject of the Sunday sermon in the church. Thus, for example, Johann Georg Gottlob Schwarz, clerical Inspector at Alsfeld in Upper-Hessen, Germany, lectured on the subject of the extraordinary weather, from 16 June until the day of the sermon (29 June 1783), on the second Sunday after Trinity:

> I see daily, for more than eight days, your attention alert. Sometimes anxious, sometimes with meditation, sometimes with a non-revealing admiration you stare at the sky and at the earth, at the, for this time of the year, unusual sunrise and sunset, and at the horizon veiled in dark exhalations. The Lord speaks daily to us and shows us in nature his omniscience, his omnipotence, his greatness - extraordinary appearances therefore speak also, and capture our attention. I praise your attention to the works of the Lord. However, I wish that this will not take an inappropriate change. I take it as my duty to direct you with all my forces, we humans and walkers to eternity, on the straight path (Schwarz, 1783).

For Milan on 29 June 1783 was written:

> The 21st of this month, a solemn procession was held to obtain from Heaven the cessation of the extraordinary rains that have not discontinued for approximately [the last] 5 weeks (*Journal Politique ou Gazette des Gazettes*, Bouillon, Première quinzaine août 1783, p.26).

In Antwerp, an account of the dry fog and the hot weather is accompanied by mention of prayers for rain:

> Antwerp: 5 August 1783. Since approximately mid June a dry and foggy exhalation has occupied the atmosphere according to the meteorological observations, so that the sun's rays fade mainly around the sunrise and sunset, and shine red-coloured through this foggy veil. This disappeared from our horizon after the showers of July 23rd. Since then the heat has been increasing, mainly from the 27th until 31st July. The 1st of August public prayers started in the divine services to obtain rain from the Almighty (*Gazette van Antwerpen*, No. 62, 5th August 1783).

A fear that the world was about to come to an end was frequently expressed:

> I am not any more surprised about it since I have seen the...terror that the people...have formed. Some are waiting and trembling [for]

the fate of Calabria; others are believing in the end of the world. That idea was so singularly substantiated that one fixed the date at July 1st (Mourgue de Montredon, 1784, p. 764).

The red colour of the sun is often described as causing fear amongst the inhabitants. Thus, for example, in Switzerland:

> Deux-Ponts [Rhineland-Palatinate]: 8 July 1783. The discouragement was extreme in some villages at the view of the sanguine colour of the sun (*Nouvelles de divers Endroits*, Berne, No. 58, 19 July 1783).

An account from Barcelona notes the belief that the fog caused the lack of rain (other accounts suggest that it caused the excess rain):

> June 1783:...and where lasted from the end of the month of June and a large part of July, such an extreme fog over the whole atmosphere, that it...nearly [obscured] the brightness of the sun, so that...many persons did believe that this fog provoked the lack of rain for a long time (Maldà, 1988).

However, some contemporaries did not agree with the general opinion on the harmful consequences of the dry fog:

> The foggy air which has given motive to so much harm, clamour and disturbances, has not been otherwise malignant, neither a fatal signal, as the people have feared, but a benign natural phenomenon of abundance, of delight, and of peace (Lapi, 1783).

Also, an account from the Upper Rhine dated 6 July 1783 notes:

> The uncertainty, the superstition and the fear...in our region was also so terrible and prophesied Plague, War with the Turks and the End of the World - the thin fog... - the fading of the sun, its red bloody face at sunset, these heavy thunderstorms without their usual black clouds, and so on, all these were probably nothing else than an excess of combustible electrical material which did not destroy but did revive (*Koblenzer Intelligenzblatt,* No. 56, 14 July 1783).

The story of the sulphur-like smelling and rumbling Gleichberg (a hill of 672m near Hildburghausen in Thüringer Wald) published and republished by nearly all newspapers describes the phenomenon as one that spread through all of Europe, thus, for example:

> Public prayers were ordered in all churches. The inhabitants of the villages of the district run away because they fear that the Gleichberg will collapse, and the physicists pretend that a new fire-spitting mouth will be opened there (*Supplément aux Nouvelles de divers Endroits*, Bern, No. 59, 23 July 1783).

On occasion, information denying such rumours was also published, usually in a much less visible way:

> Bayreuth: 17 July 1783. Travellers coming from Hildburghausen do
> not know of any changes in the Gleichberg, nor that it bursts out
> neither that it erupts (*Berlinsche Nachrichten*, No. 50, 29 July 1783).

The continuing effects of the Messina earthquake compounded the concerns of those living in the vicinity:

> The last notices from Messina announce that [on] the 19 June within
> the time of 5 hours one has felt there and in the neighbourhood four
> shocks of earthquakes. As the inhabitants are still lodging in wooden
> barracks out of the town, these commotions did not take anyone's
> life, but they have augmented their consternation and weakened their
> hopes (*Nouvelles de divers Endroits*, Berne, No. 59, 23 July 1783).

In the eighteenth-century popular mind, the link between dry fog and cold winters appears to have been known. Certainly this is suggested by a weather prediction which appeared for the autumn of 1783:

> Some very old people remember that also in the very severe winters
> of the years 1709 and 1740 such hot and foggy summers preceded. It
> is therefore necessary to [be prepared with] stocks of wood
> (Anonymous, 1784; Wiedeburg, 1784).

Benjamin Franklin (1785, 1786) has been credited as being the first scientist to have advanced the thesis of the existence of a correlation between a volcanic eruption and a consecutive cold winter. In this way, he sparked the still ongoing scientific discussion concerning relationships between volcanic eruptions and climate. Certainly, for Europe at least, there is firm evidence that explosive eruptions do tend to lead to cold summers and warm winters. A sulphate aerosol dry fog could also lead to a cold summer, but this depends on the exact composition of the aerosol. In the case of the 1783 event, this concerns the cold winter of 1783/84 as well as the hot summer of 1783. The question remains relevant in the present context of global change where it is necessary to distinguish between natural climatic variability (including the part due to volcanic events) and greenhouse-gas induced warming. The above-mentioned text of Wiedeburg, shows here that a possible connection between volcanic eruptions and climate, via the summer fogs, was a popular belief in the eighteenth century (although the link between the dry fog and vulcanism was not made until Franklin's observation).

8. DISCUSSION

The task of the eighteenth-century press was twofold. On the one hand, it fulfilled the essential role of relaying news. As such, it uses superlatives such as: "extraordinary"; "an event never seen before"; "our oldest citizens do not remember..."; "never seen as far as one can remember"; etc. On the other hand, it often disseminated reassuring explanations. An example is that published in Parisian newspapers, and discussed widely all over Europe, by the academician de la Lande (1783, p.762-763) who explains the dry fog as being an ordinary wet fog due to the frequent rain during the spring. The latter explanation was, however, challenged by many natural scientists all over Europe.

One of the main difficulties with which the eighteenth-century witnesses of the dry fog had to cope was the delay in obtaining information regarding what had really occurred in Iceland. Indeed, while the greater part of the appearance of the dry fog was seen in Europe from the middle of June until July, and had a varying intensity in August and September, the news about the *Skaftáreldar* eruption only reached Copenhagen on 1 September 1783. It was published in: Stockholm on 11 September; Wroclaw (Breslau) on 15 September; Leiden on 19 September; Vienna on 20 September; Brussels and St Petersburg on 22 September; Florence on 27 September; Paris on 30 September; Bouillon, Luxembourg, Bern and Venice on 1 October; Warsaw on 8 October; Madrid on 17 October; and finally, Barcelona on 25 October 1783. At that time, the "great dry fog" had finished flowering. Therefore, prior to the months of September and October, when the information concerning the eruption in Iceland became generally available, contemporaries were compelled to advance other hypotheses for the "great dry fog". The most obvious explanation must have been to connect it to the earthquakes of Calabria and Messina. This suggestion was fostered because, during earthquake episodes in June and July, the dry fog was not only omnipresent, but also witnessed in southern Italy.

In trying to understand the natural phenomena of the time, a few contemporary scientists did refer to previous analogous events. The *"Histoire"* of the French Academy of Sciences recalls a similar event in the year 1721 where a dry fog was noticed in Paris, Picardie, Auvergne, Milan and Persia (modern Iran). In this case, it was also suggested that this had to do with an earthquake of the town of Tauris [Tabriz, Iran] (Mourgue de Montredon, 1784). Also Lapi (1783) from Florence refers to that event, thus: "The other which happened in 1721, and from which we ignore the circumstances which preceded, which accompanied, and which followed, but which covered the whole of Italy, and France, and still other places." There was indeed an eruption of the Icelandic volcano Katla in May 1721, but this was a relatively minor eruption, and unlikely to have had as much effect in Europe as Lakagígar. It was also noticed that the disastrous Lisbon earthquake of 1 November 1755, was preceded by a fog (von Hoff, 1840/1841). This could, in fact, possibly be ascribed to the Katla eruption of October 1755 through August 1756.

Toaldo's chronology *Chronica delle oscurazioni straordinarie del Sole: e phenomeni analoghi* ("Chronicle of extraordinary obscurings of the Sun: and analogous phenomena" consulted here in a French edition, see Toaldo, 1784) contains information on natural events that are nowadays ascribed to volcanic eruptions. The first one he mentions, occurring at the time of Caesar's death, is related to the Etna eruption of the year 44 B.C. (Stothers and Rampino, 1983; Forsythe, 1988). The other one listed for the year A.D. 937 (and associated with the Eldgjá eruption in Iceland) has been dated by Stothers (1998). He suggests that c. A.D. 934 is the most plausible date for this Icelandic eruption.

9. CONCLUSIONS

The "great dry fog" of the year 1783 which was caused by the *Skaftáreldar* eruption in Iceland was, without doubt, one of the most remarkable meteorological and geophysical phenomena of the last millennium. For several months, a film of dust like a veil covered large parts of the Northern Hemisphere, obscuring the sun, and giving it the appearance of a red, bloody-looking sphere at sunrise and sunset. This image made the headlines in European newspapers for many months after its first appearance in mid June 1783. As the event appeared to be virtually unprecedented, it is not surprising that it was also at the centre of a heated scientific debate among most natural scientists of the time.

The news of the *Skaftáreldar* in Iceland did not reach Europe until it came to Copenhagen on 1 September 1783. From there it subsequently made its way through the European press in September and October. By that time, the "great dry fog" had nearly totally disappeared from the sky. The delay in reporting, as well as the state of eighteenth-century science, contributed to the difficult identification of the event. For present-day meteorologists, the links and possible impact upon the world weather of such large eruptions still remain elusive and unclear. Furthermore, the chronology of all potential sources of the "great dry fog" is not yet comprehensively established. The exact nature of the relationship between the *Skaftáreldar* event and the very warm European summer of 1783, and the extremely cold winter of 1783-1784 is not yet fully understood. Other questions to be asked are: what were its climatic and environmental impacts worldwide? What is the part of the possible *El Niño* episode 1782-1786 in the context of worldwide climatic anomalies? One interesting aspect regarding the *Lakagígar* eruption is that this was a unique type of eruption, but one from which it is likely that we can learn about the effects of tropospheric sulphate aerosols. There are some puzzling features, however. Why was it warm rather than cold in the summer? Was the hot summer just a fluctuation unrelated to the eruption event? Or, did the aerosol have a large non-sulphate absorbing component? Or, did the cloud produce an odd circulation pattern that dominated over the direct radiative effect? Further research will help to elucidate the many unanswered questions which remain regarding climate/volcanic interactions.

Acknowledgements

The authors are mindful of the invaluable help of: Mariano Barriendos (Barcelona); Hans Bergström (Uppsala); Rudolf Brázdil (Brno); Cathérine Dhérent (Dainville); Barbara Hulanicka (Wroclaw); Trausti Jónsson (Reykjavík); Eline Kadai (Tallinn); Anders Moberg (Stockholm); Elena Nieplova (Bratislava); Øyvind Nordli (Oslo); Maria Nunes de Fátima (Evora); Barbara Obrebska-Starkel (Kraków); Anne-Marie Spanoghe (Ghent); Andres Tarand (Tallinn). Astrid Ogilvie acknowledges the support of NSF grants OPP-952359 and OPP-9726510. The lively and interesting discussions with colleagues from NABO (North Atlantic Biocultural Organization) and EACH (European and Atlantic Climate Historians) at the occasion of the workshop in Reykjavík in August 1997, and the conference in Norwich in September 1998 are gratefully remembered. We thank Richard Stothers, Tom Wigley, and Trausti Jonsson for their helpful and critical comments on the manuscript.

The authors respectfully dedicate this paper to three scholars of Skaftáreldar who no longer walk this earth: Sigurdur Thórarinsson, Vilhjálmur Bjarnar, and Gísli Ágúst Gunnlaugsson.

REFERENCES

Allan, R.J., and D'Arrigo, R.D., 1999, "Persistent" ENSO sequences: how unusual was the 1990-1995 El Niño? *Holocene* 9:101-118.

Allan, R., Lindesay, J., and Parker, D., 1996, *El Niño Southern Oscillation Climatic Variability*, CSIRO Publishing, Collingwood VIC 3066, Australia.

Anonymous, 1783, *Utdrag af Wäderleks-Journalen, hållen i Åbo, år 1783.* Tidningar Utgifne Af Et Sällskap i Åbo, År 1783. Tionde Årgången. Åbo, Tryckt hos Kong. Acad. Boktryckaren J.C. Frenckels Enka.

Anonymous, 1784, *Ueber die Erdbeben und den allgemeinen Nebel von 1783.* von Joh. Ernst Basilius Wiedeburg, Kammerrath und Professor der Mathematik, in: Göttingische Anzeigen von gelehrten Sachen, Der erste Band, auf das Jahr 1784, 47. Stück, den 20. März 1784, Göttingen:470-472.

Anonymous, 1793, *Verhandelingen van de Natuur- en Geneeskundige Correspondentie-Societeit, in de Vereenigde Nederlanden, opgericht in 's Hage. Behelzende de Weer- en Ziektekundige Waarneemingen van de Jaaren 1782. tot 1790. Ingeslooten.* IV. Deel. In 's Gravenhage, By A. Van Hoogstaten, VIII+866 p.

de Bazingham, 1783, *Journal d'Abot de Bazingham*, coll. partic., Archives Départementales Pas-de-Calais, Dainville, France.

Bell, W.T., and Ogilvie, A.E.J., 1978, Weather compilations as a source of data for the reconstruction of European climate during the medieval period, *Climatic Change* 1:331-348.

Bjarnar, V., 1965, The Laki eruption and the famine of the mist, in: *Scandinavian Studies, The American-Scandinavian Foundation*, Carl Bayenschmidt and Erik J. Friis, eds, University of Washington Press:410-421.

Blondeau, R.-A., 1981, *Vander Meulen van Roesbrugge.* De IJzerbode, 1981/9, p. 58.

Bradley, R.S., and Jones, P.D., 1992, Records of explosive volcanic eruptions over the last 500 years, in: *Climate since A.D. 1500*, R.S. Bradley and P.D. Jones, eds, Routledge, London:606-622.

Briffa, K.R., Jones, P.D., Schweingruber, F.H., and Osborn, T.J., 1998, Influence of volcanic eruptions on Northern Hemisphere summer temperature over the past 600 years, *Nature* 393:450-455.

Brugmans, S.J., 1783, *Natuurkundige Verhandeling over een zwavelagtigen nevel den 24 Juni 1783 in de provincie van stad en lande en naburigen landen waargenomen*, Groningen, 58 p.

Brun, J.N., 1786, *Johan Nordahl Bruns Prædiken paa Nyt-Aars Dag 1786, i Anledning av Collection for Island*, Rasmus H. Dahl, Bergen, 16 p.

Camuffo, D., and Enzi, S., 1994, Chronology of 'dry fogs' in Italy, 1374-1891, *Theor. and Appl. Climatol.* 50:31-33.

Camuffo, D., and Enzi, S., 1995, Impact of the clouds of volcanic aerosols in Italy during the last 7 centuries, *Nat. Hazards* 11:135-161.

Chinese Academy of Meteorological Sciences, 1982, *Yearly charts of Dryness/Wetness in China for the last 500-year Period*, Cartographic Publishing House, Beijing:332 p.

Cowper, W., 1802, *Poems, by William Cowper, Of the Inner Temple, Esq.*, In two Volumes, A New Edition, London, Printed for J. Johnson, in St Paul's Church-Yard.

Crowley, T.J., 2000, Causes of climate changes over the past 1000 years, *Science* 289:270-277.

Daubrée, 1875, Chute de poussière observée sur une partie de la Suède et de la Norvège, dans la nuit du 29 au 30 mars 1875, d'après des Communications de MM. Nordenskiöld et Kjerulf, *Comptes rendus des séances de l'Académie des Sciences*, Tome LXXX:994-996 & p. 1059.

Demarée, G.R., 1996, The neo-hippocratic hypothesis - an integrated 18th century view on medicine, climate and environment, *Proceedings of the International Conference on Climate Dynamics and the Global Change Perspective, Cracow, October 17-20 1995*:515-518.

Demarée, G.R., 1997, "De grote droge nevel" van 1783 in de Zuidelijke Nederlanden: een historisch-klimatologische studie, *Tijdschrift voor Ecologische Geschiedenis* 1:27-35.

Demarée, G.R., 1999, Annus mirabilis A.D. 1783: een kroniek van "merckenweerdigste voorvallen" langs de Schreve, in: *Liber Amicorum R.A. Blondeau. Poperinge, Schoonaert*:79-94.

Demarée, G.R., 2000, Giuseppe Toaldo and his contributions to 18th century meteorology, in: *Atti dei convegno "Giuseppe Toaldo e il suo tempo*, Nel bicentenario della morte, Scienze e lumi tra Veneto e Europa", a cura di Luisa Pigatto, presentazione di Paolo Casini, Padova, 10-13 novembre 1997, Bertoncello Artigrafiche, p.645-654.

Demarée, G.R., Ogilvie, A.E.J., Zhang De'er, 1998, Comment on Stothers, R.B. "The great dry fog of 1783" (*Climatic Change* 32, 1996): Further documentary evidence of Northern Hemispheric coverage of the great dry fog of 1783, *Climatic Change* 39:727-730.

Fischer, J.B., 1791, *Versuch einer Naturgeschichte von Livland, entworfen von J.B. Fischer*. Zweite vermehrte und verbesserte Auflage. Königsberg, bey Friedrich Nicolopius.

Forsyth, P.Y., 1988, In the Wake of Etna, 44 B.C., *Classical Antiquity* Vol. 7/No. 1, University of California Press:49-57.

Franklin, B., 1785, Meteorological Imaginations and Conjectures, By Benjamin Franklin, LL.D.F.R.S. and Acad. Reg. Scient. Paris. Soc. &c. Communicated by Dr. Percival. Read December 22, 1784, *Memoirs of the Manchester Literary and Philosophical Society* ii:373-377.

Franklin, B., 1786, Idées & conjectures météorologiques, par M. B. Franklin, docteur ès loix, membre de la société royale de Londres, de l'académie royale des sciences de Paris. &c., &c., tirées des Mémoires de la Société Littéraire et Philosophique de Manchester, & récemment publiées en françois, Journal encyclopédique iv, Juin 1786:493-496.

Geikie, Sir A., 1893, *Text-Book of Geology*. Third edition, revised and enlarged. London, Macmillan and Co. and New York:1147 p.

Grattan, J.P., 1994, Acid damage to vegetation following the Laki fissure eruption in 1783 - an historical review, *The Science of the Total Environment* 151:241-247.

Grattan, J., 1998, The distal impact of Icelandic volcanic gases and aerosols in Europe: a review of the 1783 Laki Fissure eruption and environmental vulnerability in the late 20th century, in: *Geohazards in Engineering Geology*, J.G. Maund and M. Eddleston, eds, Geological Society, London, Engineering Geology Special Publications 15:97-103.

Grattan, J., and Brayshay, M., 1995, An amazing and portentous summer: environmental and social responses in Britain to the 1783 eruption of an Iceland volcano, *The Geographical Journal*,161(Part 2):125-134.

Grattan, J., and Charman, D.J., 1994, Non-climatic factors and the environmental impact of volcanic volatiles: implications of the Laki fissure eruption of AD 1783, *Holocene* 4:101-106.

Grattan, J.P., and Gilbertson, D.D., 1994, Acid-loading from Icelandic tephra falling on acidified ecosystems as a key to understanding archeological and environmental stress in northern and western Britain, *J. Archeol. Sci.* 21:851-859.

Gunnlaugsson, G.A., Guðbergsson, G.M., Thórarinsson, S., Rafnsson, S., and Einarsson, Th., eds, 1984, *Skaftáreldar 1783-1784. Ritgerdir og Heimildir*. Mál og Menning, Reykjavík.

Halpert, M.S.; and Ropelewski, C.F., 1992, Surface temperature patterns associated with the Southern Oscillation, *J. Climate* 5:577-593.

Hamilton, W., 1784, *Relation des derniers Tremblements de Terre arrivés en Calabre et en Sicile*, Envoyé à la Société Royale de Londres, Genève.

Harington, C.R., ed., 1992, *The Year Without a Summer; World Climate in 1816*, Canadian Museum of Nature, Ottawa.

Helland, A., 1882/1884, Islændingen Sveinn Pálssons beskrivelser af islandske vulkaner og bræer, *Den Norske Turistforeningens Årbok for 1882*:19-79; for 1884:27-56.

Helland, A., 1886, Lakis kratere og lavaströmme. Universitetsprogram for 2det semester 1885. Kristiania. Trykt i Centraltrykkeriet, 40 p. + 1 Kart & Planches.

Hemmer, 1785, *Vaporis anni 1783 succincta historia*. Ephemerides Societatis Meteorologicae Palatinae. Observationes Anni 1783. Manheimii, Prostant apud C.Fr. Schwan, bibliopolam aulicum:57-60.

Henderson, E., 1819, *Iceland; or the Journal of a Residence in that Island, during the Years 1814 and 1815, Containing Observations on the natural Phenomena, History, Literature, and Antiquities of the Island; and the Religion, Character, Manners, and Customs of its Inhabitants*, With an Introduction and Appendix, Edinburgh.

Hippocrates, 1800, Traité d'Hippocrate. *Des Airs, des Eaux et des Lieux;* Traduction nouvelle, Avec le texte grec collationné sur deux manuscrits, des notes critiques, historiques & médicales, un discours préliminaire, un tableau comparatif des vents anciens & modernes, une carte géographique, & les index nécessaires, Par CORAY, Paris, clxxx:170-484 p.

Holm, S.M., 1784a, *Om Jordbranden paa Island i Aaret 1783*, Peder Horrebow, København, 80 pp.

Holm, S.M., 1784b, *Vom Erdbrande auf Island im Jahr 1783*. Durch S.M. Holm, S.S. Theol. Cand. Aus den Dänischen übersetzt mit zwey Landkarten erläutert, E.G. Prost, Universitets Buchhändler, Kopenhagen, 94 p.

Holm, S.M., 1785, De incendio terrae in Islandia anno 1783, Autore S.M. Holmio, SS. Theologicae Candidato. (Monitum Traductoris Latini, Hemmer, p. 689) *Ephemerides Societatis Meteorologicae Palatinae. Observationes Anni 1783, Manheimii*, Prostant apud C.Fr. Schwan, bibliopolam aulicum:689-694.

Holm, S.M., 1799, Account of a remarkable Eruption from the Earth in Iceland, in the Year 1783. By S.M. Holm, S.S. Theol. Cand. *The Philosophical Magazine*: Comprehending the various Branches of Science, the liberal and fine Arts, Agriculture, Manufactures, and Commerce, by Alexander Tilloch. London, Vol. III, March 1799:113-120.

Hooker, W.J., 1813, *Journal of a Tour in Iceland in the Summer of 1809*, Longman *et al.*, London, 2 Volumes.

Ingram, M.J., Farmer, G. and Wigley, T.M.L., 1981, Past climates and their impact on Man: a review, in: *Climate and History*, T.M.L. Wigley, M.J. Ingram and G. Farmer, eds, Cambridge University Press:3-50

Jackson, E.L., 1982, The Laki Eruption of 1783: impacts on population and settlement in Iceland, *Geography* 67:42-50.

Jacoby, G.C., Workman, K.W., and D'Arrigo, R.D., 1999, Laki eruption of 1783, tree rings, and disaster for northwest Alaska Inuit. *Quat. Sci. Revs.* 18:1365-1371.

Jones, P.D., Briffa, K.R., and Schweingruber, F.H., 1995, Tree-ring evidence of the widespread effects of explosive volcanic eruptions, *Geophys. Res. Letts.* 22(11):1333-1336.

Kalela-Brundin, M., 1996, The narrow ring of 1784 in tree-ring series of Scots pine (*Pinus sylvestris* L.) in southwest Norway - a possible result of volcanic eruptions in Iceland, in: *Holocene Treeline Oscillations, Dendrochronology and Palaeoclimate*, Burkhard Frenzel, ed., ESF Project European Palaeoclimate and Man 13, Akademie der Wissenschaften und der Literatur Mainz. ESF, Strasbourg:107-118.

Kington, J.A., 1980, July 1783: the warmest month in the Central England temperature series, *Climate Monitor* 9(No. 3):69-73.

Kunz, K., 1998, See Steingrímsson, J. 1998.

de Lamanon, 1784, *Vues sur la nature et l'origine du brouillard qui a eu lieu cette année, Observations et Mémoires sur la Physique, sur l'Histoire naturelle, et sur les Arts et Métiers.* Janvier, 1784, Tome XXIV:8-18.

de Lamanon, 1784, *Nieuwe gedagten, over den aart en oorsprong der zeldzaamen nevels, in den jaare MDCCLXXXIII. (Door den Ridder de Lamanon.)*, Algemeene vaderlandsche letter-oefeningen, Zesde Deels, Tweede Stuk, Amsterdam:296-308.

de Lamanon, 1799, Observations on the Nature of the Fog of 1783, *The Philosophical Magazine*: Comprehending the various Branches of Science, the liberal and fine Arts, Agriculture, Manufactures, and Commerce, by Alexander Tilloch. London, Vol. V:80-89.

Lamb, H.H., 1970, Volcanic dust in the atmosphere: with a chronology and assessment of its meteorological significance, *Phil. Trans. Roy. Soc.*, Series A, 266: 425-533.

de la Lande, 1783, Lettre sur l'état actuel de l'Atmosphère. Aux auteurs du Journal. *Journal de Paris*, Numéro 182, Mardi 1er Juillet 1783:762-763.

Lapi, G., 1783, *Sulla caligine del corrente anno 1783. E sulla vigorosa vegetazione e fertilità delle piante del suddetto anno*. Congetture di Giovanni Lapi Mugellano già publico lettor di Botanica or Direttore del giardino dei Georgofili di Firenze. Firenze, XIV + 90 p.

Maldà, Rafael d'Amat i de Cortada, Barón de, 1988, Calaix de Sastre, Volum I, 1769-1791. Biblioteca Torres Amat. Selecció i edició a cura de Ramon Boixareu, Pòrtic de Jaume Sobrequés. Curial Ediciones Catalanes, Barcelona, 296 pp.

Mikami, T., 1982, Structure of famines - an example of Tenmei famine, *Chiri* 27(12):51-57 (in Japanese).

Mikami, T., and Tsukamura, Y., 1992, The climate of Japan in 1816 as compared with an extremely cool summer climate in 1783, in: *The Year without a Summer? World Climate in 1816*, C.R. Harington, ed., Canadian Museum of Nature, Ottawa:576 p.

Mohn, H., 1877, Askeregnen den 29de-30te Marts 1875, *Naturen*:152-153.

Mohn, H., 1877/78, Askeregnen den 29de-30te Marts 1875, *Forhandlinger i Videnskabelige Selskabet i Kristiania*, Kristiania 10:1-12 (med Kort).

Mourgue de Montredon, M., 1784, Recherches Sur l'origine & sur la nature des Vapeurs qui ont régné dans l'Atmosphère pendant l'été de 1783, *Histoire de l'Académie Royale des Sciences avec les Mémoires de Mathématique et de Physique pour l'Année 1781*, Imprimerie Royale, Paris:754-773.

Nordenskiöld, A.E., 1876, Distant transport of volcanic dust, *The Geological Magazine or, Monthly Journal of Geology: with which is incorporated "The Geologist"*, edited by Henry Woodward, New Series Decade II. Vol. III, January-December 1876:292-297.

Ogilvie, A.E.J., 1986, The climate of Iceland, 1701-1784, *Jökull*, 36:57-73.

Ogilvie, A.E.J., 1992, 1816 - a year without a summer in Iceland?, in: *The Year Without a Summer? World Climate in 1816*, C. R. Harington, ed., Canadian Museum of Nature, Ottawa.

de Pingré, A.G., 1784, Précis du mémoire sur l'isle qui a paru en 1783, au sud-ouest de l'Islande, lu par M. Pingré, dans la séance publique de l'académie royale des sciences de Paris, tenue le 12 Novembre dernier, *Journal encyclopédique ou universel* Tome I, Partie I, Bouillon:116-118.

de Poederlé, M. le baron, 1784, Précis des observations météorologiques faites à Bruxelles pendant l'année 1783, *L'Esprit des Journaux, françois et Ètrangers*, Tome V, Paris:326-349.

Pretorius, J.C., 1785, EXTRACTS Das observações Meteorologicas feitas em Lisboa nos annos 1783, e 1784 por Jacob Chrysostomo Pretorias, Socio da Academia Real das Sciencias. Almanach para o anno de MCCCLXXXV. Lisboa, Officina Academia Real dos Sciencias:267-276.

Salvà, F., 1783, Memoria del Dr. Salvà en las tablas meteorológicas, Año1783, Archivo Real Academia de Medicina de Barcelona.

Robock, A. 2000, Volcanic eruptions and climate, *Rev.Geophys.*, 38:191-219.

Robock, A., and Free, M., 1995, Ice cores as an index of global volcanism from 1850 to the present, *J. Geophys. Res.* 100(D6):11,549-11,567.

Robock, A., and Mao, J., 1995, The volcanic signal in surface temperature observations, *J. Climate* 8(5):1086-1103.

Santer, B.D., Wigley, T.M.L., Barnett, T.P., and Anyamba, E., 1996, Detection of climate change and attribution of causes, in: *Climate Change 1995: The Science of Climate Change, Contribution of Working Group I to the Second Assessment Report of the Intergovernmental Panel on Climate Change*, Houghton, J.T., Meira Filho, L.G., Callander, B.A., Harris, N., Kattenberg A., and Maskell, K., eds, Cambridge University Press, New York:407-443.

Schneider, S.H., 1983, Volcanic dust veils and climate. How clear is the connection ? An editorial, *Climatic Change* 5:111.

Schwarz, Johann Georg Gottlob (s.d.) Kanzelvortrag an dem II. Sonntag nach Trinitatis 1783. Vom Ausserordentlichen der Witterung (vom 16ten Junius bis auf den Tag da die Predigt gehalten wurde, nämlich den 29ten gedachten Monats) Lauterbach, 8-vo, 40 p.

Self, S., and Rampino, M.R., 1988, The relationship between volcanic eruptions and climate change: still a conundrum, *EOS* 69(6):7475, 85-86.

Self, S., Rampino, M.R., and Barbera, J.J., 1981, The possible effects of large 19th and 20th century volcanic eruptions on zonal and hemispheric surface temperatures, *J. Volcanol. and Geotherm. Res.* 11:41-60.

Steingrímsson, J., 1973, *Æfisagan og Önnur Rit.*, K. Albertsson, ed., Helgafell, Reykjavík.

Steingrímsson, J., 1998, *Fires of the Earth. The Laki Eruption 1783-1784 by the Rev. Jón Steingrímsson.* Introduction by Dr Guðundur E. Sigvaldason. English translation by Keneva Kunz. University of Iceland Press and the Nordic Volcanological Institute, Reykjavík.

Stephensen, M., 1786a, Magnus Stephensens Beschreibung des Ausbruches eines neuen Vulkans im Vestriskaftafells-Syssla im Jahre 1783, in: [C.U.D. von Eggers] *Philosophische Schilderung der gegenwärtigen Verfassung von Island, nebst Stephensens zuverlässiger Beschreibung des Erdbrandes im Jahre 1783 und anderen authentischen Beylagen*. Mit einer neuen Charte dieses Landes und Zweyen Kuphertafeln. Altona, gedruckt von J.D.A. Eckhardt, Kön. Dän. Priv. Buchdr., Leipzig in Commission bey J.S. Heinzius, p. 307-386.

Stephensen, M. 1786b, Nachricht von den schrecklichen Unglücksfällen, welche Island in dem Jahre 1783 und 1784 betroffen haben. *Hannoverisches Magazin*, 14tes Stück, c. 218-224, 15tes Stück (Schluss), c. 225-232.

Stephensen, M., 1808, *Island i det Attende Aarhundrede, historisk-politisk skildret*. *Kjøbenhavn*, Johan Rudolph Thiele:451.

Stommel, H., and Stommel, E., 1979, The year without a summer. *Sci. Amer.* 240, 6, p. 134-140.

Stothers, R.B., 1996, The great dry fog of 1783, *Climatic Change* 32:79-89.

Stothers, R.B., 1998, Far reach of the tenth century Eldgjá eruption, Iceland, *Climatic Change* 39:715-726.

Stothers, R.B., and Rampino, M.R., 1983, Volcanic eruptions in the Mediterranean before A.D. 630, from written and archeological sources. *J. Geophys. Res.* 88(B8):6357-6371.

Sydenham, Th., 1749, Thomae Sydenham med. Doct. Ac practici londinensis Celeberrimi Opera medica; in *Tomos duos divisa*. Editio Novissima ... Genevae, Apud Fratres de Tournes. Tomus primus, 711 p., Tomus secundus, 496 p.

Tett, S.F.B., Stott, P.A., Allen, M.R., Ingram, W.J., and Mitchell, J.F.B., 1999. Causes of twentieth-century temperature change near the Earth's surface, *Nature* 399:569-572.

Thórarinsson, Sigurður, 1967, *Surtsey, The New Island in the North Atlantic*, Viking Press, New York, 47pp.

Thórarinsson, S., 1969, The Lakagígar eruption of 1783, *Bull. Volcanologique* 33, Fasc. 3:910-929.

Thórarinsson, S., 1979, On the damage caused by volcanic eruptions with special reference to tephra and gases, in: *Volcanic Activity and Human Ecology*, P.D. Sheets and D.K. Grayson, eds, Academic Press:125-159.

Thórarinsson, S., 1981, Greetings from Iceland. Ash-falls and volcanic aerosols in Scandinavia, *Geografiska Annaler* 63(3-4):109-118.

Thordarson, Th., and Self, S., 1993, The Laki (Skaftár Fires) and Grímsvötn eruptions in 1783-1785, *Bull. Volcanol.* 55:233-263.

Thoroddsen, Th., 1879, De vulkanske Udbrud paa Island i Aaret 1783. (Hermed Tavle V.). Særtryk af *Geografisk Tidskrift*, Kjøbenhavn, Hoffensberg & Traps Etabl., 16 p. + Kaart.

Thoroddsen, Th., 1882, Oversigt over de islandske Vulkaners Historie. Hermed to Kort over Hekla-Partiet og Island. (Avec un résumé en français) Kjøbenhavn. Bianco Lunos, 170 p.

Thoroddsen, Th., 1925, Die Geschichte der Isländischen Vulkane (nach einem hinterlassenen Manuskript) Mit 5 Tafeln. *Det Kongelige Danske Videnskabernes Selskabs Skrifter, Naturvidensk og Mathem. Afd.* 8. Rekke, IX, Kobenhavn.

Toaldo, G., 1784, Observations météorologiques faites a Padoue au mois de Juin 1783; Avec une Dissertation sur le Brouillard extraordinaire qui a régné durant ce temps-la; Traduites de l'Italien de M. Toaldo, & accompagnées de nouvelles Vues sur l'origine de ce Brouillard, de l'Académie de Turin; Par M. le Chevallier de Lamanon, Correspondant de l'Académie des Sciences de Paris. *Observations sur la Physique*, Janvier 1784, Tome XXIV:3-18.

van Geuns, 1784, De Heerschende Persloop, &c. Recherches sur la Dysenterie Epidémique qui a regné pendant trois ans & surtout en 1783 dans la Province de Gueldre, particulièrement dans le Quartier de Veluwe. Harderwijk & Amsterdam, Nooyen & Holtrop, in: *Nouvelle Bibliothèque Belgique* Tome VI, La Haye:380-390.

Vasey, D.E., 1991, Population, agriculture, and famine: Iceland, 1784-1785, *Human Ecol.* 19(3):323-350.

Vasey, D.E., 1997, *Dear King, Please send at least six Ships each Year: Contagion, Sickness and Death Optional*, Unpublished manuscript, NABO meeting, Reykjavík, Island, August 1997, 8pp.

Vasey, D.E., 2001, Vasey, D.E.: 2001, A quantitative assessment of buffers among temperature variations, livestock and the human population of Iceland, 1784-1900, *Climatic Change* 48, in press.

von Hoff, K.E.A., 1840/1841, Geschichte der durch Überlieferung nachgewiesenen natürlichen Veränderungen der Erdoberfläche. *Chronik der Erdbeben und Vulcan-Ausbrüche. Mit vorausgehender Abhandlung, ber die Natur dieser Erscheinungen*. Erster Theil. Vom Jahre 3460 vor, bis 1759 unserer Zeitrechnung. Zweiter Theil. Vom Jahre 1760 bis 1805, und von 1821 bis 1832 n. Chr. Geburt. Gotha, Justus Perthes.

Wiedeburg, J.E.B., 1784, Ueber die Erdbeben und den allgemeinen Nebel 1783. 1) Geschichte der Erdbeben 2) Muthmassungen dar, ber 3) Vorschläge sie zu verh ten 4) Geschichte des Nebels 5) Muthmassungen, ber denselben von Johann Ernst Basilius Wiedeburg. Jena bey den Verfasser und in Commission der Eunoischen Hof, Buchhandlung, 86 p.

Wigley, T.M.L., 1996, A millennium of climate, *Earth*, December 1996:38-41.

Wigley, T.M.L., 2000, ENSO, volcanoes and record-breaking temperatures (Submitted to *Geophys. Res. Letts.*).

Wigley, T.M.L., and Santer, B.D., 2000, Differential ENSO and volcanic effects on surface and tropospheric temperatures (Submitted to *J. Climate*).

Wood, C.A., 1992, Climatic effects of the 1783 Laki eruption, in: *The Year without a Summer? World Climate in 1816*, C.R. Harington, ed., Canadian Museum of Nature, Ottawa 1992:58-77.

List of Contemporary Newspapers Consulted

Berlinsche Nachrichten. Berlin. Preußischer Kulturbesitz, Staatsbibliothek zu Berlin, Berlin, Germany
Das Wienerblättchen. Vienna. Österreichische Nationalbibliothek, Vienna, Austria
Feuilles de Flandres. Lille. Royal Library Albert I, Brussels, Belgium
Gazeta de Barcelona. Archivo historico de Barcelona, Barcelona, Catalonia, Spain
Gazeta de Madrid. Biblioteca nacional, Madrid, Spain
Gazeta Warszawska, Warsaw. Biblioteka Jagiellonska, Kraków, Poland
Gazette de Cologne. Köln. Royal Library Albert I, Brussels, Belgium
Gazette de France. Paris. Royal Library Albert I, Brussels, Belgium
Gazette van Antwerpen. Royal Library Albert I, Brussels, Belgium & Stadsbibliotheek Antwerpen, Antwerp, Belgium
Gazette van Gent. Royal Library Albert I, Brussels, Belgium
Gazety Wilenskie. Vilnius. Biblicteka Jagiellonska, Kraków, Poland
Gazzetta universale. Florence. Biblioteca Nazionale Centrale, Florence, Italy
Journal historique et politique de Genève. Bibliothèque publique et universitaire, Ville de Genève, Switzerland
Journal historique et littéraire. Luxemburg, Royal Library Albert I, Brussels, Belgium & Stadsbibliotheek Antwerpen, Antwerp, Belgium
Journal politique, ou Gazette des Gazettes. Bouillon. Royal Library Albert I, Brussels, Belgium
Koblenzer Intelligentzblatt. Stadtarchiv Koblenz, Germany
Notizie del mondo ossia Gazzetta. Venezia, Biblioteca universitaria di Padova, Padova, Italy
Nouvelles de divers Endroits. Berne. Bibliothèque publique et universitaire, Ville de Genève, Switzerland
Nouvelles extraordinaires de divers endroits. Gazette de Leyde. Leiden, Royal Library Albert I, Brussels, Belgium
Pressburger Zeitung, Bratislava. University Library, Bratislava, Slovak Republic
Schlesische privilegirte Zeitung. University Library, Wroclaw, Poland
The continental journal and weekly advertiser. Boston. Helsinki University Library. The National Library of Finland, Helsinki, Finland
Wiener Zeitung. Österreichische Nationalbibliothek, Vienna, Austria

THE EFFECT OF CLIMATIC VARIATION ON PELAGIC FISH AND FISHERIES

Jürgen Alheit and Eberhard Hagen

Baltic Sea Research Institute
Seestr. 15
18119 Warnemünde, Germany

1. INTRODUCTION

Evidence is accumulating that marine ecosystems undergo decadal-scale fluctuations which appear to be driven by climate variability (e.g. Beamish, 1995; Bakun, 1996). Climatic variations affect marine communities and trophodynamic relationships, and may induce regime shifts where the dominant species replace each other on decadal time scales. One way to predict how marine ecosystems will react to future climate variability or to climatic change is to search for causal relationships of past patterns of natural variability and to draw conclusions on the basis of retrospective studies. Long-term biological time-series are essential for such retrospective analyses of climate impact on marine ecosystems; however, they are not readily available. Because of their economical importance, fish populations usually provide longer records than other biological components of marine ecosystems. The dynamics of exploited fish populations are affected by both environmental variability and man-made activities (fishing, habitat alteration) and retrospective studies will help to distinguish between the two. Earlier summaries on climate and fisheries have been published by e.g. Cushing (1982), Wyatt and Larraneta (1988), and Laevastu (1993). More recently, further studies have been stimulated by: the world-wide public awareness of global changes and the predicted greenhouse effect; the initiation of global international research programmes such as the World Climate Research Programme (WCRP) and the International Geosphere Biosphere Programme (IGBP); vastly improved co-operation across disciplinary boundaries; and accumulating knowledge on climate variability, particularly on the decadal scale.

2. SMALL PELAGIC FISHES

Small pelagic fishes such as the sardine, anchovy, herring, and others, represent about 20 – 25 % of the total annual world fisheries catch. They are widespread and occur in all oceans. They support important fisheries all over the world and the economies of many

History and Climate: Memories of the Future?
Edited by Jones *et al.*, Kluwer Academic/Plenum Publishers, 2001

247

countries depend on those fisheries. They respond dramatically and quickly to changes in ocean climate. Most are highly mobile, have short, plankton-based food chains, and some even feed directly on phytoplankton. They are short-lived (3-7 years), highly fecund, and some can spawn all year round. These biological characteristics make them highly sensitive to environmental forcing and extremely variable in their abundance (Hunter and Alheit, 1995). Thousandfold changes in abundance over a few decades are characteristic for small pelagics, and well-known examples include the Japanese sardine, sardines in the California Current, anchovies in the Humboldt Current, sardines in the Benguela Current, and herring in European waters. Their drastic stock fluctuations often caused dramatic consequences for fishing communities, entire regions and even whole countries. Their dynamics have important economic consequences as well as ecological ones. They are important food sources for larger fish, seabirds, and marine mammals. The collapse of small pelagic fish populations is often accompanied by sharp declines in marine bird and mammal populations that depend on them for food (Hunter and Alheit, 1995). Major changes in abundance of small pelagic fishes may also be accompanied by marked changes in ecosystem structure, for example, in abundance and species composition of zooplankton. The great potential plasticity in the growth, survival and other life-history characteristics of small pelagic fishes is the key to their dynamics and makes them ideal targets for testing the impact of climate variability on marine ecosystems and fish populations. This review will therefore focus on the impact of climatic variability on small pelagic fish stocks.

3.   ENSO

El Niño – Southern Oscillation (ENSO) is the strongest climate signal of global impact which also affects marine ecosystems, fish stocks and fisheries (Glantz, 1996) on time scales of 3-7 years. However, the negative impacts of ENSO on the dynamics of anchovies and sardines are of rather short duration and, usually, fish populations and fisheries have bounced back after a few years, as observed, for example, after the strong ENSO event in 1982/83. The Peruvian anchovy fishery was the largest fishery for one single species in history. In 1970, 12.5 million metric tonnes (MT) were landed (Tsukayama, 1983; Alheit and Bernal, 1993; Fig. 1). This was about one sixth of the total world catch of marine fishes which amounted to about 70 million MT in 1970. According to Castillo and Mendo (1987), the real catch in 1970 was 15 million MT as under-reporting of the real catches was frequent. The spectacular crash of the Peruvian anchovy fishery has often been ascribed to a combined negative impact of overfishing and the 1972/73 ENSO, and still serves as a classic text book example. However, as will be shown below, the Peruvian anchovy stock had already entered a decreasing phase before the onset of the 1972/73 ENSO (Alheit and Bernal, 1993). This decrease appears to have been environmentally induced (Alheit and Bernal, 1993), but was likely to have been accelerated and aggravated by the extremely high fishing mortality. Clearly, long-term dynamics of the anchovy and sardine populations in the Humboldt Current seem to be governed on a longer timescale than on the typical ENSO scale. Single ENSO events can dramatically affect these populations. However, this tends to be only for a rather short duration (unless individual ENSO events trigger long-lasting changes in food chain relationships, for example, a crash in predator populations). Nevertheless, if there are changes in the frequency and intensity of ENSO occurrences, these will certainly influence the fish-population dynamics in the Humboldt Current.

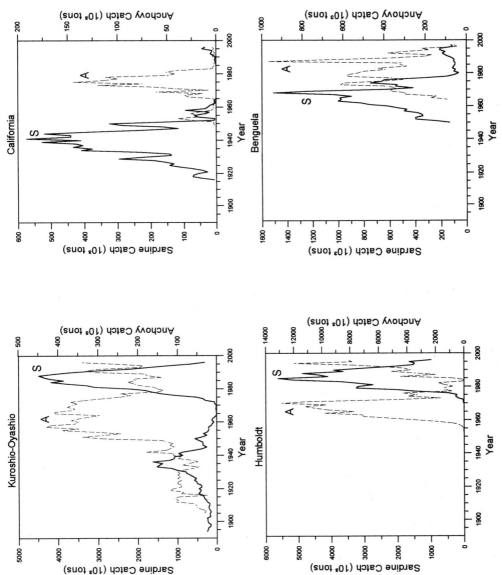

Figure 1. Variations in the annual catch of sardine and anchovy in four major current systems during the twentieth century (modified after Lluch-Belda *et al.*, 1989). (A = anchovy; B = sardine)

4.   DECADAL-SCALE REGIME SHIFTS

Huge populations of sardines and anchovies live in the upwelling ecosystems of the eastern boundary currents (California, Humboldt, Canary, and Benguela Currents) and in the waters around Japan. They support important fisheries, mainly for fish meal, and the well-being of the economy of the riparian countries of upwelling systems depends heavily on these fisheries. The dynamics of these anchovy and sardine populations are characterised by their inverse relationships. When one species is doing well and supports a large biomass and high production, the other species usually sustains a rather low biomass. These shifts between sardine-dominated and anchovy-dominated states seem to restructure the entire ecosystem, as concomitant qualitative and quantitative changes in ecosystem components other than the sardine and anchovy populations have been observed. Because of their dramatic and long-lasting nature, these switches have been termed "regime shifts" (Lluch-Belda *et al.*, 1989, 1992). The term "regime" has been phrased by Isaacs (1976) to describe distinct environmental or climatic states and regime shifts are transitions between different regimes (Lluch-Belda *et al.*, 1989, 1992; MacCall, 1996). The bulk of the data on fish population dynamics stems from the fisheries. Catch data may be a measure of fish abundance, however, albeit a crude one, as they usually give an acceptable signal of the trends in population dynamics. Additional data arise from research surveys of fisheries management institutions.

4.1 Humboldt Current

The famous fishery for the Peruvian anchovy (*Engraulis ringens*) was initiated in the early 1950s (Fig. 1). From 1960 until 1972, its average biomass was 15 million MT (Tsukayama, 1983; Alheit and Bernal, 1993). Biomass peaked in 1967 at about 22-23 million MT (Pauly and Palomares, 1989). The stock then dropped rather rapidly to 12 million MT in 1969 and decreased dramatically to 9 million MT in early 1972 and further to 4 million MT at the end of 1972. Within one single year, the stock lost about 5 million MT because of excessive fishing. Subsequently, catches remained between 1 and 6 million MT until 1982, and then decreased to an extremely low level during the ENSO of 1982/83, probably far below 1 million MT (Alheit and Bernal, 1993). This extremely low stock virtually exploded to several million MT in 1985 and 1986 (probably well above 6 million MT, according to Csirke *et al.*, 1996), when 0.7 and 3 million MT, respectively, were caught. Whatever caused this surprising recovery of the Peruvian anchovy, it is obvious that this species can increase its population size rapidly and drastically when environmental conditions become favourable. Although the anchovy population is reduced by the effect of ENSO, this climate signal does not seem to have such a long lasting impact on the anchovy, nor are its effects necessarily as drastic and deleterious in the long term as often assumed (Alheit and Bernal 1993). After 1986, the anchovy stock steadily recovered and reached a biomass of about 10 million MT by the early 1990s (Csirke *et al.*, 1996). In 1994, almost 10 million MT of anchovy were caught (Schwartzlose *et al.*,. 1999). Sardine (*Sardinops sagax*) catches were insignificant during the 1950s and 1960s. From 1964 to 1971, the only distinct spawning areas were in northern Peru and northern Chile (Bernal *et al.*, 1983). After 1971, the sardine showed considerable expansion to the North and the South and after the 1972/73 ENSO, spawning increased strongly and the spawning area was extended considerably (Zuta *et al.*, 1983; Bernal *et al.*, 1983; Alheit and Bernal, 1993). Catches of sardines in the Humboldt Current started to increase in the early 1970s (Fig. 1) and did so particularly in the second half of the 1970s. They rose steadily until 1985 when they reached a peak of 5.6 million MT (Schwartzlose *et al.*, 1999). Thereafter, sardine catches decreased steadily and were below 1 million MT in 1996 (Schwartzlose *et al.*, 1999). This decrease was particularly dramatic in the Chilean sardine fishery where catches dropped from a high of 2.7 million MT in 1985 to 1.4 million MT in 1989, to 0.8 million MT in 1990 and to 0.02

million MT in 1994. As indicated above, the time series of catches and biomass clearly shows an inverse relationship for the dynamics of anchovy and sardine populations in the Humboldt Current. When anchovies have a high biomass and support a strong fishery, sardine biomass and yield are low. As pelagic fisheries were only initiated in the 1950s, there are no catch data from earlier periods. However, such information may be acquired from disciplines other than fisheries research. For example, analysis of fish scale accumulations in varved anaerobic sediments facilitates retrospective studies on fish abundances (see below). Recent preliminary investigations by Baumgartner and colleagues of sediment cores from the upwelling waters off Peru revealed a sardine period from the 1920s to about 1940 (Baumgartner, pers. comm.; Schwartzlose *et al.*, 1999). Studies on the dynamics of sea birds and their diet indicate that there were high abundances of anchovies in the Humboldt system during the 1920s to the early 1930s, as the sea birds which rely heavily on an anchovy diet had flourishing populations during this period (Crawford and Jahnke, 1999). The change from an anchovy-dominated Humboldt system to a sardine-dominated one, initiated in the late 1960s, and continued throughout the 1970s, is also documented by time-series data from zoo- and ichthyoplankton (Alheit and Bernal, 1993). Zooplankton biomass as collected with plankton nets of 300 micron mesh size in two independent studies, one off Peru and the other one off Chile, exhibited a dramatic decrease from the late 1960s to the early 1970s (Bernal *et al.*, 1983; Carrasco and Lozano, 1989). Other strong evidence for fundamental biological changes in the Humboldt system from 1969 – 1970 is the marked shift in the relative abundance of larval fish of non-fished mesopelagic species (Loeb and Rojas, 1988), at the time when the anchovy population initiated its drastic decline. A summary of the regime shifts between anchovies and sardines in the Humboldt Current system is given in Fig. 2. By using data from different disciplines, we have been able to reconstruct a total of 3 anchovy and 2 sardine periods from 1920.

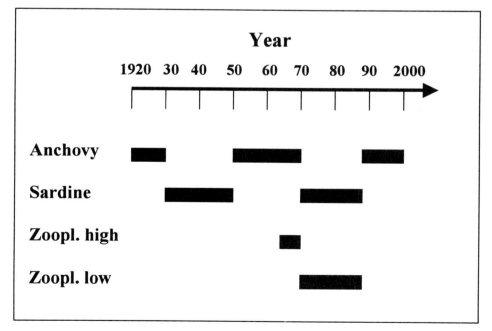

Figure 2. Schematic presentation of the periods of anchovy and sardine regimes in the Humboldt Current, including periods of high and low zooplankton biomass.

4.2 California Current

Surveys on larval fish showed that the Californian anchovy (*E. mordax*) had reached a very low population in the early 1950s, but started to increase in the mid-1960s (MacCall, 1996). Biomass estimates were between 0.2 and 0.6 million MT from 1963 to 1972. These fish had a brief period of high abundance in the mid-1970s with a peak of 1.6 million MT in 1973. Thereafter, biomass decreased and was particularly low between 1990 and 1993, around 0.15 million MT. In 1994, it had increased again to 0.4 million MT (Schwartzlose *et al.*, 1999). Californian sardine (*S. caerulea*) fisheries were very strong from the mid-1920s to 1951 (Fig. 1). The annual catch maximum was recorded in 1937 with 0.7 million MT. Biomass peaked in 1934 with 4 million MT and then declined, particularly after 1944 and, then, again, after 1959, to below 0.1 million MT. A moratorium was exercised in the early 1970s when it was thought that the northern stock was almost extinct. The first signs of recovery were observed in the early 1980s. The stock has also recolonised the northern part of its range up to Vancouver Island where it had been commercially extinct for more than 50 years (MacCall, 1996; Schwartzlose *et al.*, 1999). In 1997, the fishery yield had increased to more than 0.1 million MT, however, this was nowhere near the peak of the 1930s. Biomass had reached 0.5 million MT in 1996. Some authors report a major recovery of the Californian sardine since the mid-seventies (Lluch-Belda *et al.*, 1989; 1992). However, they have erroneously included catch data from the Gulf of California, an area which is not part of the California Current, and from which we do not have quantitative records on sardines before 1969. Over the past 70 years, the California Current has also undergone switches from a sardine-dominated system in the 1930s and 1940s to an anchovy-dominated system in the 1970s and, possibly, back to a sardine system in the late 1980s.

4.3 Japanese Waters

Catches of the Japanese anchovy (*E. japonica*) were between 0.02 and 0.18 million MT from 1905 to 1949 (Fig. 1). Then, they increased abruptly to 0.03 million MT and fluctuated between 0.2 and 0.4 million MT until 1977. Catches ranged between 0.1 to 0.2 million MT from 1978 to 1989. Thereafter, they increased again up to 0.34 million MT in 1996 (Schwartzlose *et al.*, 1999). Catch records for the Japanese sardine (*S. melanosticta*) have been kept since at least 1894 (Fig. 1) and were between 0.1 and 0.6 million MT up to 1927. From this time, they increased sharply until 1942, with a peak of 1.6 million MT in 1936. Thereafter, they decreased again to former levels until the late 1950s when they dropped dramatically to 0.02 million MT, and then dropped even more during the 1960s. They started to recover again around 1971, and reached levels well above 1 million MT from 1976 to 1994, with a peak of 4.5 million MT in 1988 when they started to decrease steadily again. It must be noted that the bulk of the catches during the peak period in the 1930s came from waters east of Japan, whereas the record catches in the 1980s were made mainly in waters west of Japan. Japanese waters were dominated by sardines from the late 1920s to the early 1950s. Then the anchovy became more important until the mid-1970s when the system shifted again to sardine dominance. When the sardine catch decreased in 1989/90, the anchovy gained in importance again, however, its catches were still below those of sardines.

4.4 Benguela Current

The turbulent upwelling cell off Lüderitz separates the anchovy (*E. capensis*) and the sardine (*S. ocellata*) into separate northern (Namibia) and southern (South Africa) stocks. As the fish population dynamics are slightly different in these two areas, they are treated separately here.

Southern Benguela: Anchovies were not fished before the mid-1960s. According to fishermen´s reports and studies of stomach contents of sea birds, they were probably at low levels in the 1950s, but started to increase around the mid-1960s. Anchovy catches dominated the fishery from 1966 to 1995 (Fig. 1). Catches were between 0.14 to 0.60 million MT, with the peak catch in 1988. Thereafter, catches dropped dramatically to an average of 0.05 million MT in 1996 and 1997. Sardine was abundant in the 1950s and, particularly, in the 1960s. Biomass peaked in 1961 with 0.6 million MT, and catches in 1962 with 0.4 million MT (Fig. 1). The fishery and biomass collapsed in the mid-1960s and catches were well below 0.1 million MT (exceptions: 1972 and 1976) until 1991 with an extreme low catch of 0.02 million MT in 1974. Since 1991, annual catches were around 0.1 million MT, except in 1993. The southern Benguela Current was a sardine-dominated system from the 1950s until the mid-1960s. The regime shifted then to an anchovy-dominated one until the late 1980s when the anchovy stock started to collapse and gave way to a sardine regime from the mid-1990s.

Northern Benguela: The anchovy catch data from this region may not be suitable as an indicator of natural population fluctuations because of the management strategy, which was different from that in the southern Benguela. This northern Benguelan fishery only commenced in the mid-1960s. After the collapse of the sardine fishery, the anchovy was subjected to exceptionally heavy fishing pressure in the belief that the removal of a competitor would enhance the recovery of the sardine resource. The highest catches were from 1968 to 1983, well above 0.1 million MT (with the exception of 1976 and 1982) with a peak of 0.36 million MT in 1978. Thereafter, catches declined, particularly after 1983. They then fluctuated between 0.02 and 0.08 million MT (with the exception of 1987 and 1988 with 0.4 and 0.1 million MT respectively). The sardine fishery had high catches from the early 1950s to 1977. The highest yields were in the 1960s, with peaks of 1.4 and 1.8 million MT in 1968 and 1969 respectively. The biomass peaked from 1963 to 1965 between 10.3 and 11.1 million MT. The stock declined thereafter, collapsed in the mid-1970s, and entered a period of catches below 0.1 million MT until 1997, with the exception of 1993 and 1994 when catches were above 0.1 million MT. Although there seemed to be a recovery during the late 1980s and early 1990s, the catches were at extremely low levels in 1996 and 1997. Conclusions concerning regime shifts are problematic because of the very biased anchovy catch data. However, it seems that a sardine-dominated period terminated between the late 1960s and the mid-1970s, and was followed by an anchovy-dominated period. Between the mid- to late 1980s, the system shifted back to sardine dominance.

4.5 Salmon Fisheries in Northeast Pacific

Yields of the salmon fisheries in the subarctic northeast Pacific showed clear interdecadal fluctuations during this century (Francis and Hare, 1994; Hare and Francis, 1995). Total Alaskan salmon catch increased from the mid-1920s and peaked in the mid-1930s. Thereafter, it decreased and the average yield from the early 1950s to the mid-1970s was comparatively low. Catches increased again from the early/mid-1970s and were rather high from the late 1970s. Most of the salmonid production in Alaskan waters comes from four populations which comprise 80% of the total salmon catches in Alaska (by number, during the period of investigation). These are Western and Central Alaska sockeye salmon (*Oncorhynchus nerka*) and Central and Southeast Alaska pink salmon (*Oncorhynchus gorbuscha*). The use of intervention analysis, a time-series analysis technique, demonstrated that the production of these four stocks alternates between periods of low and high production, and that the timing of the transitions from one period to another is almost synchronous for the four different species as well as across a large part of their spatial range in Alaska (Francis and Hare, 1994). Intervention analysis also showed that there was a high production regime from the early 1920s to the late 1940s/early 1950s followed by a low production regime to the mid-1970s. This gave way to a high production regime up to the present (1998). Zooplankton in the northeast Pacific also underwent regime changes from low biomass values in the late 1950s/early 1960s to rather high values in the 1980s (Brodeur and Ware, 1992).

5.   CENTENNIAL-SCALE VARIABILITY

5.1   European Herring and Sardine Fisheries

Bohuslän Herring. The Bohuslän region is a coastal stretch between the North Sea and the Baltic Sea, at the eastern side of the Skagerrak (Fig. 3). Periodically, large amounts of spent herring (*Clupea harengus*) migrated to this coast in the autumn and overwintered in the skerries and fjords. These migrations occurred for periods of decades and supported considerable fisheries (Devold, 1963; Höglund, 1978; Cushing, 1982; Lindquist, 1983; Sahrhage and Lundbeck, 1992; Alheit and Hagen, 1997). These Bohuslän herring periods can be traced back for about 1000 years and nine such periods have been identified (Fig. 4; Table 1). Naturally, the further one goes back, the less precise are the data.

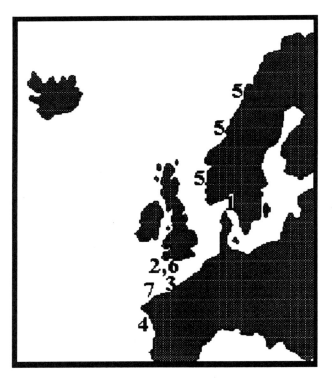

Figure 3. Geographical locations of different European herring and sardine fisheries:
1, Bohuslän herring; 2, Devon and Cornwall herring; 3, French catches of herring
in the English Channel; 4, Bay of Biscay herring; 5, Norwegian spring spawning
herring; 6, Devon and Cornwall sardine; 7, northern Brittany sardine (modified
after Alheit and Hagen, 1997).

Norwegian Spring Herring Fisheries. The Norwegian spring herring forms the largest stock of the Atlanto-Scandian herring and its total biomass may have ranged between 15 to 29 million MT. Its main habitat is the Norwegian Sea (Fig. 3). Historical records indicate large fluctuations of the fishery during the last 500 years. Periods of large catches have alternated with periods of extreme scarcity. The fishing periods of the Norwegian spring-spawners and the Bohuslän periods seem to alternate with each other (Ljungman, 1879; Devold, 1963; Beverton and Lee, 1965; Skjoldal *et al.*, 1993; Alheit and Hagen, 1997; Fig. 4).

Herring and Sardine Fisheries of Southwest England. The English Channel is roughly the geographical boundary between the areas of distribution of the rather cold-water-preferring herring, and the more warm-water-adapted sardine (*Sardina pilchardus*) (Fig. 3). Fishing for both species off the southwestern tip of England, off Cornwall and Devon, has been reported since at least the sixteenth century (Southward *et al.*, 1988). The geographical boundary between the two species seems to shift to and fro on a decadal scale and, consequently, periods of the herring fishery have alternated with those of the sardine fishery (Southward, 1980; Alheit and Hagen, 1997). These periods seem to be linked to the Russell Cycle. This is a periodic and synchronous alternation of appearance and disappearance of a large number of pelagic species, zooplankton and fish, including fish eggs and larvae, in the western Channel, and has been recorded since 1924 (Russell, 1973; Cushing and Dickson, 1976; Southward, 1980; Cushing, 1982).

French Herring and Sardine Fisheries in the English Channel. There are only a few data available on the herring and sardine fisheries of French fishermen in the English Channel from the eighteenth century onwards (Fig. 3). This herring was caught in the eastern Channel off Normandy and Picardy and the periods of this fishery were from about 1750 to 1810 and from 1880 to 1910 (Binet, 1988; Alheit and Hagen, 1997; Fig. 4). A sardine fishery occurred sporadically north of Brest, off northern Brittany (Fig. 3). Various reports on this exist from the period between 1726 and 1764. A notable decrease of sardine abundance was then reported in the first years of the nineteenth century, but the fishery resumed in the 1860s and 1870s (Binet, 1988; Alheit and Hagen, 1997; Fig. 4).

Table 1. Bohuslän herring fisheries periods.

Period	Duration
1	end of 10th century - early 11th century (ca. 970-1020)
2	end of 11th century - early 12th century (ca. 1110-1130)
3	end of 12th century - mid 13th century
4	end of 13th century – 1307/1330
5	mid 15th century
6	1556 - 1589
7	1660 - 1680
8	1747 – 1809
9	1877 – 1906

Bay of Biscay Herring Fisheries. A small herring population exists in the Bay of Biscay off southern Brittany (Fig. 3). This was first mentioned in 1728, however, due to its relative insignificance, reports are rather sporadic, but indicate, nevertheless, alternating periods of presence and absence (Binet, 1988; Alheit and Hagen, 1997; Fig. 4).

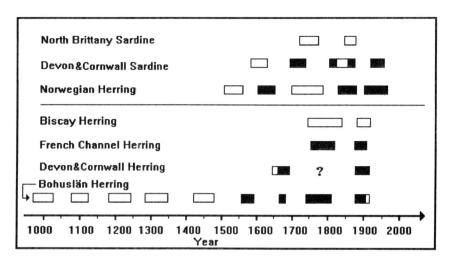

Figure 4. Historical periods of European herring and sardine fisheries. Open rectangles depict periods where the extensions are not known precisely (modified after Alheit and Hagen, 1997).

All these herring and sardine fisheries periods are reported historically as the persistent presence or absence of fish and fisheries at certain locations. The drastic changes between fish-rich and fish-poor periods ensured the maintenance of historical records because the regional economies and the well-being of the local populations were heavily affected. The fluctuations of these fisheries could be the result of real fluctuations of biomass, or of decadal changes in migration routes moving the fish to areas not accessible to the limited range of the fisheries methods employed. However, fluctuations in biomass and migration routes are likely to be concomitant phenomena. These herring and sardine fisheries fall into two groups with alternating periods of occurrence (Fig. 4). The Group 1 species comprise the Bohuslän herring, the herring off south-western England, the herring caught by the French fleet in the eastern Channel off Normandy and Picardy and the Bay of Biscay herring. The Group 2 species consist of the Norwegian spring spawning herring, the sardines off southwest England, and the sardines caught by the French fleet in the English Channel. The records collected over several centuries demonstrate that periods of abundance in Group 1 fisheries alternated with those of Group 2 fisheries (Alheit and Hagen, 1997).

Japanese Sardine Fishery. The Japanese sardine fishery has a long history starting at the time of the beginning of the Tokugawa era (1600 – 1867). Favourable and unfavourable periods of sardine fisheries have been compiled by Kikuchi (1959; as reported by Schwartzlose *et al.*, 1999) and Tsuboi (1987; as reported by Yasuda *et al.*, 1999. See Table 2). However, it can be very difficult to determine when a fishing period begins or ends because no common criteria have been developed. Consequently, authors have used their own judgement and this explains why some of the starting points and end points of the periods given by the two authors differ. However, the important issue is that the peaks of the periods coincide and there is excellent agreement for periods 1, 2, 5 and 6. (Periods 3 and 4 are listed only by Tsuboi)

Table 2. Japanese sardine fisheries periods.

Period	Duration	
	after Kikuchi (1959)	after Tsuboi (1987)
1	1633 - 1660	1638 - 1657
2	1673 - 1725	1678 - 1722
3		1748 – 1754
4		1790 – 1803
5	1817 – 1843	1818 – 1831
6	1858 – 1882	1858 – 1867
7	1920 – 1945 (after Schwartzlose *et al.*, 1999)	
8	1975 – 1995 (after Schwartzlose *et al.*, 1999)	

Paleosedimentary Records of Fluctuations of Californian Sardines and Anchovies

Some ecosystems, particularly the upwelling systems, have varved anoxic sediments which preserve fish scales. Accumulation rates of fish scales in aged sediment cores indicate relative historical fish abundances, and can be calibrated against recent biomass values to infer historical biomass data. Pioneering work has been done by Soutar and Isaacs (1969; 1974) on sediment samples from the Santa Barbara Basin in the California Current. Baumgartner *et al.* (1992) extended the work of Soutar and Isaacs and succeeded in reconstructing a 1750-year time series proxy of anchovy and sardine abundance from A.D. 270 to 1970 (Fig. 5). It clearly demonstrates that anchovy and sardine populations have undergone dramatic natural fluctuations over the last 2000 years, even during periods when fisheries were absent or insignificant.

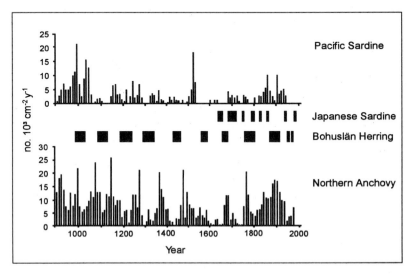

Figure 5. Coincidence of Bohuslän herring periods (rectangles) and Japanese sardine periods (rectangles) (modified after Alheit and Hagen, 2000) with fluctuations in paleosedimentary sardine and anchovy scale deposition in the Santa Barbara Basin, California, (bar graphs; modified after Baumgartner *et al.*, 1992).

6. GLOBAL SYNCHRONY OF FISH POPULATION DYNAMICS AND TELECONNECTION PATTERNS

At the "FAO Expert Consultation to Examine Changes in Abundance and Species Composition of Neritic Fish Resources" in Costa Rica in 1983, Kawasaki (1983) compared fluctuation trends of the three large Pacific sardine stocks (Japan, Humboldt Current, California Current) and demonstrated clearly their synchronous population swings with high biomass values during the 1930s and 1940s (except Peru, as no fishery was developed then), low biomass during the late 1950s and 1960s and increasing catches from the early 1970s to 1980 when the time series ended (at the time of the conference). Not much attention was given to Kawasaki's study in 1983, but now, 17 years later, research teams in Africa, Europe and the Americas, are intensively studying the possible causes of synchronous fish population fluctuations (see for example, Lluch-Belda *et al.*, 1989; Lluch-Belda *et al.*, 1992; Schwartzlose *et al.*, 1999) and additional data can now be added to the time series presented by Kawasaki. With his work on sediment analysis, albeit preliminary, Baumgartner (in: Schwartzlose *et al.*, 1999) was able to prove that the Humboldt sardine also had a period of rather high biomass values during the 1930s and early 1940s similar to that of the sardine populations off Japan and California. Monitoring of the fish populations since 1983 also showed that sardine abundances off Japan and in the Humboldt Current peaked during the second half of the 1980s and dramatically decreased after that time (Fig. 6; and see, e.g., Lluch-Belda *et al.*, 1992; Schwartzlose *et al.*, 1999). The Californian sardine, however, seems to be running out of phase with its congeneric Pacific populations. Also, it showed the first signs of recovery in the early 1980s and this recovery is still continuing at a very slow rate. Schwartzlose *et al.* (1999) assume that the Californian sardine had been fished to such a low level that the re-building of the population takes a very long time. Interestingly, examination of sardine fluctuations in the Benguela Current shows an inverse trend to the Pacific stocks. Benguela sardines had high abundances during the 1960s when sardine catches were very low in the Pacific (Fig. 6). Anchovy present the same scenario, with inverse relationships between the two oceans. Also, as shown above, their population swings are inverse to those of sardines (Lluch-Belda *et al.*, 1992; Schwartzlose *et al.*, 1999). Other fish species also fluctuate in phase with Pacific sardines such as the salmons in the subarctic Northeast Pacific. Similar in-phase relationships in abundance fluctuations have been shown for the European herring and sardine populations (Alheit and Hagen 1997; Fig. 4) on a longer, centennial scale.

The central issue here is that fish populations which are separated by thousands of kilometres and which do not have any opportunities for interacting with each other are fluctuating synchronously (Alheit and Hagen 1997, Bakun 1998). These teleconnection patterns are found within ocean basins (e.g. sardines in Pacific ocean) and between ocean basins (e.g. sardines in North Pacific and in South Atlantic). Even more compelling is a comparison of the historical paleoecological time-series data. There is a surprisingly good match between the Bohuslän fishing periods (Table 1) with the periods when sardine fishing was poor off Japan (Table 2) (Fig. 5). In turn, favourable periods for European sardines match with those of Japanese sardines. Similarly, when the periods of the Swedish Bohuslän herring fishery are superimposed on the paleochronology of sardine and anchovy scales from the Santa Barbara Basin, California, they seem to coincide with high abundances of anchovies and low abundances of sardines off California (Schwartzlose *et al.* 1999; Fig. 5). These teleconnection patterns between fish populations seem to be widespread, and may indicate a global synchrony of fluctuations of many important fish populations around the world (Bakun 1998a). It is difficult to assume that the surprising match of these population fluctuations is just pure coincidence, and the question arises; what forces fish populations on a global scale to fluctuate synchronously?

7.   THE SEARCH FOR CORRELATIONS AND CAUSAL RELATIONSHIPS

Although the synchronous population swings of anchovies and sardines in the Pacific and the South Atlantic have been reported for several years, there is no satisfying theory or hypothesis which explains these teleconnection patterns, nor is there any clear explanation of the causes of regime shifts, the switch from an anchovy-dominated ecosystem to a sardine-dominated one. It is highly unlikely that the forcing signal travels through the ocean as the fish populations are too widely separated for such a signal to initiate population changes at the same time in e.g. the North Pacific and the South Pacific. Consequently, environmental forcing for long-term population dynamics of anchovies and sardines must be brought about by atmospheric processes which are not linearly coupled with ocean dynamics. Because of the decadal nature of the fish-population dynamics it is highly likely that climate variability is the driving force.

Yasuda *et al.* (1999) correlated the 400-year-long time-series on high and low sardine catches in Japanese waters (Table 2) with a temperature series from the northwestern coast of North America reconstructed from tree rings (Fritts, 1991). High catches coincided with periods of positive temperature anomalies and vice versa. However, this temperature series is negatively correlated with sea-surface temperature to the east of Japan. This relationship depends on the variability in location and intensity of the Aleutian Low (AL) system. In periods when the AL is intensified and shifts to the southeast, westerly winds east of Japan are strong, resulting in low sea surface temperatures off Japan, and warm southwesterly winds on the west coast of North America. So, the Japanese sardine seems to flourish with negative temperature anomalies in its environment (Yasuda *et al.*, 1999).

Francis *et al.* (1998) studied a major reorganisation of the northeast Pacific biota caused by a "regime shift" in the mid-1970s. During the winter 1976/77, the Aleutian Low intensified and moved eastward. Prior to this change, sea surface temperatures exhibited cold anomalies along the North American coast and the central North Pacific. This pattern was reversed after the shift of the AL and resulted in population increases of North Pacific salmon and California Current sardines (Fig. 6).

According to Bakun (1998 b) the synchronous variations of fish populations seem to be generated by some type of very large external forcing, most probably through climatic teleconnections acting through the atmosphere. He excludes temperature as the key environmental variable as temperature trends have not been consistent in the various regions. For example, during the period in the 1970s to the mid-1980s when the sardines stocks off Japan and in the Gulf of California (Lluch-Belda *et al.*, 1992) dramatically increased, and when the sardines of both sides of the North Pacific vastly expanded, the northwestern Pacific was in a warm phase whereas the northeastern Pacific clearly had a cool period (Bakun 1996). Bakun (1998b) suggests that the most likely external driving force must be a mechanical one. Wind stress acting on the sea surface is the predominant mechanism for transfer of mechanical energy and momentum between the atmosphere and the ocean. The resulting input of kinetic energy works against stratification of oceanic water. The oceanic stratification regionally depends not only on net heat fluxes between ocean and atmosphere but also on lateral advection processes acting on different oceanic scales not only in time, but also in space. Oceanic currents redistribute heat from low latitudes to high latitudes and influence atmospheric forcing fields around the globe. Consequently, it would be expected that the causal mechanism for the observed regime changes of sardines and anchovies might be a sum of processes driven by the action of the wind on the sea surface depending on regional anomalies in the heat transport of oceanic current systems.

The causes of the teleconnections of the synchronous reactions of fish populations in the northern and southern Pacific are largely obscure. There is some statistical evidence that a negative SOI (Southern Oscillation Index; standardised sea level air pressure gradient

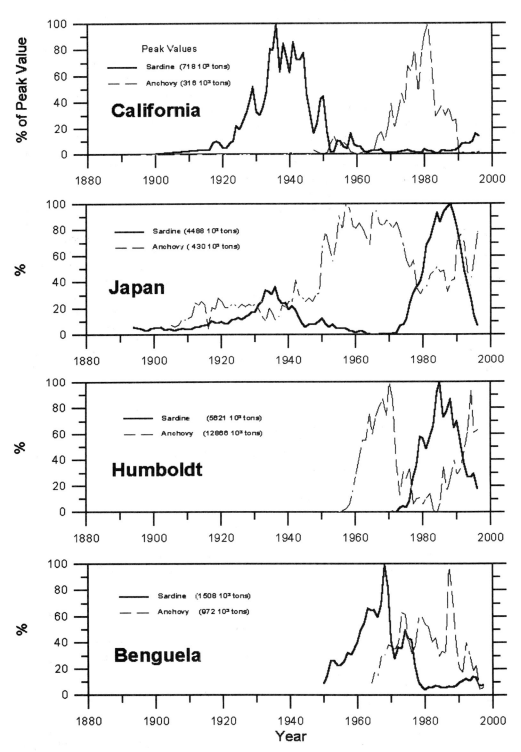

Figure 6. Synchrony of catch records of sardine and anchovy in four major current systems. Catches plotted as percentages of maximum historical annual values (modified after Alheit and Hagen 2000).

between Tahiti and Darwin) correlates with a positive anomaly in the Aleutian Low Pressure index and vice versa. Since ENSO is known to be the dominant mode of interannual-scale climatic variability throughout the world, the ENSO phenomenon seems a possible candidate for implication in the global-scale "regime shifts" due to its pronounced extra-tropical consequences. ENSO-like changes have been reported from areas around the whole globe. Unstable air-sea interactions and the subsurface "thermal memory" of the ocean which is governed by its large heat capacity control such ENSO quasi-cycles (Schwartzlose *et al.*, 1999). The driving forces might not be single ENSO events, but changes in frequency and intensity of ENSO. So, the resulting (very) low frequency changes in the non-linearly coupled ocean-atmosphere system have a strong oceanic component. Regional anomalies in the net meridional heat transport from equatorial latitudes to subarctic and arctic latitudes transform oceanic time scales into the atmospheric circulation. Decadal-scale changes are also observed in the intensity of westerlies at mid-latitudes.

Over the North Atlantic Ocean, the strength of the westerlies is well described by sea level air pressure gradients between the Azores and Iceland. Related fluctuations are expressed by the index of the North Atlantic Oscillation (NAO). It exhibits a clear decadal variability and is frequently attributed to variations in the thermohaline circulation (Weisse and Mikolajewicz, 1994). Such variations are forced by density anomalies at the sea surface which modify the spatial pattern and the strength of oceanic deep water convection which, in turn, leads to large-scale variations in the oceanic general circulation. Released anomalies in net heat fluxes between ocean and atmosphere modify atmospheric forcing fields. Alheit and Hagen (1997) correlated the different herring and sardine populations in Europe with the NAO index. They showed that the Bohuslän fishing periods coincided with periods of severe winters as determined by the winter severity index of Lamb (1972; Fig. 7). Whenever this index showed a dramatic decrease, there was a strong herring fishery in the Bohuslän region. Such periods of severe winters occur when the NAO index is negative. Thus, there were strong fisheries for Group 1 species (herrings) (Fig. 4) under the following conditions: negative NAO index; cold air temperatures at mid- and higher latitudes of the eastern North Atlantic; negative sea-surface anomalies; increased ice cover off Iceland and in the northern Baltic Sea; reduced westerly winds over the North Atlantic; minimum frequency of southwesterly winds over England; and cold water in the North Sea, English Channel and in the Skagerrak. Fisheries for Group 2 species (sardines and Norwegian spring spawning herring) were well developed during periods with a persistent positive NAO index with concomitant environmental conditions contrasting to the scenario described above (Alheit and Hagen, 1997). These findings agree well with the experience that herrings prefer colder environments than sardines and that fluctuations in herring and sardine fisheries are related to temperature regimes (Southward *et al.*, 1988). The Norwegian spring spawning herring might be an exception as its habitat is in the far North and a cooling of the eastern North Atlantic might have caused a deterioration of its environment.

A very tentative explanation for possible teleconnections between fish populations in the Atlantic and the Pacific might be the following scenario. When the subtropical Pacific gyre circulation intensifies, more warm tropical water will be transported poleward by western boundary currents and their extensions. This leads to positive anomalies in the sea-surface temperature in the North Pacific. The atmospheric response was explained by Latif and Barnett (1996). It is expressed by anomalies in the Pacific North America (PNA) air pressure pattern. In this way, the impact of the ENSO phenomenon on atmospheric circulation is not confined to the Pacific/North American region but may also be recognised in the Atlantic/European sector. Here, the response is much weaker than in the Pacific region (Fraedrich, 1994). The decadal nature of fish-population fluctuations might arise from time scales of coupled ocean/atmosphere processes. According to Grötzner *et al.* (1998), the long-term memory of the ocean circulation in the North Atlantic can be conceptually separated into a wind-driven and a thermohaline component. The former is

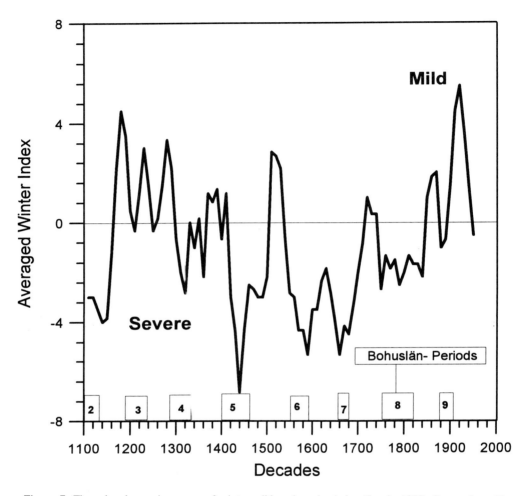

Figure 7. Three-decade running mean of winter mildness/severity index (Lamb, 1972). Rectangles with numbers denote Bohuslän herring periods which correspond to negative peak values of the index (modified after Alheit and Hagen, 1997).

associated with horizontal circulations in the oceanic "warm water sphere" while the latter is related to a meridional circulation and deep convection in subpolar regions. They concluded from simulations with a coupled ocean-atmosphere model that the adjustment time of the gyre circulation can be expected to be between 10 and 20 years, while that of the thermohaline circulation indicates time scales of several decades. Observations analysed by Deser and Blackmon (1993) and Kushnir (1994) support this conclusion.

All these correlation studies indicate more or less strongly that fish-population fluctuations and regime shifts are the consequence of climate variability, however, they do not definitively prove this. The existence of the physical forcing source (the coupled ocean/atmosphere system) and the reaction of fish-populations is well established, but the exact processes leading to changes in fish population dynamics are not well understood (Baumann, 1998).

A further crucial issue is whether effects of climate variability work their way up through the food chain before fish stocks are affected, or whether there is a more direct impact on fish populations (Bakun, 1998a). Francis *et al.* (1998) present a conceptual model of physical-biological linkages via the food web to explain the impact of climate variability on marine ecosystems. Based on a modelling exercise of phytoplankton response to physical forcing, Polovina *et al.* (1995) suggest that changes in the mixed layer depth caused by the shift and intensification of the AL have resulted in changes in biological production regimes from primary production to fish production. In contrast, Bakun (1998a) rejects the idea that climatic effects work their way through complex planktonic food webs because trophic interactions in marine ecosystems are characterised by chaotic behaviour. He concludes that the biological dynamics involved must be very simple, such as a rather direct effect of the external forcing acting either on the fish themselves at some sensitive stage in their complex life cycles, or directly on a primary food source.

8. IMPLICATIONS FOR FISHERIES MANAGEMENT

The evidence that fluctuations of exploited fish populations are to a large extent governed by climate variability certainly does not mean that no fisheries management will be required. In addition to the impact of natural variability, the dynamics of exploited fish populations are strongly influenced by fishing mortality. The growing body of environmental information will serve as a valuable tool for appropriate fisheries management. Thus, for example, fishing activities could be adapted to whether fish populations are on the ascending or the descending slope in the course of their decadal population variability. The crucial issue, of course, is to establish which regime and which phase of the regime are prevailing at any given time. This is the task of current research projects such as the "Small Pelagic Fishes and Climate Change" (SPACC) project (Hunter and Alheit, 1995) of the IGBP Global Change Core Programme "Global Ocean Ecosystem Dynamics" (GLOBEC) programme (IGBP 1997).

Acknowledgement

This study was partially supported by the ENVIFISH-Project (MA53-CT96-0058) funded by the European Union.

REFERENCES

Alheit, J., and Bernal, P., 1993, Effects of physical and biological changes on the biomass yield of the Humboldt Current ecosystem, in: *Large Marine Ecosystems – Stress, Mitigation and Sustainability.* K. Sherman, L.M. Alexander and B.D. Gold, eds., American Association for the Advancement of Science, Washington.

Alheit, J., and Hagen, E., 1997, Long-term climate forcing of European herring and sardine populations, *Fish. Oceanogr.* 6:130.

Alheit, J., and Hagen, E., 2000, Climate variability and historical NW European Fisheries, in: *Past Climate and its Significance for Human History in NW Europe*, the Last 10 000 Years, in press.

Bakun, A., 1996, *Patterns in the Ocean: Ocean Processes and Marine Population Dynamics*, University of California Sea Grant, San Diego.

Bakun, A., 1998a, Radical interdecadal stock variability and the triad concept: a window of opportunity for fishery management science? in: *Reinventing Fisheries Management*, T.J. Pitcher, P.J.B. Hart and D. Pauly, eds., Chapman and Hall, London.

Bakun, A., 1998b, Global climatic teleconnections and fisheries ecology. A perspective (with comments on the Benguela system), Paper distributed at International Symposium on Environmental Variability in the South-East Atlantic, Swakopmund, Namibia, 30 March – 1 April 1998.

Baumann, M., 1998, The fallacy cf the missing middle: physics → .... → fisheries, *Fish. Oceanogr.* 7:63.

Baumgartner, T.R., Soutar, A., and Ferreira-Bartrina, V., 1992, Reconstruction of the history of Pacific sardine and northern anchovy populations over the past two millennia from sediments of the Santa Barbara Basin, California. Calif. Coop. Oceanic Fish. Invest. Rep. 33:24.

Beamish, R.J., ed., 1995, *Climate Change and Northern Fish Populations, Can. Spec. Publ. Fish. Aqat. Sci.* 121.

Bernal, P.A., Robles, F., and Rojas, O., 1983, Variabilidad fisica y biologica en la region meridional del sistema de corrientes Chile-Peru. FAO Fish. Rep. 29:683.

Beverton, R.J.H., and Lee, A.J., 1965, Hydrodynamic fluctuations in the North Atlantic Ocean and some biological consequences, in: *The Biological Significance of Climatic Changes in Britain,* C.G. Johnson and L.P. Smith, eds, Academic Press, New York.

Brodeur, R.D., and Ware, D.M., 1992, Long-term variability in zooplankton biomass in the subarctic Pacific Ocean, *Fish. Oceanogr.* 1:32.

Carrasco, S., and Lozano, O., 1989, Seasonal and long-term variations of zooplankton volumes in the Peruvian Sea, 1964-1987, in: *The Peruvian Upwelling Ecosystem: Dynamics and Interactions,* D Pauly, P. Muck, J. Mendo and I. Tsukayama, eds, ICLARM Conference Proceedings 18:82.

Castillo, S., and Mendo, J., 1987, Estimation of unregistered Peruvian anchoveta (*Engraulis ringens*) in official catch statistics, 1951 to 1982, in: *The Peruvian Anchoveta and its Upwelling Ecosystem: Three Decades of Change,* D. Pauly and I. Tsukayama, eds, ICLARM Studies and Reviews 15:109.

Crawford, R.J.M., Shannon, L.V., and Shelton, P.A., 1988, Characteristics and management of the Benguela as a large marine ecosystem, in: *Biomass Yields and Geography of Large Marine Ecosystems,* K. Sherman and L.M. Alexander, eds, American Association for the Advancement of Science, Washington.

Crawford, R.J.M., and Jahnke, J., 1999, Comparison of trends in abundance of guano-producing seabirds in Peru and southern Africa, *S. Afr. J. mar. Sci.* 21:145.

Csirke, J., Guevara-Carrasco, R., Cárdenas, G., Niquen, M., and Cipollini, A., 1996, Situacion de los recursos anchoveta (*Engraulis ringens*) y sardina (*Sardinops sagax*) a principios de 1994 y perspectivas para la pesca en el Peru, con particular referencia a las regiones norte y centro de la costa Peruana, Bol. Instituto del Mar del Peru 15:1.

Cushing, D., 1982, *Climate and Fisheries,* Academic Press, London.

Cushing, D., and Dickson, R.R., 1976, The biological response in the sea to climatic changes, *Adv. Mar. Biol.* 14:1.

Deser, C., and Blackmon, M. L., 1993, Surface climate variations over the North Atlantic ocean during winter: 1900-1989, *J. Climate* 6:1743.

Devold, F., 1963, The life history of the Atlanto-Scandian herring, Rapp. *Proc.-verb. Réun. Cons. Int. Explor. Mer* 154:98.

Fraedrich, K., 1994, An ENSO impact over Europe? *Tellus* 46A:541.

Francis, R.C., and Hare, S.R., 1994, Decadal-scale regime shifts in the large marine ecosystems of the Northeast Pacific: a case for historical science, *Fish. Oceanogr.* 3:279.

Francis, R.C., Hare, S.R., Hollowed, A.B., and Wooster, W.S., 1998, Effects of interdecadal climate variability on the oceanic ecosystems of the NE Pacific, *Fish. Oceanogr.* 7:1.

Fritts, H.C., 1991, *Reconstructing large-scale Climate Patterns from Tree-ring data,* The University of Arizona Press, Tucson.

Glantz, M.H., 1996, *Currents of Change – El Niño's impact on climate and society,* University Press, Cambridge.

Grötzner, A., Latif, M., Timmermann A., and Voss R., 1998, Internal to decadal predictability in a coupled ocean-atmosphere general circulation model, Max-Planck-Institut für Meteorologie Hamburg, Report No. 262.

Hare, S.R. and Francis, R.C., 1995, Climate change and salmon production in the Northeast Pacific Ocean, in: *Climate Change and Northern Fish Populations,* R.J. Beamish, ed., Can. Spec. Publ. Fish. Aqat. Sci. 121:357.

Höglund, H., 1978, Long-term variations in the Swedish herring fishery off Bohuslän and their relation to North Sea herring, *Rapp. Proc.-verb. Réun. Cons. Int. Explor. Mer* 172:175.

Hunter, J.R. and Alheit, J., 1995, International GLOBEC Small Pelagic Fishes and Climate Change program. *GLOBEC Report No. 8.*

IGBP, 1997, Global Ocean Ecosystem Dynamics (GLOBEC) – *Science Plan. IGBP Report* 40.

Isaacs, J.D., 1976, Some ideas and frustrations about fishery science, *Cal. Coop. Oceanic Fish. Invest. Rep.* 18:34.

Kawasaki, T., 1983, Why do some pelagic fishes have wide fluctuations in their numbers? – Biological basis of fluctuation from the viewpoint of evolutionary ecology, *FAO Fish. Rep.* 291:1065.

Kikuchi, T., 1959, A relation between the alternation between good and poor catches of sardine and the establishment of Shinden and Naya villages, *Memorial Works dedicated to Professor K. Uchida*:84.

Kushnir, Y., 1994, Interdecadal variations in the North Atlantic sea surface temperature and associated atmospheric conditions, *J. Climate* 7:141.

Laevastu, T., 1993, *Marine Climate, Weather and Fisheries,* Fishing News Books, Blackwell Scientific Publications, Oxford.

Lamb, H.H., 1972, *Climate: Past, Present and Future. I. Fundamentals and Climate Now.* Methuen, London.

Latif, M. and Barnett, T.P., 1996, Decadal variability over the North Pacific and North America: dynamics and predictability, *J. Climate* 9:2407.

Lindquist, A., 1983, Herring and sprat: fishery independent variations in abundance. *FAO Fish. Rep.* 291:813.

Ljungman, A., 1879, Contribution towards solving the question of the secular periodicity of the great herring fisheries, *US Commission Fish Fisheries* 7 (7):497.

Lluch-Belda, D., Crawford, R.J.M., Kawasaki, T., MacCall, A.D., Parrish, R.H., Schwartzlose, R.A., and Smith, P.E., 1989, World-wide fluctuations of sardine and anchovy stocks: the regime problem, *S. Afr. J. mar. Sci.* 8:195.

Lluch-Belda, D., Schwartzlose, R.A., Serra, R., Parrish, R., Kawasaki, T., Hedgecock, D., and Crawford, R.J.M., 1992, Sardine and anchovy regime fluctuations of abundance in four regions of the world oceans: a workshop report, *Fish. Oceanogr.* 1:339.

Loeb, V.J., and Rojas, O., 1988, Interannual variation of ichthyoplankton composition and abundance relations off northern Chile, 1964-83, *Fish. Bull., U.S.* 86:1.

MacCall, A.D., 1996, Patterns of low-frequency variability in fish populations of the California Current, *Cal. Coop. Oceanic Fish. Invest. Rep.* 37:100.

Pauly, D., and Palomares, J.L., 1989, New estimates of monthly biomass, recruitment, and related statistics of anchoveta (*Engraulis ringens*) off Peru (4^0-14^0S), 1953-1985, in: *The Peruvian Upwelling Ecosystem: Dynamics and Interactions.* D. Pauly, P. Muck, J. Mendo and I. Tsukayama, eds, ICLARM Conference Proceedings 18:189.

Russell, F.S., 1973, A summary of the observations on the occurrence of the planktonic stages of fish off Plymouth 1924-1972, *J. mar. Biol. Ass. U.K.* 53:347.

Sahrhage, D., and Lundbeck, J., 1992, *A History of Fishing,* Springer-Verlag, Berlin.

Schwartzlose, R.A., Alheit, J., Bakun, A., Baumgartner, T., Cloete, R., Crawford, R.J.M., Fletcher, W.J., Green-Ruiz, Y., Hagen, E., Kawasaki, T., Lluch-Belda, D., Lluch-Cota, S.E., MacCall, A.D., Matsuura, Y., Nevarez-Martinez, M.O., Parrish, R.H., Roy, C., Serra, R., Shust, K.V., Ward, N.M. and Zuzunaga, J.Z. 1999, Worldwide large-scale fluctuations of sardine and anchovy populations, *S. Afr. J. mar. Sci.* 21:289.

Skjoldal, H.R., Noji, T.T., Giske, J., Fossa, J.H., Blindheim, J. and Sundby, S., 1993, *MARE COGNITUM – Science Plan for Research on Marine Ecology of the Nordic Seas,* Inst. Mar. Res., Bergen.

Soutar, A., and Isaacs, J.D., 1969, History of fish populations inferred from fish scales in anaerobic sediments off California, *Calif. Coop. Oceanic Fish. Invest. Rep.* 13: 63.

Soutar, A., and Isaacs, J.D., 1974, Abundance of pelagic fish during the 19th and 20th centuries as recorded in anaerobic sediment off the Californias, *Fish. Bull., U.S.* 72:257.

Southward, A.J., 1980, The Western English Channel – an inconstant ecosystem, *Nature* 285:361.

Southward, A.J., Boalch, G.T., and Maddock, L., 1988, Fluctuations in the herring and pilchard fisheries of Devon and Cornwall linked to change in climate since the 16th century, *J. mar. Biol. Ass.* 68:423.

Tsuboi, M., 1987, Japanese sardine spawning grounds circuiting around Honshu, Shikoku and Kyushu (1-3), Sakana (*Bull. Tokai Reg. Fish. Lab.*), 38-40 (in Japanese).

Tsukayama, I., 1983, Recursos pelagicos y sus pesquerias en el Peru, Rev. Com. Perm. *Pacifico Sur* 13:25.

Weisse, R. B., and Mikolajewicz, U., 1994, Decadal variability of the North Atlantic in ocean general circulation model, *J. Geophys. Res.* 99:12411.

Wyatt, T., and Larraneta, M.G., 1988, *Long Term Changes in Marine Fish Population,* Instituto de Investigaciones Marinas, Vigo.

Yasuda, I., Sugisaki, H., Watanabe, Y., Minobe, S.-S., and Oozeki, Y., 1999, Interdecadal variations in Japanese sardine and ocean/climate. *Fish. Oceanogr.* 8:18.

Zuta, S., Tsukayama, I., and Villanueva, R., 1983, El ambiente marino y las fluctuaciones de las principales poblaciones pelagicas de la costa peruana, *FAO Fish. Rep.* 291:179.

CLIMATE AND HUMAN HEALTH LINKAGES ON MULTIPLE TIMESCALES

Henry F. Diaz[1], R. Sari Kovats[2], Anthony J. McMichael[2], and Neville Nicholls[3]

[1]Climate Diagnostics Center, Environmental Research
Laboratories, NOAA, Boulder, CO 80303, USA
[2]London School of Hygiene and Tropical Medicine,
Keppel Street, London, WC1E 7HT, UK
[3]Bureau of Meteorology Research Centre,
PO Box 1289K, Melbourne 3001, Australia

1. INTRODUCTION

Interest in public health aspects of climatic variability has grown greatly in the past decade. This has come about as knowledge about climatic changes and its impacts on society in general has increased, leading to greater efforts being made toward documenting and establishing possible links between climatic variability and changes in human health and well-being.

The idea that human health and disease are linked with the climate probably predates written history. The Greek physician Hippocrates (about 400 BC) related epidemics to seasonal weather changes. Robert Plot, Secretary to the Royal Society, took weather observations in 1683–84 and noted that if the same observations were made "in many foreign and remote parts at the same time" we would "probably in time thereby learn to be forewarned certainly of divers emergencies (such as heats, colds, dearths, plagues, and other epidemical distempers)." (Quoted in Symons 1900.) Public interest in the climate-health relationship strengthened during the 18th and 19th centuries after many studies indicated that certain diseases occurred more often in particular seasons. This close climate-health relationship is apparent in the titles of many books published during the 19th century, for instance, *Climate and health in Australasia*, by James Bonwick, published in London in 1886, and James Kilgour's 1855 book *Effect of the climate of Australia upon the European Constitution in Health and Disease*. These books were among many arguing whether Australia's climate was beneficial or harmful to human health (Nicholls 1997a).

The interest in climate as a precondition for disease probably reached its peak towards the end of the 19th century. For instance, Latham (1900) read a major review paper relating climate conditions to plague to the Royal Meteorological Society, in December 1899.

History and Climate: Memories of the Future?
Edited by Jones *et al.*, Kluwer Academic/Plenum Publishers, 2001

267

Latham, who was President of the Society in 1890–92, noted that "while it is admitted that plague is due to a specific microbe, it cannot spread except under certain meteorological conditions associated with the conditions of the ground, which must be in such a state as to exhale what is necessary for the propagation and spread of this particular disease". The general acceptance that germs were the proximate cause of most diseases, and the increasing ability to treat these diseases with pharmaceuticals, led to a decline in interest in the climate-health relationship until late in the 20th century. Major epidemics associated with the 1982/83 El Niño event, encouraged studies on the impact of climate on health, and concerns about possible impacts of climate change on health inspired further studies during the 1990s. Some of these studies (e.g., Nicholls 1986) seemed to confirm that Robert Plot had been correct, in that climate observations (of the El Niño–Southern Oscillation) could be used to "forewarn" of some epidemics.

In this chapter, we document some key possible connections between aspects of climatic variability and variations in the incidence of some human diseases which appear to be modulated, at least partly, by climatic variations. Since the large-scale ocean-atmosphere phenomenon known as El Niño/Southern Oscillation (ENSO) represents the largest source of interannual variability in the modern climate system, ENSO will be a main focus of attention here. We also examine some aspects of climatic variability that lead to changes in human well-being, such as drought and floods, heat and cold waves, and other extreme events.

As this paper represents a contribution to the proceedings volume of the Second International Climate and History Conference, we try here to develop aspects of our subject matter related to their historical context. For a thorough treatment of the subject of climate change related to aspects of human health, we refer the reader to McMichael *et al.* (1996), and references therein, which is a state-of-the-art assessment prepared for several United Nations organizations. Other useful treatments of this subject may be found in Martens *et al.* (1999), Patz *et al.* (2000), and McMichael and Githeko (2001).

2.   CLIMATIC VARIABILITY FACTORS IMPORTANT TO HUMAN HEALTH

Climate varies across all space and time scales. However, locally, natural ecosystems, including humans become adapted to the prevailing climate conditions, as these normally change only slowly over time. Interannual and decadal-scale variability are intrinsic parts of current as well as past climates. However, for a variety of reasons, including competition for resources among human populations and with other species, the effects of shorter-term climatic fluctuations—for example, those occurring on human time scales of 100 years or less—can sometimes lead to environmental changes that affect the ability of animal and human species to adapt, locally or regionally, to these naturally occurring climatic changes.

In the following, we first present a review of ENSO and its teleconnections, followed by a brief overview of the main findings of previous studies regarding possible connections between climatic changes, vector-borne diseases, chronic ailments and others linkages. We have made an effort to point out some of the intervening circumstances, which are often present, that ameliorate or exacerbate the effect of climatic variability on humans.

2.1 El Niño and the Southern Oscillation

The ENSO phenomenon is one the best examples of quasi-periodic internal climate variability on interannual time scales (Glantz *et al.* 1991; Diaz and Kiladis 1995; Anderson *et al.* 1998). It is not possible, at the present time, to state unequivocally whether external climate forcing mechanisms, such as solar influences, or increasing atmospheric greenhouse gas concentrations, can alter—or indeed *have* altered—the basic ENSO system. The few studies made of the subject (e.g., Enfield and Cid 1991; Tett 1995; Latif *et. al.* 1997; Rajagopalan *et al.* 1997) are not conclusive, and the observational climate record of the last century, suggests a substantial degree of decadal-scale variability in the Pacific Ocean that resembles, in both spatial and temporal characteristics, the variations in sea surface temperature associated with the interannual ENSO phenomenon (Latif *et al.* 1997; Zhang *et al.* 1997; Broccoli *et al.* 1998).

The alternating phases of the ENSO system are associated with marked deviations from the mean atmospheric and oceanic conditions. The development of ENSO events tends to peak from late Northern Hemisphere summer to early winter. Notable exceptions occurred in the warm (El Niño) events of 1982, 1986, and 1991, which developed late in the calendar year and peaked during northern winter-spring. The temporal variability of various indices of the SO contain broad spectral peaks in the range of 3 to 7 years, with little power at 2 years but a secondary peak at about 28 months (Rasmusson and Carpenter 1982; Trenberth and Shea 1987).

The most pronounced signals occur during the year that an ENSO extreme first develops (referred to as "Year 0," following the convention of Rasmusson and Carpenter 1982) and into the following year ("Year+1"). Figures 1–3, modified and extended from Kiladis and Diaz (1989), show regions of statistically significant differences in precipitation (Figs. 1 and 2) and temperature (Fig. 3) between cold and warm events for selected seasons, and will be used here as a basis for the following discussions. Only the western hemisphere temperature signal is illustrated in Fig. 3.

2.2 Effect of ENSO on Precipitation and Temperature

ENSO appears closely tied to the Asian and Australian monsoon circulations, which are notably weaker during warm events (Webster and Yang 1992). Figures 1 and 2 illustrate the effect of ENSO on precipitation in the region of India and around northern Australia. While precipitation increases in the central and equatorial Pacific during warm events it becomes markedly drier over regions bordering the western Pacific and eastern Indian Ocean such as Indonesia and Australia (Allan 1988; 1991; Nicholls 1992; Ropelewski and Halpert 1987), India (Rasmusson and Carpenter 1983), and southeast Asia (Fig. 1a). Failure of monsoon rains during warm events can have catastrophic impacts (Kiladis and Sinha 1991). Conversely, cold events are often associated with flooding in India. The Australasian signal is more evident during the northern fall of year 0, and persists over the monsoon region of northern Australia into the December through February rainy season (Fig. 1b). However, during strong ENSO events, the entire continent of Australia can be affected by severe drought conditions (Allan 1991).

As warm events evolve towards their "mature" phase in the northern winter season (DJF+1), drier than normal conditions continue in the region of the western Pacific Intertropical Covergence Zone from the Philippines eastward, and east of Australia in the normal position of the South Pacific Convergence Zone (Kiladis and van Loon 1988). Similarly, a large region of southeastern Africa, including parts of Zimbabwe, Mozambique, and South Africa, has a marked tendency for drought during DJF+1 of warm events. Precipitation in this region is highly seasonal, so this signal can have an especially large impact, as it occurs during the normal southern summer rainy season.

Farther north in Africa, there are indications that warm events favor drought conditions from the Sahel eastward to the highlands of Ethiopia during the normal summer rainy season of JJA 0. It should be noted, however, that the Sahel region experiences high amplitude rainfall variability at lower frequency than that associated with ENSO variability (Lamb and Peppler 1991).

While coastal Ecuador and northern Peru often experience flooding during warm events, other regions of the Americas often register large rainfall anomalies. A consistent

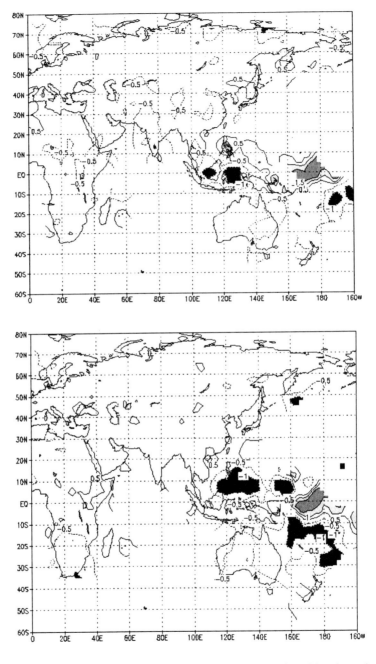

Figure 1. ENSO precipitation signal in the Eastern Hemisphere. Map shows the precipitation difference in standardized units for the past century, between El Niño and La Niña years for the three months seasons of (top panel) June–August of "Year 0," and (bottom panel) December–January of "Year +1." Solid (dashed) contours indicate positive (negative) deviations during the El Niño phase. The shading indicates difference anomalies exceeding ±1 (local) standard deviations. Modified after Kiladis and Diaz (1989).

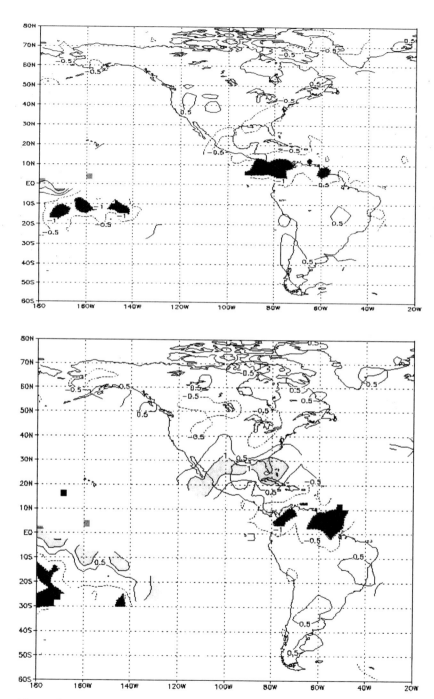

Figure 2. ENSO precipitation signal in the Western Hemisphere. Map shows the standardized temperature difference for the past century between El Niño and La Niña years for the three months seasons of (top panel) June–August of "Year 0," and (bottom panel) December–January of "Year +1." Solid (dashed) contours indicate positive (negative) deviations during the El Niño phase. The shading indicates difference anomalies exceeding ±1 (local) standard deviations. Modified after Kiladis and Diaz (1989).

dry signal is found from JJA 0 through DJF+1 over northern South America (see Figure 2; also, Rogers 1988). Over northeast Brazil, the periodic occurrence of severe drought in the agriculturally rich "Nordeste" region in connection with El Niño events has resulted in severe economic hardship and occasional famines in this region (see a review by Hastenrath 1995). Widespread and severe famine was reported during the great El Niño of 1877–78 (Kiladis and Diaz 1986), while other severe El Niño episodes, such as the ones that occurred in 1982–83, and. most recently in 1997–98, have led to great suffering.

The northern Amazon basin is also affected, while, in contrast, much of southern South America is wet during JJA 0 (Fig. 2a), and this signal persists into SON 0 in Uruguay, and central Chile and Argentina (not shown; see Diaz and Kiladis 1995). A remnant of this signal is still present in DJF+1 (Fig. 2b), when the southeastern United States and northern Mexico also show above normal precipitation. In the heavy rainfall areas of the Paraná and Paraguay River Basins in South America, MAM+1 precipitation tends to be above normal, and this contrasts with the relatively dry rainy season over Central America and Colombia. These signals are highly reliable from event to event.

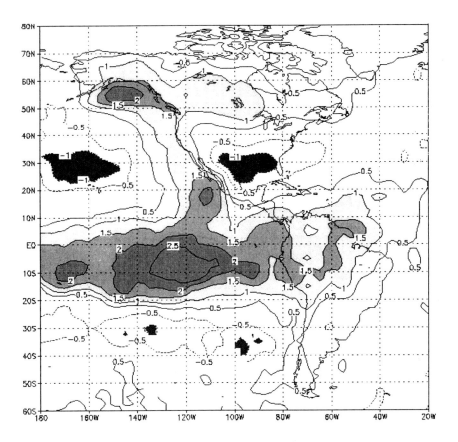

Figure 3. ENSO temperature signal in the North and South American sector. Map shows the standardized surface temperature difference for the past century between El Niño and La Niña years for the three months season of December–January of "Year +1." Solid (dashed) contours indicate positive (negative) deviations during the El Niño phase. The shading indicates difference anomalies exceeding ±1 (local) standard deviations. Modified after Kiladis and Diaz (1989).

Although the best correlations between ENSO and temperature are observed in the tropics and subtropics, the Americas can experience large midlatitude temperature anomalies during ENSO events. Strong and relatively mild westerly flow from the Pacific into North America during warm events is responsible for a large region of positive temperature departures from southern California northwards along the west coast to Alaska, then inland across western and central Canada (Figure 3). This signal over northwest North America is one of the most reliable in the extratropics from event to event (see Kiladis and Diaz 1989; Diaz and Kiladis 1992). Cold (La Niña) events have been equally reliable in their being associated with anomalously cold winter seasons in this region. In those years, a tendency towards a weak jetstream over the central Pacific leads to atmospheric "blocking" patterns in the Gulf of Alaska, which in turn are associated with anomalous northerly flow over northwestern North America.

We should note, however, that the most recent La Niña episode, extending from mid-1998 to the time of this writing, in mid-2000, has departed considerably with respect to its expected surface temperature signal in much of North America. Whether this represents a random chance occurrence is not possible to tell at present. However, recent studies have shown that regional warming over North America since the late 1960s (Livezey and Smith 1999) is associated with rising sea surface temperatures (SST) in much of the globe (see also Levitus *et al.* 2000). The impact of this long-term rise in regional US surface temperatures may have been large enough to reduce the expected probability of occurrence of "below average" winter and spring temperatures during La Niña events in parts of North America.

Precipitation patterns during the cold La Niña phase of the Southern Oscillation (SO) also exhibit highly coherent anomalous signals throughout the globe, as they do during warm El Niño events (Ropelewski and Halpert 1989). While in general, the climatic anomalies arising from the cold SO phase are opposite those of the El Niño phase, some areas experience greater or lesser impacts from one or the other phase of the phenomenon, and in some regions La Niña conditions lead to a more consistent climatic response (less inter-event variability) than during the El Niño (see Hoerling and Kumar 1997; Hoerling et al 1997). Changes in precipitation patterns related to ENSO phases associated with the large-scale pattern of rising worldwide temperatures have not been detected at present.

2.3 Historical changes in ENSO

Most investigators have concluded that there has been no statistically significant change in the frequency of El Niño itself between the period known as the "Little Ice Age[1]" and recent times (since the late 19th century, see Diaz and Markgraf 1992). There are some indications of century-scale oscillation in the frequency of El Niño over the past few centuries, based on a variety of historical records, suggesting that El Niño was relatively more frequent during the periods 1680 to 1770 and 1820-1930 (Diaz and Pulwarty 1992, 1994). This is also consistent with proxy records of tree ring width over the southwestern United States and northern Mexico (Michaelsen 1989), and other climate indicators in regions sensitive to ENSO.

Work on historical reconstruction of El Niño events was pioneered by W. H. Quinn (Quinn *et al.* 1978, 1987; Quinn and Neal 1992; Quinn 1992, 1993). More recent work has extended that work and major El Niño episodes have been further substantiated from

---

[1]   There has been a great deal of research on the so-called Little Ice Age (LIA), and debate continues as to the exact timing of this supposedly global period of cooler climate. Nominally, the LIA has been taken to run from about the mid-16th through the early part of the 19th century. Some argue for an earlier start—into the 15th century, or even earlier, while evidence from the Southern Hemisphere, suggests a somewhat different timing for the LIA there. The reader is referred to Bradley (2000) for a short review of the controversy and a list of applicable references.

archival sources of major climatic anomalies (i.e., severe drought and widespread famine, or major floods) in India (Grove 1998), and South America (Ortlieb 2000). Major El Niño events affecting India were documented in the years around 1685–1688, 1789–1793, and 1877–1878 (Grove 1998; Kiladis and Diaz 1986; Ortlieb 2000).

The ENSO teleconnection patterns as described above, may also vary substantially over time. For example, Kiladis and Diaz (1989) and Diaz and Pulwarty (1992) noted that warm events were associated with drier than normal conditions at some stations in Ethiopia and southern Sudan, near the headwaters of the Nile. Quinn (1992) and Diaz and Pulwarty (1992) show a reasonably good correspondence between the stronger El Niño events and deficient Nile flow for the 1824-1973 period. The most recent portion of the Nile data reflects persistent drought conditions which has affected the Sahel region of Africa since 1972.

Since the mid-1970s, the ENSO system has favored the warm (El Niño) phase of the oscillation (Trenberth and Hurrell 1994; Trenberth and Hoar 1996). The result has been generally warmer than average tropical SSTs and a pattern of mid-latitude atmospheric circulation since that time that tends to resemble the mean atmospheric response to individual El Niño events (Zhang *et al.* 1997; Hoerling and Kumar 2000).

3.   CRITICAL HEALTH LINKS TO CLIMATIC VARIABILITY

In the past 10 years or so, increasing attention has been focused within the climate and medical communities on the issue of climatic change and human health impacts (e.g., WHO 1990; NHMRC 1991; CGCP 1995; Epstein 1995; Unninayar and Sprigg 1995; Martens *et al.* 1995a,b; McMichael *et al.* 1996; Curson *et al.* 1997; Martens *et al.* 1997; McMichael and Haines 1997). We will note, at the outset, that the emphasis of this study, is aimed more on aspects of the natural variability of the climate system, as it may be related to variations in human disease and human health.

McMichael *et al.* (1996) pointed out that research in this relatively new area of climate impacts in the health sciences area is rapidly developing, and that empirical information upon which to base various types of mechanistic climate and health models is generally inadequate. Environmental health concerns usually focus on toxicological or microbiological risks to health from factors within the local environment. The following subsection illustrates the need to understand the health impact of climate variability within a population-level, and predominantly ecological, framework.

3.1 Long-term Climate Variations, Food Production, Famines, and Other Disasters

A fundamental influence on human health is the effect of a change in climate upon food production. Such effects occur on two different timescales. First, long-term changes in climatic conditions can alter the extent and viability of a civilisation. Second, short-term climatic fluctuations can cause disruption of food supplies, famine, mortality and social unrest. Indeed, acute famines have been a longstanding characteristic of pre-industrial agriculturally-based societies everywhere.

Over recent millennia, various civilisations and settlements have struggled to feed themselves as background climatic conditions have changed. One such example is the mysterious demise of the Viking settlements in Greenland in the fourteenth and fifteenth centuries (see elsewhere in this volume), as temperatures in and around Europe began to fall. Established during the so-called Medieval Warm Period (MWP) around A.D. 1000 (see Hughes and Diaz 1994), these culturally conservative livestock-dependent settlements could not cope with the progressive deterioration in climate. Food production declined, food importation became more difficult as sea-ice persisted, and the native Inuit population in

Greenland was pressing southwards, probably in response to the ongoing climate change. The settlements eventually died out or were abandoned (Pringle 1997).

Historical accounts of acute famine episodes, occurring in response to climatic fluctuations, abound (Bryson and Murray 1977). Throughout pre-industrial Europe, diets were marginal over many centuries, and the mass of people survived on monotonous diets of vegetables, grain gruel and bread. As late as 1870, for two-thirds of the French population, diet consisted of bread and potatoes. Yields of animal foods were low, and meat and fish were dietary luxuries. Over these centuries, average daily intakes were less than 2000 calories, falling to around 1800 in the poorer regions of Europe. This permanent state of dietary insufficiency led to widespread malnutrition, susceptibility to infectious disease, and low life expectancy. The superimposed frequent famines inevitably culled the populations, often drastically. The famine of 1696–97 in Finland killed around 30% of the population. In the late eighteenth and nineteenth centuries, elsewhere in the world, similar proportions died from major famines in Bengal and Ethiopia, respectively.

In Tuscany, between the fourteenth and eighteenth centuries there were over one hundred years of recorded famine. France experienced frequent famines between the tenth and eighteenth century, including 26 famines that affected the whole country during the eleventh century and 16 such famines in the eighteenth century. Meanwhile in China, where the mass rural diet of vegetables and rice accounted for an estimated 98% of caloric intake, between 108 BC and A.D. 1910 there were famines that involved at least one province in over 90% of years (Bryson and Murray 1977).

A particularly dramatic example in Europe was the great medieval famine of 1315–17. Climatic conditions deteriorated over much of the decade, and the cold and soggy conditions led to widespread crop failures, food price rises, hunger, and death. Social unrest increased, robberies multiplied, and bands of desperate peasants swarmed over the countryside. Reports of cannibalism abounded from Ireland to the Baltic. Animal diseases proliferated, contributing to the die-off of over half the sheep and oxen in Europe. This tumultuous event, and the Black Death which followed thirty years later, is deemed to have contributed to the weakening and dissolution of feudalism in Europe.

The later decades of the sixteenth century were a time of increasing cold, occurring at a time when the European population had already reached the limit that the agricultural system could support (Ponting 1991; Post 1985). The growing season contracted by around a month, and the altitude at which crops could be grown declined by around 200 metres. A succession of four bad harvests in 1594–97 resulted in a desperate return to cannibalism and to the eating of cats and dogs. The parish register of the late 1590s in Orlosa, western Sweden, has recorded the local experience:

> "The soil was sick for three years, so that it could bear no harvest. After these inflictions it happened that even those who had good farms turned their young people away, and many even their own children, because they were not able to watch the misery of them starving to death in the homes of their mothers and fathers. Afterwards the parents left their house and home going whither they were able, till they lay dead of hunger and starvation. . . . At times these and other inflictions came and also the bloody flux [dysentery] which put people in such a plight that countless died of it."

The climate remained extremely cold for the next few years. Indeed, in 1602–03, temperatures were sufficiently cold for the Guadalquivir river in Seville, Spain, to have frozen in winter. This Europe-wide downturn in food production around the turn of the century lead to widespread malnutrition, starvation and death. This, in turn, is likely to have contributed to the unusually acute period of political instability within Europe at that time (Ladurie 1972; Ponting 1991).

Similar worsening periods of famine, disease and death occurred in Europe in the late seventeenth and eighteenth centuries during periods of intensified cold weather. Great famines occurred in France in 1693 and 1709, late in the reign of Louis XIV. The last severe famine to affect the whole of Europe came in 1816–17. While exacerbated by the Napoleonic wars and the depredations of the Grand Armée, the underlying cause was the "year without summer" in 1815 following the atmospheric clouding caused by the massive volcanic eruption of Tambora in Indonesia. Crop failure was widespread, food riots (including food-export protests) broke out across most of Europe in 1816 and 1817, the number of infectious disease epidemics increased, and death rates rose (Hufton 1985).

The last major, and devastating, famine in Europe occurred in Ireland in the 1840s. The climate in Europe had deteriorated during the 1830s, again apparently contributed to by volcanic eruptions in the early 1830s. Famines occurred in Norway in 1836 and 1838 as that country struggled through the seven bad harvest years of 1835–1841. During 1837–39, the death rate doubled in the Dombas district, associated with starvation (O. Nordli, pers. comm.).

Meanwhile, in Ireland, population pressure on land had increased enormously. The population had grown ten-fold since 1500, and, by the 1840s, around 650,000 landless labourers were living in squalid and destitute conditions. The pressure to produce food from tiny plots of land led to an increasing reliance on the potato, such that in the early nineteenth century potato farming accounted for 40% of total crop area in Ireland. Potatoes, having originated in the Andes, were not well adapted to the wet climate of northwest Europe (Ireland was particularly wet), and were prone to outbreaks of disease. By the 1830s, poor potato harvests were becoming common in Ireland, perhaps exacerbated by the deterioration in climate during that decade. In 1845, catastrophe occurred when a fungal disease was introduced from America. The ensuing potato blight wiped out crops in Ireland, partially in 1845, and totally in 1846 (Ponting 1991). The blight spread elsewhere in northern Europe. The Irish catastrophe was heightened by the British Government's repeal of its own Corn Laws. This allowed the import of grain to relieve the concurrent shortages due to bad harvests in England. The impoverished Irish peasantry had no money to buy imported grain; besides, Irish grain was now being exported, often under armed guard, to Britain. Overall, an estimated one million Irish died from the potato blight famine. Many deaths were due to starvation; many were due to the outbreaks of infectious diseases, especially typhus, that accompanied the famine.

3.2 ENSO, famines, and other disasters

Worldwide, rates of persons affected by drought/famine account for about half of all disaster victims, and these show a significant association with the ENSO cycle. El Niño is important because it is associated with drought in many vulnerable regions at the same time (see discussion in earlier section). This aggregate effect has even led to several world food crises (Dyson 1996). For example, in 1972, drought hit several of the major grain-producing regions of the world, including northern and west parts of South Asia and north-east China. The most effected populations were in Ethiopia, the Sahel and in parts of India and China. The world food crisis of 1982–83 was also linked to the El Niño, when famines struck populations in Ethiopia and the Sahel, which were also badly affected by the civil war.

Perhaps not unexpectedly, due to the pervasive influence of the ENSO phenomenon on atmospheric and oceanic phenomena, a number of studies have found significant asoociations worldwide between ENSO and increased risks of natural disasters (see, e.g., Dilley and Heyman 1995; Bouma *et al.* 1997). Data on natural disasters (1964–1994) obtained from the EM-Dat Disaster Events based at the Centre for Epidemiology of Disasters at the Catholic University of Louvain, Belgium (Sapir and Misson 1992), show that in nearly all El Niño years (with the exception of 1976), there was an increase in the number of people affected by natural disasters. The data suggest an El Niño disaster cycle, with low risk in the years before an El Niño and high risk years during and after an El Niño (years 0, +1). This cycle of risk should be seen in the context of the trend in the increased vulnerability of populations to natural disasters. Regional analyses show that the impact of ENSO on the number of persons affected by natural disasters is strongest in South Asia, which contributes more than 50% of all disaster victims, due to its high population density and high absolute population level.

Earlier ENSO events also resulted in disasters (Nicholls 1991). About 10 million people perished in North China due to famine in the 1877 El Niño, with over 8 million deaths in India. In many districts a quarter of the population died. Perhaps as many as a million also died in northeast Brazil in that event, and droughts caused food shortages in famine through Africa, while Tahiti was struck by a "terrible hurricane." The 1888 event again affected India and Brazil, but the greatest effect was on Ethiopia where about one-third of the population died. In some areas up to 80% of the population was lost, leading to the total desertion of some previously well-populated areas.

Southern Africa is subject to recurrent droughts which cause severe food shortages. The 1991–92 El Niño event was accompanied by the worst drought in southern Africa this century, affecting nearly 100 million people (Dyson 1996). In 1992, 20 million people needed food relief in the Southern African Development Community (SADC) region, 15% of the total African population. The worst hit countries were Zimbabwe, Mozambique and northern parts of South Africa. These problems were exacerbated by the war in Mozambique and related refugees in Zimbabwe. In 1991, Zimbabwe had been forced to sell some of its grain reserves by the World Bank. At this time, an El Niño had been forecast but this forecast was not communicated to the government (Dyson 1996). Clearly, if the Zimbabwean government had been aware of the forecast, they might have been able to resist the pressure from the World Bank.

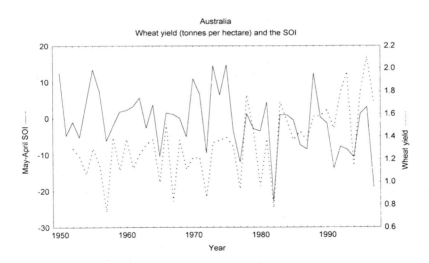

Figure 4. Australian crop yields and ENSO. Time series of the SOI (solid curve) and average wheat yields (dashed curve).

Several studies have shown a relationship between ENSO and food production in southern Africa. Cane *et al.* (1994) found a strong relationship between SOI and maize yields in Zimbabwe. SOI and SST parameters are both related to seasonal rainfall in the SADC region (Matariria and Unganai 1995; Mason *et al.* 1994). A comparison of ENSO-related SST anomalies and NDVI (a satellite-derived vegetation index) shows that NDVI can be used to identify specific locations in Africa where ENSO activity has an important effect on climate (Anyamba and Eastman 1996).

Australian crop yields are closely related to (and predictable with) ENSO indices (Nicholls 1985, 1988; Rimmington and Nicholls 1993). Figure 4 shows time series of Australian average wheat yield and the Southern Oscillation Index (or SOI, the standardised pressure difference between Tahiti and Darwin). Wheat yield shows an increase due to other factors including recent increases in minimum temperatures (Nicholls 1997b), but the interannual variations are clearly related to ENSO, with low yields tending to occur when the SOI is negative (i.e., during El Niño events).

The foregoing discussion is meant to provide a broad brushstroke overview of some critical events in world history (though admittedly with some European bias), where the intersection of human choices and natural phenomena related to climatic variations helped to produce catastrophic results. In what follows, we examine some specific climate–health linkages, starting with heatwaves and cold spells, followed by a discussion of some recent studies of possible physical connections between ENSO-mediated changes in seasonal aspects of climatic variability and the incidence of disease.

3.3 Infectious Diseases and Climate Variability

Climate plays a dominant role in determining the distribution and abundance of insects and tick species, either directly or indirectly through its effects on host plants and animals. The direct effects of climatic variability on vector biology occur through changes in temperature, precipitation, humidity and winds. Vector-borne disease transmission is sensitive to temperature fluctuations. Increases in temperature can accelerate vector life cycles or allow the vector to colonise areas that were previously too cold. Increases in temperature also decrease the intrinsic incubation period of the parasite or virus (e.g. malaria parasite, dengue or yellow fever virus) and so the vectors become infectious more quickly.

There is often an increased risk of a disease following an extreme weather event. Several mechanisms can explain the association between rainfall anomalies associated with ENSO (drought, heavy rain, flooding) and disease outcome (see Kovats *et al.* 1999). There is a well-studied relationship between rainfall and diseases spread by insect vectors which breed in water and are therefore dependent on water availability (Gilles 1993).

The main species of interest are mosquitoes which spread malaria and viral diseases such as dengue, Rift Valley fever and yellow fever. Mosquito-borne disease transmission is also sensitive to temperature fluctuations. Increases in temperature can accelerate vector life cycles or allow the vector to colonise areas that were previously too cold. Increases in temperature also decrease the intrinsic incubation period of the parasite or virus and so the vectors become infectious more quickly (MacDonald 1957).

One of the most severe outbreaks of yellow fever, a viral disease transmitted by the *Aedes aegypti* mosquito, affected the southern United States in the summer of 1878. The economic and human toll was enormous, and the city of Memphis, Tennessee was one of the most affected. Estimates placed the death toll as high as 20,000 people, out of about 100,000 cases of the disease, with attendant economic losses of as much as $200 million (Bloom 1993). Diaz and McCabe (1999) carried out a study of possible connections of this outbreak of yellow fever to climatic conditions prevailing in the southern United States during and immediately preceding the 1877–78 El Niño event. As noted above, this was one of the strongest El Niño episodes on record, and it led to the development of exceptional climate anomalies throughout the world. In the United States, large and persistent temperature and precipitation anomalies were recorded from mid-1877 to mid-1878. Diaz and McCabe (1999) suggested that the unusual mildness and humid conditions prevalent during this period, may have been partly responsible for the widespread nature and severity of the 1878 yellow fever outbreak.

Diaz and McCabe (1999) also noted that other years with major outbreaks of yellow fever in the 18th and 19th centuries also occurred during the course of El Niño episodes in the eastern United States, a fact which appears not to have been noted before in the literature.

The relationships between rainfall and water-washed or water-ingested diseases have been less well-studied. These are principally diseases that are spread via water or food that is contaminated with faecal material due to flooding or drought. Examples of such diseases include cholera, typhoid, and other diarrhoeal diseases. Outbreaks of such diseases can occur after flooding. Note that outbreaks of infectious disease are often associated with catastrophic events not only due to the initial cause (e.g. flooding) but also due to population displacement and overcrowding. Many health consequences have been recorded after flooding in South America associated with the El Niño of 1982–83. Increases in the incidence of acute diarrhoeal diseases and acute respiratory diseases were recorded in Bolivia (Telleria 1986), and in Peru (Gueri *et al.* 1986). Flooding in the Horn of Africa associated with the El Niño of 1997 was linked with an upsurge of deaths due to cholera due to affected sanitation and contaminated water supplies (WHO 1998). Other authors have suggested an association between dengue fever in the South Pacific and ENSO (Hales *et al.* 1996).

The majority of deaths and disease that are associated with El Niño can be attributed to disasters. This is illustrated by the impacts of torrential rain, landslides and flooding in Peru during El Niño. A report on the impacts of the 1982/83 El Niño in Peru lists the following impacts in 1983 (percentages refer to increase compared to 1982):

- 39.79% increase in all-cause mortality
- 103% increase in infant mortality;
- 176% increase in infant mortality due to intestinal disease;
- 284% increase in infant mortality due to respiratory infection;
- 70.36% increase in mortality due to malnutrition.

In 1998, very heavy rainfall was experienced along the coastal regions of Ecuador and Northern Peru. In the Piura region of Peru, there were 12 separate days with at least half its annual rainfall and in Talara, Peru, five times the normal annual rainfall fell in a single day. It is not surprising that severe health consequences followed these events. The Peru-UNICEF programme reported that 22 of the 24 departments in the country were badly affected. It was estimated that 67,068 families were affected with over 20,000 homes destroyed.

3.4 Malaria

In areas where malaria may be marginally endemic, transmission is particularly sensitive to climate variations (Bouma 1995). Further, in these areas of "unstable" malaria, populations lack protective immunity and are prone to epidemics when weather conditions facilitate transmission. Many such areas across the globe experience drought or excessive rainfall due to ENSO teleconnections.

Climate variability may also affect disease transmission in areas with 'stable' or endemic malaria (Macdonald 1957). Work by Colombian investigators have shown that the

ENSO affects the number of reported malaria cases in that country (see Anon 1996; Poveda and Rojas 1996, 1997; Poveda *et al.* 1999, 2000). The work by Poveda and colleagues demonstrates the need for careful analysis of both the epidemiological data in conjunction with careful study of the effects of large-scale patterns of climatic variability, such as the ENSO phenomenon. In Colombia, for example, the effects of El Niño are felt as a reduction of the normal high rainfall regime in much of the country (Poveda and Mesa 1997; Poveda *et al.* 1997; 2000). It is hypothesized that reduced high runoff and streamflow associated with the lower rainfall, actually benefits the reproduction of the mosquito vector, by providing more slack water areas in which to breed. Temperatures also tend to be higher during El Niño episodes, also favoring mosquito activity and disease transmission (Poveda et al 2000).

The relationship between ENSO and malaria has also been examined in Venezuela (Bouma and Dye 1997). This country also experiences below average rainfall during El Niño. Figure 5 illustrates that malaria rates increased on average by 37% in the post-Niño year (+1). However, this analysis used total country data and does not reflect the dramatic localized surges of malaria that can occur during post-Niño years.

The link between malaria and anomalous climatic events has long been the subject of study in places like the Indian subcontinent. Earlier this century, the Punjab region of India experienced periodic epidemics of malaria. This geographic plain region, irrigated by the five rivers that give the province its name, borders the Thar desert. Excessive monsoon rainfall has been firmly identified as a major epidemic factor since 1908 through its effect on the vector (increased breeding and a longer lifespan due to higher rainfall-related humidity) (Christophers 1911). Since 1921, forecasts of malaria epidemics in the districts of Punjab have been issued based on established relationships between rainfall and malaria mortality (Gill 1920, 1921, 1923). This system was probably the first mathematically-supported malaria early warning system ever used (Swaroop 1946). The ability to forecast the socially and economically important annual monsoon was a major incentive for the

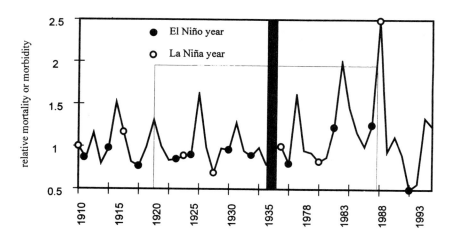

Figure 5. Time series of the malaria cases and deaths in Venezuela. El Niño years are marked by black dots. Source: Bouma and Dye (1997).

careful recording of accurate climate data. In addition, good records of the malaria epidemics in the Punjab (1868–1943) (now north-east Pakistan and north-west India) remain. Historical analyses have shown that the risk of a malaria epidemic increased five-fold during the year following an El Niño (+1 year) in this region (Bouma and van der Kaay 1994, 1996). The risk of an epidemic is greater in a year with excess rain in critical months.

The Punjab no longer experiences malaria epidemics due to economic and ecological changes. Epidemic malaria is still a serious problem in the more arid areas in Western Rajasthan and Gujarat in India, and Pakistan. Malaria epidemics in these areas are linked to excessive rainfall which is strongly associated with ENSO. A strong correlation is found between annual rainfall on the one hand and the number of rainy days on the other and malaria incidence in most districts of Rajasthan and some districts in Gujarat (Akhtar and McMichael 1997). Risk years, with excessive monsoon rainfall can be expected particularly in years following an El Niño (+1 year) and during La Niña years (Bouma *et al.* 1994). Recent studies have shown promising advances in the use of remote sensing as an aid in forecasting vector borne illnesses, such as Rift Valley fever (Linthicum *et al.* 1999). Together with modelling efforts, and the availability of other ecological and meteorological data, this technique offers hope that prediction capabilities with a few months lead time, could soon be in use.

3.5 Arboviruses in Australia

Epidemic polyarthritis is caused by infection with several arboviruses, of which Ross River virus is the most important. Ross River virus is transmitted by a wide range of mosquito species in complex transmission cycles with several possible mammal and bird intermediate hosts. The disease is distributed throughout Australia and elsewhere in the South Pacific. In northern and central Queensland, south-east Australia, cases are reported throughout the year. Tong *et al.* (1998) demonstrated that notified cases of epidemic polyarthritis (1986–95) in one district in Queensland, were positively associated with monthly SOI. The number of cases were also positively associated with temperature, rainfall and humidity.

In other regions of Australia, virus activity tends to be epidemic following spring and summer rains (Mackenzie and Smith 1996). In arid areas of Australia, the virus is thought to persist in mosquito eggs for considerable time (Lindsay *et al.* 1993). When environmental conditions become favourable, such as with heavy rain or flooding, the eggs hatch into infected mosquitoes and a localised outbreak of disease may occur. In coastal regions with salt marsh habitats (principally south-western Australia), sporadic cases may occur at anytime of year. In these areas, sea level rise and tidal effects have been shown to more important than rainfall patterns for local vector abundance.

Outbreaks of Ross River virus disease are linked to discrete vector/virus cycles in different parts of the country. Therefore, local studies are needed to determine a consistent relationship with the ENSO cycle. Harley and Weinstein (1996) found no relationship between La Niña events and epidemic polyarthritis outbreaks "years" or notified cases in Australia as a whole. La Niña is associated with above average rainfall in parts of Australia. Infection with Ross River virus confers life-long immunity, and time since last outbreak is an important factors affecting the risk. Major outbreaks have occurred every 3-4 years despite years of high rainfall in-between.

Murray Valley Encephalitis (MVE), also known as Australian Encephalitis, is another arboviral disease in Australia. Interannual rainfall variability in eastern and northern Australia are closely related to the Southern Oscillation. Frequent small epidemics of MVE occur in tropical Australia. However, infrequent but severe epidemics of MVE have occurred in temperate south-east Australia, after well above average rainfall and flooding associated with La Niña episodes. Thus, years with MVE cases in south-east Australia are

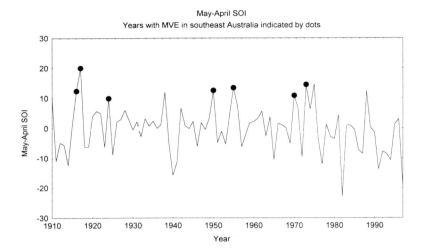

Figure 6. Time series of the Tahiti–Darwin Southern Oscillation Index averaged from May–April and years with outbreaks of Murray Valley Encephalitis (MVE) marked by large black dots.

positively correlated with the SOI (Nicholls 1986, 1988, 1993). Figure 6 shows time series of the SOI with MVE epidemics in south-east Australia indicated. These only occur when the SOI is positive, ie during La Niña events.

3.6 Heatwaves and Cold Spells

Within certain ranges of tolerance human biology can handle most variations in climate, whether these relate to rate of change or degree of change. But marked short-term fluctuations in weather can cause "acute" adverse effects. These are often indicated by increased death rates, upswings in hospital admissions, and increases in the number of individuals complaining of mental stress such as depression.

A U-shaped relationship has been widely observed between daily temperature and mortality in populations temperate regions. Mortality is lowest within an intermediate comfortable temperature range which depends on the local climate. For example, lowest mortality rates are observed when mean daily temperature is approximately 23°C in Greece, but 16°C in the Netherlands (Katsouyanni 1998; Kunst *et al.* 1993). Mortality appears to be more strongly associated with temperature for heat-related deaths than for cold-related deaths, although the latter occur over a much greater temperature range. The rate of increase in deaths as daily winter temperature decreases appears to be considerably less steep than that accompanying increasing temperatures in summer. Thus, mortality appears to be more strongly associated with temperature for heat-related deaths than for cold-related deaths, although the latter occur over a much greater temperature range.

For temperatures above that intermediate comfortable range, there is sometimes a "threshold" above which mortality increases markedly (Kalkstein and Tan 1995). In a typical year, relatively few people die from heat stroke. However, it is now becoming clear that hot weather can also increase the likelihood of dying from other causes. During heatwaves in the USA and in Europe, deaths from all causes are shown to increase. In July 1995 in Chicago, 514 deaths were certified as heat-related (primarily cardiovascular deaths) but a total of 696 excess deaths were attributed to the severe heatwave (Whitman *et al.* 1997). The immediately following heatwave in London, UK, was associated with a 15% increase in daily mortality and approximately 520 excess deaths (Rooney *et al.*, 1998). Physiological acclimatisation to hot weather can occur over a few days. The impact of the first heatwave on mortality is often greater than the impact of subsequent heatwaves in a single summer (see McMichael *et al.* 1996). This effect can also be explained in part by the accumulating deaths of susceptible individuals. i.e. towards the end of the summer there are fewer susceptible people alive to die in a heatwave. A major heatwave in July 1987 in Athens, Greece, was associated with 2000 excess deaths (Katsouyanni *et al.* 1988). In all urban areas in Greece, excluding Athens, a 32.5% increase in mortality was observed in this month compared to the average July mortality for 1981-1986 (Katsouyanni *et al.* 1993). A subsequent Athens heatwave in July 1988, was associated with 28 heat-related deaths (Katsouyanni *et al.* 1988). Although this heatwave was shorter than the one in 1987, it is likely that the both the general population and medical personnel were better prepared.

Cold spells are also associated with short-term increases in mortality. Although death rates during the winter season are often 10-25% higher than those in the summer (Kilbourne 1992), an important cause of this excess mortality is deaths from respiratory disease and influenza which are not directly related to temperature. Further, behavioural responses such as cold avoidance are an important thermoregulatory process at very low temperatures in populations well-accustomed to extreme cold. Other influences include seasonal patterns of respiratory infections (such as influenza) and diet. Thus, the extent of winter-associated mortality that is directly attributable to stressful weather is difficult to determine. Some evidence exists however that, in extreme climates, stormy rather than very cold weather is responsible for some wintertime excess mortality (Rogot and Padgett 1976; Glass and Zack 1979).

People can and do acclimate to a large range of climatic conditions—after all, humans are distributed just about everywhere over the land surface of the earth. As noted above, marked short-term fluctuations in temperature, as well a those that persist for several days, especially those which are accompanied by high humidty can result in a large excess in mortality compared to seasonal norms. The literature on these effects of climatic variations on human health and well-being is extensive. We refer the reader to Chapter 3 of McMichael *et al.* (1996) for a more thorough review and bibliographic reference list.

3.7 Complex Climate-Health Relationships

The social, ecological, and economic impacts of climate anomalies such as droughts can be mediated or aggravated by interactions between the climate system, human actions, and the ecosystem. These interactions can even result in climate impacts unforeseeable and unpredictable from a simple consideration of the climate system alone.

As noted earlier, the very strong 1877–78 ENSO episode had severe impacts in many parts of the world, with over eight million deaths in India which suffered severe drought during this El Niño. The immediate cause of many of the deaths was malaria, but most of these deaths occurred after the breaking of the drought (Whitcombe 1993). Malaria incidence increased because of several factors. The drought had probably reduced the numbers of predators of mosquitoes, so that when mosquito breeding increased after the drought-breaking rains there was nothing to restrict the population. The drought would have weakened the human population, making them more vulnerable to malaria. Perhaps as important, however, large numbers of people had assembled in relief centres and towns where food was available, encouraged by the Government. The concentration of people increased the pressure on water supplies, further weakening the population and increasing their vulnerability to malaria. As well, the concentration of population supplied the breeding mosquitoes with an adequate food supply for breeding. Government actions to alleviate the effects of the drought and famine, therefore, ultimately led to increased loss of life after the end of the drought. Such a consequence could only have been avoided if the interactions between crowded labour camps, mosquitoes, and malaria were understood. Even today, relief camps are used in response to famines. Such complex interactions between human action and climate can mask relationships between climate and health consequences, or can exacerbate the apparent effect of climate variations on human health.

4. CONCLUDING REMARKS

While the interactions between climatic factors and human health may be varied and complex, there are a few key elements of this emerging science that are likely to have fundamental importance. These factors may be classified into three broad categories, although this is not meant to imply that they are independent of one another.

We might call these climate-human health interactions "mediating processes." One set of climatic processes, such as thermal extremes, and increased frequency of storms, can modulate the rates of heat- and cold-related illnesses and death, either directly (through fatal cold and heat exposure, or from the violent effects of a storm), or indirectly through psychological disorders and through the impacts associated with a diminished public health system. These two factors—increased morbidity due to the personal after-effects of natural disasters (post-traumatic stress disorder, reduced economic resiliency), as well as greater risks of disease incidence due to a compromised public health infrastructure—often lead to enhanced human impacts, particularly in developing countries.

Another factor arises through the interactions of humans with their environment. Ecological disturbances caused by an ever expanding human population affect the range and activity of disease vectors and infective parasites. These could occur via changes in the geographic ranges of the vector carriers, either from human activities such as land-clearing , or through the effects of climatic changes (whether natural of anthropogenic), whereby, regions that were at one time beyond the range of a particular infective agent, become accessible to, say a mosquito carrier, as a consequence of increases in temperature. Changes caused by humans and/or climate can also change the distribution of pests (e.g., rats, fungal agents, molds, etc.), which can impact food production, as well as rates of transmission of infectious diseases).

A third component, is related to the second one, but it's considered here in its own right, because it tends to operate on more global scales, compared to the more regional impacts associated with the second category of climate-human health interactions, described above. Global climate change, including ozone depletion, may ultimately have very major public health consequences, through, *inter alia*, changes in the distribution and transmission of vector borne diseases, global and/or regional food production, air and water quality and water quantity, changes in storminess and other extremes, etc.

Although we believe that improving the public health systems of nations across the globe is likely to yield the greatest immediate and long-term health benefits, we believe that research into the mechanisms associated with climatic influences on human and ecosystem health is also important. In particular, studies aimed at prevention of outbreaks of vector-borne diseases, which are often tied to drought and flood cycles associated with ENSO and other climate mechanisms are possible. Our current ability to forecast the ENSO cycles well in advance, and to use numerical models of the coupled ocean-atmosphere system to make predictions about likely outcomes in monthly or seasonal temperature and precipitation for many regions of the globe may ultimately lead to mitigation, and in some cases, possible prevention of climate-senstive illnesses.

The confluence of the globalization of human society, and the potential for significant climatic change in the next century resulting from anthropogenic causes, indicates that increasing our knowledge of their possible interactions will be critical to the well being of humankind. Indeed, we find ourselves happily in agreement with Hippocrates, when he states in the very first sentence of his classic book, *On Airs, Waters, and Places*, "Whoever wishes to investigate medicine properly, should proceed thus: in the first place to consider the seasons of the year, and what effects each of them produces for they are not at all alike, but differ much from themselves in regard to their changes."

REFERENCES

Anon, 1996, El Niño-Oscillacion del sur y su relación con la incidencia de malaria en Colombia, *Informe, Quincenal Epidemiológica Nacional* 1:29-33

Akhtar, R., and McMichael, A.J., 1996, Rainfall and malaria outbreaks in western Rajasthan, *Lancet* 348:1457–1458.

Allan, R.J., 1988, El Niño–Southern Oscillation influences in the Australasian region, *Progress in Physical Geography* 12:4-40.

Allan, R.J., 1991, Australasia. in: *Teleconnections Linking Worldwide Climate Anomalies*, M.H. Glantz, R.W. Katz, and N. Nicholls, eds., Cambridge University Press, pp.73-120.

Anderson, D.L.T., Sarachik, E.S., Webster, P.J., and Rothstein, L.M, 1998, The TOGA Decade, Reviewing the Progress of El Niño Research and Prediction, *Journal of Geophysical Research* 103:14,167–14,510.

Anyamba, A., and Eastman, J.R., 1996, Interannual variability of NDVI over Africa and its relation to El Niño/Southern Oscillation, *International Journal of Remote Sensing* 17:2533-2548.

Bloom, K. J., 1993, *The Mississippi Valley's Great Yellow Fever Epidemic of 1878*, Louisiana State University Press, 290pp.

Bouma, M.J., 1995, *Epidemiology and Control of Malaria in Northern Pakistan. With Reference to the Afghan Refugees, Climate Change, and El Niño Southern Oscillation*, ICG Printing, Dordrecht.

Bouma, M.J., and Dye, C., 1997, Cycles of malaria associated with El Niño in Venezuela, *Journal of the American Medical Association* 278:1772–1774.

Bouma, M.J., Kovats, S., Goubet, S.A., Cox, J., and Haines, A., 1997, Global assessment of El Niño's disaster burden, *Lancet* 350:1435–1438.

Bouma, M.J., Sondorp, H.E., and van der Kaay, H.J., 1994, Climate changes and periodic epidemic malaria. *Lancet* 343:1440.

Bouma, M.J., and van der Kaay, H.J., 1994, Epidemic malaria in India and El Niño Southern Oscillation. *Lancet* 344:1638–1639.

Bouma, M.J., and van der Kaay, H.J., 1996, The El Niño Southern Oscillation and the historic malaria epidemics on the Indian subcontinent and Sri Lanka: an early warning system for future epidemics? *Tropical Medicine and International Health* 1: 86–96.

Bradley, R.S., 2000, Climate paradigms for the last millennium, *PAGES Newsletter* 8(No. 1):2–3.

Broccoli, A.J., Lau, N.-C., and Nath, M.J., 1998, The Cold Ocean–Warm Land pattern: Model simulation and relevance to climate change, *Journal of Climate* 11:2743–2763.

Bryson, R.E., and Murray, T.J., 1977, *Climates of Hunger: Mankind and the World's Changing Weather*, Madison, Wisc.: University of Wisconsin Press.

Canadian Global Change Program (CGCP), 1995, *Implications of Global Change and Human Health: Final Report of the Health Issues Panel to the Canadian Global Change Program*. Ottawa, The Royal Society of Canada (CGCP Technical Series).

Cane M.A., and Eshel G., and Buckland, R.W., 1994, Forecasting Zimbabwean maize yields using eastern Equatorial Pacific sea surface temperature, *Nature* 370:204–205.

Christophers, R., 1911, *Malaria in the Punjab*, Scientific Memoir, Med. and San Dept, Government of India, Superintendent Government Printing, Calcutta.

Curson, P., Guest, C., and Jackson, E., eds., 1997, *Climate Change and Human Health in the Asia-Pacific Region*, Canberra: Australian Medical Association and Greenpeace International.

Diaz, H.F., and V. Markgraf, eds., 1992, *El Niño: Historical and Paleoclimatic Aspects of the Southern Oscillation*, Cambridge: Cambridge University Press, 476pp.

Diaz, H.F., and Kiladis, G.N., 1992, Atmospheric teleconnections associated with the extremes phases of the Southern Oscillation, in: H.F. Diaz and V. Markgraf, eds., *El Niño: Historical and Paleoclimatic Aspects of the Southern Oscillation*, Cambridge University Press, pp. 7–28.

Diaz, H.F., and Kiladis, G.N., 1995, Climatic Variability on Decadal to Century Time Scales, in: A. Henderson-Sellers, ed., *Future climates of the world: a modelling perspective*, World Survey of Climatology, Elsevier Publ. Co.:191–244.

Diaz, H.F., and Pulwarty, R.S., 1992, A comparison of Southern Oscillation signals in the tropics, in: H.F. Diaz and V. Markgraf, eds., *El Niño: Historical and Paleoclimatic Aspects of the Southern Oscillation*, Cambridge University Press:175–192.

Diaz, H.F., and Pulwarty, R.S., 1994, An analysis of the time scales of variability in centuries-long ENSO-sensitive records in the last 1000 years, *Climatic Change* 26:317–342.

Diaz, H.F., and McCabe, G.J., 1999, A possible connection between the 1878 yellow fever epidemic in the southern United States and the 1877–78 El Niño episode, *Bull. Amer. Meteor. Soc.* 80:21–27.

Dilley, M., and Heyman, B.N., 1995, ENSO and disaster: droughts, floods and El Niño/Southern Oscillation warm events, *Disasters* 19:181–193.

Dyson, T., 1996, *Population and Food: Global Trends and Future Prospects*, London: Routledge.

Enfield, D.B., and Cid, S.L., 1991, Low-frequency changes in El Niño–Southern Oscillation, *Journal of Climate* 4:1137–1146.

Epstein, P.R., 1995, Emerging diseases and ecosystem instability: New threats to public health, *American Journal of Public Health* 85:168–172.

Gill, C.A., 1920, The relationship between malaria and rainfall, *Indian Journal of Medical Research* 7:618–632.

Gill, C.A., 1921, The role of meteorology and malaria, *Indian Journal of Medical Research* 8:633–693.

Gill, C.A., 1923, The prediction of malaria epidemics, *Indian Journal of Medical Research* 10:1136–1143.

Gilles, H.M., 1993, Epidemiology of malaria, in: H.M. Gilles and D.A., Warrell, eds., *Bruce-Chwatt's Essential Malariology*, 3rd ed., London, Edward Arnold, pp.124–163.

Glantz, M.H., Katz, R.W., and Nicholls, N., eds., 1991, *Teleconnections Linking Worldwide Climate Anomalies*, Cambridge, Cambridge University Press, 535 pp.

Glass, R.T., and Zack, M.M.J., 1979, Increase in deaths from ischaemic heart disease after blizzards, *Lancet* i:485–487.

Grove, R.H., 1998, Global impact of the 1789–93 El Niño, *Nature* 393:318–319.

Gueri, M., Gonzalez, C., and Morin, V., 1986, The effect of the floods caused by El Niño on health, *Disasters* 10:118–124.

Hales, S., Weinstein, P., and Woodward, A., 1996, Dengue fever in the South Pacific: Driven by El Niño Southern Oscillation? *Lancet* 348:1664–1665.

Harley, D.O., and Weinstein, P., 1996, The Southern Oscillation Index and Ross River virus outbreaks [letter]. *Medical Journal of Australia* 165:531–532.

Hastenrath, S., 1995, Recent advances in tropical climate prediction, *Journal of Climate* 8:1519–1532.

Hoerling, M.P., and Kumar, A., 1997, Why do North America climate anomalies differ from one El Niño event to another? *Geophysical Research Letters* 24:1059–1062.

Hoerling, M.P., and Kumar, A., 2000, Understanding and predicting extratropical teleconnections related to ENSO, in: H.F. Diaz and V. Markgraf, eds., *El Niño and the Southern Oscillation, Multiscale Variability and Global and Regional Impacts*, Cambridge: Cambridge University Press:57–88.

Hoerling, M.P., Kumar, A., and Zhong, M., 1997, El Niño, La Niña, and the nonlinearity of their teleconnections, *Journal of Climate* 10:1769–1786.

Hufton, O., 1985, Social Conflict and the Grain Supply in Eighteenth Century France, in: R.I., Rotberg and T.K., Rabb, eds., *Hunger and History: The Impact of Changing Food Production and Consumption Patterns on Society*, Cambridge, Cambridge University Press, pp. 105-133.

Hughes, M.K. and Diaz, H.F., 1994, Was there a "Medieval Warm Period", and if so, where and when? *Climatic Change* 26:109-142.

Kalkstein, L.S. and Tan, G., 1995, Human health, in: K.M. Strzepek and J.B. Smith, eds., *As Climate Changes: International Impacts and Implications*, Cambridge University Press, New York, pp. 124–145.

Katsouyanni, K., 1998, *Heat waves in Southern Europe: is there any evidence of an early effect of climate change?* WHO European Centre for Environment and Health, Rome, Working Group Paper EHRO 020502/11.

Katsouyanni, K., Pantazopoulu, A., and Touloumi, G., 1993, Evidence for interaction between air pollution and high temperature in the causation of excess mortality, *Arch. Environ. Health* 48:321–324.

Katsouyanni, K., Tricholpoulos, D, Zavitsanos, X. and Touloumi, G., 1988, The 1987 heatwave in Athens. *Lancet* 2:573.

Kiladis, G.N. and Diaz, H.F., 1986, An analysis of the 1877–78 ENSO episode and comparison with 1982-83, *Monthly Weather Review* 114:1035–1047.

Kiladis, G.N. and Diaz, H.F., 1989, Global climatic anomalies associated with extremes of the Southern Oscillation, *Journal of Climate* 2:1069–1090.

Kiladis, G.N. and Sinha, S.K., 1991, ENSO, monsoon and drought in India, in: M.H. Glantz, R.W. Katz and N.Nicholls, eds., *Teleconnections Linking Worldwide Climate Anomalies*, Cambridge University Press:431–458.

Kiladis, G.N. and van Loon, H., 1988, The Southern Oscillation. Part VIII: Meteorological anomalies over the Indian and Pacific sectors associated with the extremes of the oscillation, *Monthly Weather Review* 116:120–136.

Kilbourne, E.M., 1992, Illness due to thermal extremes, in: J.M. Last and R.B. Wallace, eds., *Public Health and Preventive Medicine*, 13th ed. Norwalk, Appleton Lange:491–501.

Kovats, R.S., Bouma, M.J., and Haines, A., 1999, *El Niño and Health*, Technical Report, World Health Organization, Geneva (WHO/SDE/PHE/99.4).

Kunst, A.E., Looman, C.W.N., and Mackenbach, J.P., 1993, Outdoor air temperature and mortality in the Netherlands: a time series analysis, *Amer. J. Epidemiol.* 137:331–341.

Ladurie, L.R., 1972, *Times of Feast, Times of Famine: A History of Climate Since the Year 1000*, London: Allen and Unwin.

Lamb, P.J. and Peppler, R.A., 1991, *West Africa*, in: *Teleconnections Linking Worldwide Climate Anomalies*, M.H. Glantz, R.W. Katz and N. Nicholls, eds., Cambridge University Press:121-189.

Latham, B., 1900, The climatic conditions necessary for the propagation and spread of plague, *Q. J. Roy. Meteor. Soc.* 26:37–94.

Latif, M., Kleeman, R., and Eckert, C., 1997, Greenhouse warming, decadal variability, or El Niño? An attempt to understand the anomalous 1990s, *Journal of Climate* 10:2221–2239.

Levitus, S., Antonov, J.I., Boyer, T.P., and Stephens, C., 2000, Warming of the world ocean, *Science* 287:2225–2229.

Lindsay, M., Broom, A.K., Wright, A.E., *et al.*, 1993, Ross River virus isolations from mosquitoes in arid regions of Western Australia: Implications of vertical transmission, *American Journal of Tropical Medicine and Hygiene* 49:686–696.

Linthicum, K.J., Anyamba, A., Tucker, C.J., Kelley, P.W., Myers, M.F., and Peters, C.J., 1999, Climate and satellite indicators to forecast Rift Valley fever epidemics in Kenya, *Science* 285:397–400.

Livezey, R. E., and T. M. Smith, 1999, Covariability of aspects of North American climate with global sea surface temperatures on interannual to interdecadal timescales, *J. Climate* 12:289–302.

Macdonald, G., 1957, *The Epidemiology and Control of Malaria*, Oxford University Press, London, U.K., 201pp.

MacKenzie, J.S., and Smith, D.W., 1996, Mosquito-borne disease and epidemic polyarthritis, *Medical Journal of Australia* 164:90-93.

Martens, W.J.M., Jetten, T.H., Rotmans, J., and Niessen, L.W., 1995a, Climate change and vector-borne diseases: A global modelling perspective, *Global Environmental Change* 5:195–209.

Martens, W.J.M., Niessen, L.W., Rotmans, J., Jetten, T.H., and McMichael, A.J., 1995b, Potential impact of global climate change on malaria risk, *Environmental Health Perspectives* 103:458-464.

Martens, W.J.M., Jetten, T.H., and Focks, D.A., 1997, Sensitivity of malaria, schistosomiasis and dengue to global warming, *Climatic Change* 35:145–156.

Martens, P., Kovats R.S., Nijhof, S., de Vries P., Livermore M.T.J., Bradley D., Cox J., McMichael, A.J., 1999, Climate change and future populations at risk of malaria, *Global Environmental Change* 9:special issue, S89-S107.

Mason, S.J., Lindesay, J.A. and Tyson, P.D., 1994, Simulating drought in southern Africa using sea surface temperature variations, *Water South Africa* 20:15–22.

Matariria, C.H. and Unganai, L.S., 1995, *A rainfall prediction model for southern Africa based on the Southern Oscillation phenomena*, FAO/SADC Regional Early Warning System Project. Harare, Zimbabwe.

McMichael, A.J., and Haines, A., 1997, Global climate change: The potential effects on health, *British Medical Journal* 315:805–809.

McMichael, A.J., Haines, A., Slooff, R., and Kovats, S., eds., 1996, *Climate Change and Human Health*, World Health Organization, Geneva, 297pp.

McMichael, A.J., and Githeko, A., 2001, Human Health, in: The IPCC Third Assessment Report, Working Group II (*in press*).

Michaelsen, J., 1989, Long-period fluctuations in El Niño amplitude and frequency reconstructed from tree-rings, in: D.H. Peterson, ed., *Aspects of Climate Variability in the Pacific and the Western Americas*, Geophysical Monograph 55, American Geophysical Union, Washington, D.C.:69–74.

National Health and Medical Research Council (NHMRC), 1991, *Health implications of long-term climatic change*, Canberra, Australian Govt. Publishing Serv.

Nicholls, N., 1985, Impact of the Southern Oscillation on Australian crops, *J. Climatology* 5:553–560.

Nicholls, N., 1986, A method for predicting Murray Valley Encephalitis in southeast Australia using the Southern Oscillation, *Aust. J. Exp. Biol. Med. Sci.* 64:587–594.

Nicholls, N., 1988, El Niño–Southern Oscillation impact prediction, *Bull. Amer. Meteor. Soc.* 69:173–176.

Nicholls, N., 1991, Teleconnections and health, in: *Teleconnections Linking Worldwide Climate Anomalies*, M., Glantz, R.W. Katz, and N. Nicholls, eds., Cambridge University Press:493–510.

Nicholls, N., 1992, Historical El Niño/Southern Oscillation variability in the Australasian region, in, H.F. Diaz and V. Markgraf, eds., *El Niño: Historical and Paleoclimatic Aspects of the Southern Oscillation*, Cambridge University Press:151–173.

Nicholls, N., 1993, El Niño-Southern Oscillation and vector-borne disease, *The Lancet* 342:1284–1285.

Nicholls, N., 1997a, "A healthy climate"? in: E.K. Webb, ed., *Windows on Meteorology*, CSIRO publishing, Collingwood, Australia:105–117.

Nicholls, N., 1997b, Increased Australian wheat yield due to recent climate trends, *Nature* 387:484-485.

Ortlieb, L., 2000, The documentary historical record of El Niño events in Peru: An update of the Quinn records (sixteenth through nineteenth centuries), in: H.F. Diaz and V. Markgraf, eds, *El Niño and the Southern Oscillation, Multiscale Variability and Global and Regional Impacts*, Cambridge: Cambridge University Press:207–295.

Patz, J.A., McGeehin, M.A., Bernard, S.M., Ebi, K.L., Epstein, P.R., Grambsche, A., Gubler, D.J., Reiter, P., Romieu, I., Rose, J.B., Samet, J.M., and Trtanj, J., 2000, The potential health impacts of climate variability and climate change for the United States, *Executive Summary*, Report of the health sector of the U.S. National Assessment. *Environmental Health Perspectives* 108:367–376.

Ponting, C., 1991, *A Green History of the World*, London: Penguin.

Post, J.D., 1985, *Food Shortage, Climatic Variability and Epidemic Disease in Pre-Industrial Europe*, Ithaca, NY:Cornell University Press.

Poveda, G., and Mesa, O.J., 1997, Feedbacks between hydrological processes in tropical South America and large scale oceanic-atmospheric phenomena *Journal of Climate* 10:2690–2702.

Poveda, G., and Rojas, W., 1996, Impact of El Niño phenomenon on malaria outbreaks in Colombia (in Spanish), *Proceedings XII Colombian Hydrological Meeting,* Colombian Society of Engineers, Bogotá:647–654.

Poveda, G., and Rojas, W., 1997, Evidences of the association between malaria outbreaks in Colombia and the El Niño-Southern Oscillation (in Spanish), *Revista Academia Colombiana de Ciencias* XXI(81):421-429.

Poveda, G., Gil, M.M., and Quiceno, N., 1997, Impact of ENSO and NAO on the annual cycle of the Colombian hydrology (in Spanish), *Proc. International Workshop on the Climatic Impacts of ENSO at Regional and Local Scales*, ORSTOM-INHAMI, Quito, Ecuador.

Poveda, G., Gil, M.M., and Quiceno, N., 1999, Associations between ENSO and the annual cycle of Colombia´s hydro-climatology, *Proceedings 10th Symposium on Global Change*, 79th American Meteorological Society Meeting, Dallas, Texas, 1999b.

Poveda, G., Graham, N.E., Epstein, P.R. Rojas, W., Quiñonez, M.L., Vélez, I.D., and Martens, W.J.M., 2000, Climate and ENSO Variability Associated with Vector-Borne Diseases in Colombia, in: H.F. Diaz and V. Markgraf, eds, *El Niño and the Southern Oscillation, Multiscale Variability and Global and Regional Impacts*, Cambridge: Cambridge University Press:183-204.

Pringle, H., 1997, Death in Norse Greenland, *Science* 275:924-926.

Quinn, W.H., 1992, A study of Southern Oscillation-related climatic activity for A.D. 622-1990 incorporating Nile River flood data, in: H.F. Diaz and V. Markgraf, eds, *El Niño: Historical and Paleoclimatic Aspects of the Southern Oscillation*, Cambridge: Cambridge University Press:119–149.

Quinn, W.H., 1993: The large-scale ENSO event, the El Niño, and other important features. *Bull. de l'Institut Français d'Etudes Andines*, Lima, 22:13–34.

Quinn, W. H. and Neal, V. T., 1992, The historical record of El Niño events, in Bradley, R. S. and Jones, P. D. (eds.), *Climate Since A. D. 1500*. Routledge, 623-648.

Quinn, W. H., Neal, V. T. and Antunez de Mayolo, S. E., 1987, El Niño occurrences over the past four and a half centuries. *Journal of Geophysical Research* 92:14,449-14,461.

Quinn, W. H., Zopf, D. O., Short, K. S. and Kuo Yang, R. T. W., 1978, Historical trends and statistics of the Southern Oscillation, El Niño. and Indonesian droughts, *Fisheries Bulletin* 76:663-678.

Rajagopalan, B., Lall, U., and Cane, M A., 1997, Anomalous ENSO occurrence: An alternate view, *Journal of Climate* 10:2351–2357.

Rasmusson, R.M., and Carpenter, T.H., 1982, Variations in tropical sea surface temperature and surface wind fields associated with the Southern Oscillation, *Monthly Weather Review* 110:354–384.

Rasmusson, R. M., and Carpenter, T. H., 1983, The relationship between eastern equatorial Pacific sea surface temperatures and rainfall over India and Sri Lanka, *Monthly Weather Review* 111:517-528.

Rimmington, G. M., and Nicholls, N., 1993, Forecasting wheat yields in Australia with the Southern Oscillation Index, *Aust. J. Agric. Res.* 44:625–632.

Rogers, J.C., 1988, Precipitation variability over the Caribbean and tropical Americas associated with the Southern Oscillation, *Journal of Climate* 1:172–182.

Rogot, E., and Padgett, S.J., 1976, Associations of coronary and stroke mortality with temperature and snowfall in selected areas of the United States, 1962–1966, *Amer. J. Epidemiol.* 103:565–575.

Rooney, C., McMichael, A.J., Kovats, R.S., and Coleman, M., 1998, Excess mortality in England and Wales, and in Greater London, during the 1995 heatwave, *Journal of Epidemiology and Community Health* 52:482–486.

Ropelewski, C.F. and Halpert, M.S., 1987, Global and regional scale precipitation patterns associated with El Niño/Southern Oscillation, *Monthly Weather Review* 115:1606–1626.

Ropelewski, C.F. and Halpert, M.S., 1989, Precipitation patterns associated with the high-index phase of the Southern Oscillation, *Journal of Climate* 2:268–284.

Swaroop, S., 1946, Forecasting of epidemic malaria in the Punjab region, India, *American Journal of Tropical Medicine* 29:1-17.

Sapir, D.G. and Misson, C., 1992, The development of a database on disasters, *Disasters* 16:74–80.

Symons, G. J., 1900, Jubilee Address, *Q. J. Roy. Meteor. Soc.* 26:176–181.

Telleria, A.V., 1986, Health consequences of floods in Bolivia in 1982, *Disasters* 10:88–106.

Tett, S., 1995, Simulation of El Niño–Southern Oscillation-like variability in a global AOGCM and its response to CO_2 increase, *Journal of Climate* 8:1473–1502.

Tong, S., Peng, B., Parton, K., Hobbs, J. and McMichael, A.J., 1998, Climate variability and transmission of epidemic polyarthritis, *Lancet* 351:1100.

Trenberth, K.E. and Hoar, T.J., 1996, The 1990–1995 El Niño-Southern Oscillation event: Longest on record, *Geophys. Res. Lett.* 23:57–60.

Trenberth, K.E. and Hurrell, J.W., 1994, Decadal atmosphere-ocean variations in the Pacific, *Clim. Dyn.* 9:303–319.

Trenberth, K. E. and Shea, D. J., 1987, On the evolution of the Southern Oscillation, *Monthly Weather Review* 115:3078–3096.

Unninayar, S.S. and Sprigg, W., 1995, Climate and the emergence and spread of infectious diseases, *EOS* 76(47):p. 478.

Whitcombe, E., 1993, Famine mortality, *Economic and Political Weekly* 28:1169–1179.

Whitman, S., Good, G., Donoghue, E.R., Benbow, N., Shou, W. and Mou, S. 1997, Mortality in Chicago attributed to the July 1995 heat wave, *American Journal of Public Health* 87:1515-1518.

World Health Organization (WHO), 1990, Potential health effects of climate change: report of a WHO task group, Geneva, (unpublished document WHO/PEP/90.10).

World Health Organization (WHO), 1998, Choiera in 1997, *Weekly Epidemiological Record* 73:201–208.

Zhang, Y., Wallace, J.M., and Battisti, D.S., 1997, ENSO-like interdecadal variability: 1900-93, *Journal of Climate* 10:1004–1020.

INDEX